Defensive (anti-herbivory) Coloration in Land Plants

Simcha Lev-Yadun

Defensive (anti-herbivory) Coloration in Land Plants

Anti-Herbivory Plant Coloration and Morphology

Springer

Simcha Lev-Yadun
Department of Biology & Environment
Faculty of Natural Sciences
University of Haifa – Oranim
Tivon, Israel

ISBN 978-3-319-42094-3 ISBN 978-3-319-42096-7 (eBook)
DOI 10.1007/978-3-319-42096-7

Library of Congress Control Number: 2016954234

Printed on acid-free paper

This Springer imprint is published by Springer Nature
The registered company is Springer International Publishing AG Switzerland

This book is dedicated to my school biology teachers, the late Hanna Kedari and Levi Ovadia, my many university teachers and mentors, colleagues and students, and especially to my family.

Foreword

This book is unashamedly a provocation. Simcha has long believed that a significant proportion of the coloration, patterning, and even structure of plants can be explained by selection pressures associated with influencing the sensory and cognitive systems of herbivores to reduce the damage they impose on plants. That is, Simcha believes that just as camouflage, mimicry, and aposematism are widespread among animals, so they will be in plants. Simcha has been publishing on this for 15 years, but this book represents by far the fullest exposition of his arguments.

Bluntly, I think Simcha's ideas remain (as he admits himself) only a little proven but logically plausible. There is no doubt that herbivores certainly impose selection pressures on plants and that there should be potential for the appearance of plants to be selected because of the effect that this has on the sensory and cognitive systems of herbivores. Further, some defenses open to animals against would-be attackers are not available to plants. Animals can often flee when detecting a predator, can gang together to mob the predator and drive it away, or can adapt their behavior to avoid places where they have seen predators. None of these options are open to plants; if a plant is detected and recognized by a herbivore, then there is very little the plant can do to stop the herbivore from attacking. Thus you might argue that plants might have particular need of the type of sensory defenses that Simcha champions (to hide from herbivores or otherwise dissuade them from attacking).

Alternatively, it may be that herbivores simply do not apply sufficiently strong selection pressures compared to an animal's predators. When a rabbit is captured by a fox, all its chances of further contributing to its fitness are at an end, but a plant attacked by caterpillars can still flower and set seed. It might also be that exploitation of herbivore vision is restricted by the demands of exposing large areas of chlorophyll to sunlight and/or by the lack of mobility to influence the lighting regime and microhabitat that together form the viewing environment. I don't think we can address the importance of herbivore senses in shaping plant traits with argument; we need careful experiment. One way or another, it would be a disservice to science if Simcha's provocative book is ignored; we should strive for an understanding of how important sensory and cognitive manipulation of herbivores has been in shaping plant traits – at the moment we simply don't know. Simcha's book argues powerfully that searching for an answer to this question should be a valuable exercise, and he points us to some very interesting case studies that would benefit from experimental manipulation now.

The good news is that the extensive experimental effort that we have seen expended on the investigation of attraction of pollinators should transfer readily to repulsion of herbivores, as should much of the methodology associated with sensory aspects of animal predator-prey interactions. More good news is that these experiments generally require relatively low-tech and inexpensive approaches; we really could make a lot of progress with short-term, small-scale projects like those that typify the research projects often carried out by final-year undergraduates.

There has been one conspicuous exception to the general neglect of how the senses of herbivores might have influenced plant traits. As Simcha covers in Chaps. 45, 46, 47, 48, 49, 50 and 51, there has been substantial interest in the last 15 years or so on the involvement of herbivores as a selective force in the coloration of autumn leaves. This was originally triggered by a set of papers arguing that at least some autumn leaf coloration might be a signal of individual plant "quality" selected because of its effect on autumn-flying aphids. This interest soon made clear that there was a question to answer: why is there so much variation in autumn leaf coloration between species, between years, between individuals, and even within individual plants. It became clear that we really could not explain this, and the aphid idea was a novel, plausible, and testable hypothesis. This triggered a great deal of activity testing not just that hypothesis but generating a plethora of alternative hypotheses for mechanisms influencing leaf coloration. The upshot of this endeavor is that we are beginning to have a much better idea of what the key mechanisms are that underlie variation in autumn leaf coloration. As it transpires, there is still (in my view) no strong evidence that autumn-flying aphids are an important component of this, but we are much further forward than we were in understanding what the important components are. The fact that some of these have nothing to do with the senses of any herbivores in no way detracts from the vital role that the original hypothesis had in stimulating activity. On a broader scale, even if Simcha is ultimately proved wrong, if this book is the trigger for causing us to investigate the impact that herbivore sensory systems have on which plants they exploit, then that will be an important stride forward not just for pure but also applied plant science.

The seminal work in animal camouflage was that of Hugh Cott around the time of the Second World War. Cott's insights where sharpened both by his military experience and by his considerable talent as an artist. It is probably no coincidence that Simcha's background features photography and military experience very strongly. Cott and Simcha were and are extraordinary careful and thoughtful observers of the natural world. Science needs a small number of such people that are fecund generators of original ideas, but it also needs rigorous and imaginative developers and testers of these ideas. We had to wait almost 50 years before there was a thriving community of scientists developing and testing Cott's ideas; it would be a great shame if the same fate befell Simcha's. Both Simcha and I would be happy if this book falls rapidly out of date. I urge all those interested in plant-herbivore interactions and all those interested in sensory ecology to read this fascinating, thoughtful, and stimulating book and then join Simcha in understanding an unjustly neglected and important aspect of the natural world.

School of Biology
University of St Andrews,
Fife, Scotland

Graeme D. Ruxton

Preface

This book about various types of visual defense (coloration, morphology, movement) of plants against herbivores is not just a list of facts and hypotheses, with a reference source, but mainly an ideological one, a manifesto. The book aims to establish visual anti-herbivory defense as an integral organ of botany or plant science as it is commonly called today by people with insufficient knowledge of botany. It is intended to be the modern and updated botanical parallel to Cott's (1940) book *Adaptive Coloration in Animals*, in order to intrigue and stimulate students of botany/plant science and plant/animal interactions for a very long time. This book is tailored to a readership of biologists and naturalists of all kinds and levels and more specifically for botanists, ecologists, evolutionists, and those interested in plant/animal interactions. It is written from the point of view of a naturalist, ecologist, and evolutionary biologist that I hold, considering natural selection as the main although not the only drive for evolution. According to this perspective, factors such as chance, founder effects, genetic drift, and various stochastic processes that may and do influence characters found in specific genotypes are not comparable in their power and influence to the common outcomes of natural selection, especially manifested when very many species belonging to different plant families, with very different and separate evolutionary histories, arrive at the same adaptation (convergent evolution), something that characterizes many of the visual patterns and proposed adaptations described and discussed in this book. I think that when the same apparent adaptive solution is found in many species with independent evolutionary histories, this can point more strongly to both the evolution and function of a character than many meticulous experiments conducted on a specific species or genotype and under specific experimental conditions. Many of the visual defensive mechanisms I propose, describe, or discuss are aimed at operating before the plants are damaged, i.e., to be their first line of defense. In this respect, I think that the name of the book by Ruxton et al. (2004) *Avoiding Attack* is an excellent phrase for the assembly of the best types of defensive tactics. While anti-herbivory is the theme of this book, I do remember, study, and teach physiological and developmental aspects of some of the discussed patterns of coloration and morphology, and I am fully aware of the simultaneous and diverse functions of many plant characters. The complexity of many of the discussed and proposed adaptations and the many and even contrasting selection directions of the various selective agents involved in the evolution of these characters makes things difficult to understand, model, and test in the field or in the laboratory.

Since the mid-nineteenth century, in parallel with the emergence of Darwinism and certainly after its meteoric establishment in the year 1859, the evolution of defensive coloration and morphology in animals has been and will probably continue to be the bread and butter of many zoologists, including in recent decades some molecular ones, and in an absurd way, an unknown or in the best cases considered an anecdotal issue for botanists till the beginning of the twenty-first century. I am trying clearly, sharply, and frankly to put an end to this absurdity. To make the absurdity even greater, many animal species acquire both their defensive coloration and their associated defensive chemistry directly from plants or indirectly from prey herbivores that got them directly from plants, to establish their chemically based visual aposematism. However, the plants that serve as the source for these defensive and signaling molecules were almost never, except for some of the pioneering works of Miriam Rothschild and a few others, considered to be aposematic.

After the introduction chapters that outline several essential theoretical issues related to coloration, visuality in general, and attack and defense, the main part of the book begins with camouflage and related defenses, followed by aposematism (warning coloration), and proceeds with the still only partly understood spectacular phenomenon of autumn leaf coloration of temperate and boreal trees and shrubs. The book ends with a review of defensive visual animal and animal action mimicry by plants. Defensive olfactory issues, especially of an aposematic nature, are described and discussed only briefly, and they are mentioned when it illuminates or complements visual issues.

I have had a special, complicated, and variable career background that attracted me to and allows me to explore nature and visual (as well as other) defenses as a naturalist. I grew as a nature boy in the pre-urban central coastal plain of Israel when it was still dominated by partly natural noncultivated grazing lands and by agricultural areas. As a child who used to walk barefoot in the summer, I had many opportunities to encounter and suffer from thorns, spines, and prickles in the environs of my home, some ending with extracting spines or thorns from my foot with metal plyers usually used to extract nails from wood, followed by burning iodine application and sometimes by an anti-tetanus injection. In my youth, I did not consider the defensive signaling of the defended plants and suffered again and again because of this, but I got a first-class understanding of their defensive functions. I also had a 15-year-long career as a military and civil photographer, something that sharpened my sensitivity and awareness of visual issues. As a military photographer, I lost my whole left leg and suffered several additional wounds in the 1973 war, but this did not block my military or any other career. My pre-1973 and later diverse military experiences gave me a special point of view about risks, damage, and recovery and a realistic view of defense and attack, a view that cannot result from working only in the ivory tower of the academy. My great interest in military history, which began at age 14 and resulted in some 1000 books that I have bought and read on this subject, further helped me in my studies of biological defense and attack mechanisms. My university studies were long and diverse [B.Sc. in biology (1980), M.Sc. in botany-ecology (1986), and M.A. in archaeology (1991) – including 3 years of full complementary studies for a B.A.; I received my Ph.D. in developmental botany (1992) and did a postdoc in molecular genetics and cell biology (1992–1994)]. During my M.Sc. studies, I decided that since the academic system pressed young people to focus on a single narrow issue, I would do the opposite and train myself to deal with multidisciplinary issues. Throughout my studies, the official program was only a minor part of what I did. I read much more than I had to officially and conducted research outside my official theses in both related and unrelated subjects (i.e., reproductive plant biology, ethnobotany, plant architecture, etc.). After my postdoc, 21 years after losing my leg in war, I declined a civilian tenure-track scientist position, passed an officer course, and served for several tense, hectic, and educational years as a captain and a major in the IDF (Intelligence). This gave me an even broader and deeper understanding of defense and attack. Originally, when I began to study biology at Tel Aviv University, I wished to be a zoologist, but there were several factors that operating together caused me to change my major interest from zoology to botany only several months after I began my studies. However, my love of zoology since my early youth serves me very well and is fulfilled when I study plant/animal interactions. I think that this combination of diverse education and my various other experiences not mentioned here allows me to deal better with multidisciplinary issues in general and with visual plant defenses in particular.

Interestingly, my career background is similar in salient points to that of the American artist Abbott H. Thayer with his critical contributions for understanding countershading and animal and military camouflage; the British naval painter and marine officer Norman Wilkinson, with his critical contribution of inventing military naval dazzle painting; and the British officer Hugh B. Cott, the great contributor to the area of defensive animal coloration, who was also a wildlife artist and a photographer.

My theoretical interest in the biological significance of plant coloration not related to photosynthesis or reproduction began when I was still an M.Sc. student in botany. I started to read papers and gather data on nonreproductive and nonphotosynthetic plant coloration in the year 1982, although with no specific theoretical framework or clear direction. I just thought that there is no basic ecological and evolutionary hypothesis concerning nonreproductive and nonphysiological plant coloration paralleling the many areas of research into defensive coloration in zoology and that this issue should be studied and shaped into an organized scientific botanical discipline. My long and vast experience as a professional photographer and my excellent visual memory helped me to notice and document various pigmentation patterns that either went unnoticed or were not considered important by many others. As with many other scientific ideas, I decided that one day I would try to understand what is going on with non-photosynthetic and nonreproductive plant colors. During my massive fieldwork in the 1980s and 1990s, it became obvious to me that in addition to several coloration mutants that I found in various wild plant populations and species, there is a huge array of types and patterns of vegetative

Fig. 1 A typical zebra like leaf of *Silybum marianum*. My determination to understand the evolution of this unique leaf coloration initiated my targeted field and theoretical work on defensive plant coloration. After considering it just to mimic tunneling damage mimicry for defense from herbivory, it took a year and a half to understand the potential aposematic function of this type of coloration. Additional defensive types by this coloration emerged later.

coloration in plants that was not described in floras and plant identification field handbooks and that were not considered to be functional or important by almost any botanist or zoologist.

Concerning defensive plant coloration, I made my significant and target-oriented move only in the early winter of 1994/1995. Then, I decided that I must understand why some local wild plant species look like green zebras (Fig. 1). In February 1996, I understood following my fieldwork that like many dangerous animals, such plants visually advertise their defensive qualities, and I began a very broad study of defensive plant coloration, knowing that visually oriented animals were the selective agent for such types, because they attacked differently defended and non-defended genotypes.

Even when animal coloration is considered, there are many only partly solved and totally unsolved questions (e.g., Kemp et al. 2015), let alone in plants. The classic works of Bates (1862), Müller (1879), Poulton (1890), Thayer (1918), Cott (1940), and Kettlewell (1973) and their hypotheses are still studied and examined today, some after more than 150 years. The situation concerning defensive plant coloration will probably not be different 100 years from now. I am fully aware of the meager amount of direct experimentation concerning defensive plant coloration and that many of the proposed hypotheses (including those proposed by me) need both rigorous experimentation (see Ruxton 2014) and, even more important and naturally, a much more complicated proof by documentation of their various functions in the wild. However, I am not afraid of failure in the shape of hypotheses turning out to be wrong or if they only partly explain reality. I do not claim to always be right; I just do my best to understand, intrigue, and stimulate. Many of the defensive mechanisms I proposed should currently be considered only as working hypotheses, something that will allow progress, and they should not be considered as the ultimate possible level of understanding. I know scientists who say that they have all the answers concerning the issues they study, but I freely admit, although with great frustration, that I don't even have all the good or essential questions. Since my goal in writing this book is not only technical but also educational, I will in some chapters describe my personal road to formulating a hypothesis, because this may help young scientists not to give up interesting questions and to expect that wonderful moments of discovery may arrive by surprise. At this initial stage of the study of defensive plant coloration, I am certain that many additional types and aspects of defensive plant coloration remained unrecognized, notwithstanding experimentally explored and refined. I am currently studying several such issues, but they are not sufficiently ripe to be presented. Moreover, I am certain that these are not the only yet undescribed types of visual defensive plant systems. Consistent with Grubb's (1992) view that defense systems are not simple and with Diamond's (2005) view that single-factor explanations can fail when complex environmental issues are being discussed, I consider that the evolution of various types of plant coloration reflects adaptations both to physiological pressures and to relations with other organisms but not only defensive ones. Such synergistic gains may cause the evolution of plant color traits to be quicker and more frequent. Another important issue is that practically nothing has been done concerning defensive plant coloration with the aid of modern molecular techniques, including studying the epigenetic components that are probably involved in some of the visual defenses. The time is ripe to do it, and the first ones to do so may gain highly valuable results and reputation.

My frequent contributions to defensive plant coloration are also partly related to geography. I did most of my fieldwork in the Near East, a region with a very long history of strong mammalian herbivory impact on the vegetation during the Pleistocene or possibly already in the Miocene, which increased dramatically after the establishment of herding as an important lifestyle in the Near East several millennia ago. This has resulted in both the evolution and the spread of many plant taxa with various types of anti-herbivory mechanisms. Hundreds of thorny and spiny species, as well as many poisonous ones, are very common in the landscape. Fieldwork in an ecosystem that has a millennia-long history of large-scale grazing, such as the land of Israel, clearly and "sharply" indicates the ecological benefit of being spiny. A continuous blanket of thorny, spiny, and prickly trees, shrubs, and thistles covers large tracts of the land, and other parts of the landscape are just rich with up to dozens of such plant species per km^2 that comprise a significant part of the vegetation. This dominance clearly indicates the adaptive value of being thorny or spiny when grazing pressure is high. It does not just slow the rate of feeding but gives a considerable advantage to such plants. For instance, spiny plants, such as *Echinops* sp. (Asteraceae), which usually grow as individuals or in small groups, sometimes become the most common perennial plant over many acres in grazing lands, and the thorny low shrub *Sarcopoterium spinosum* may cover large tracts of land in a continuous blanket (Ronel et al. 2007). A wealth of various nonreproductive (and non-documented) color patterns exists in our regional flora, and they deserve to be explained. Fieldwork in Canada, the eastern USA, Estonia, Finland, Germany, Hungary, northern Italy, Norway, Russia, Sweden, Switzerland, and the UK clearly indicated to me that the level of mammalian herbivory there is very much lower than in the Near East, as reflected by the low levels of grazing and browsing damage to the vegetation and therefore by only a few thorny, spiny, or prickly plants at both the flora and landscape levels.

Tivon, Israel Simcha Lev-Yadun

Acknowledgments

With a great pleasure, I thank the people who significantly influenced me in my studies of defensive plant coloration and related issues, by doing research together, meetings, fieldwork, correspondence, and conversations: Shahal Abbo, John A. Allen, Marco Archetti, Iris Aviezer, Frantisek Baluska, Pua Bar (Kutiel), Guy Bar-Oz, Amots Dafni, Thomas Döring, Adi Dror, Moshe Flaishman, Yoram Gerchman, T. Glasser, Kevin Gould, Mario Gutman, Snorre Hagen, Malka Halpern, Jarmo Holopainen, Niky Hughes, Ido Izhaki, Ofir Katz, Tamar Keasar, Said Khateeb, Avi Koplovich, Nicoletta La Rocca, Simon Leather, David Lee, Itamar Lupo, Haviva Malkiel, Yiannis Manetas, Johanna Mappes, Nitsa Mirsky, Omran Nassar, Pekka Niemelä, Helen Ougham, Avi Perevolotsky, Dina Raats, Michal Ronel, Graeme Ruxton, Paul Schaberg, Martin Schaefer, Ron Sederoff, Uri Shanas, Nissim Silanikove, Aki Sinkkonen, Howard Thomas, Pilli Urbas, Avivit Waissler, and Kazuo Yamazaki. Special thanks for collaboration and discussions are given to Malka Halpern, Jarmo Holopainen, Moshe Inbar, Gadi Katzir, and Gidi Ne'eman. I also thank the many reviewers and editors of my papers on plant defenses published in many journals (Advances in Applied Microbiology, Arthropod-Plant Interactions, BioEssays, Biological Journal of the Linnean Society, Botanical Journal of the Linnean Society, Botanical Review, Chemoecology, Communicative and Integrative Biology, Environmental and Experimental Botany, Environmental Microbiology, Flora, Functional Plant Science and Biotechnology, Israel Journal of Entomology, Israel Journal of Plant Sciences, Journal of Arid Environments, Journal of Natural History, Journal of Theoretical Biology, Journal of the Torrey Botanical Society, Naturwissenschaften, New Phytologist, Plant Signaling & Behavior, Proceedings of the National Academy of Sciences USA, Progress in Botany, Symbiosis Journal,

The Biochemist, Trends in Ecology & Evolution, Trends in Plant Sciences) for illuminating comments and very helpful suggestions that improved the published papers and by this the text of this book. I thank the Botanical Gardens of Tel Aviv University and Hudi Benayahu for allowing me to study their cacti and succulent plant collections. I thank my wife Susan Lev-Yadun for her comments on the manuscript and especially Graeme D. Ruxton for writing the foreword. Special thanks are given to Zuzana Bernhart and Mariska van der Stigchel and to other Springer people that enabled to actually turn my ideas and computer files into a book.

Contents

Part I
General Background

Chapter 1
Introduction

Defensive plant coloration (camouflage, aposematism, various types of mimicry, undermining herbivorous insect camouflage, masquerade, dazzle effects, trickery coloration, exploiting animals' perceptual biases, various types of signaling) and related visual aspects has received very limited attention till the year 2000 (e.g., Hinton 1973; Wiens 1978; Rothschild 1986; Smith 1986; Givnish 1990) compared to defensive animal coloration (e.g., Bates 1862; Müller 1879; Wallace 1889; Poulton 1890; Thayer 1918; Cott 1940; Kettlewell 1973; Edmunds 1974; Majerus 1998), but gained much more attention in the twenty-first century (e.g., Archetti 2000, 2009a, b; Lev-Yadun 2001, 2006a, 2009a, 2014a, b, c; Lev-Yadun et al. 2004a; Ruxton et al. 2004; Lee 2007; Lev-Yadun and Gould 2007, 2009; Archetti et al. 2009a; Fadzly et al. 2009; Klooster et al. 2009; Lev-Yadun and Holopainen 2009; Schaefer and Ruxton 2009, 2011; Burns 2010; Cooney et al. 2012; Lev-Yadun and Ne'eman 2012, 2013; Farmer 2014; Hughes and Lev-Yadun 2015). The surprisingly small number of papers in botany related to defensive coloration or defensive mimicry as compared to zoology is clearly reflected in the annotated bibliography by Komárek (1998) with thousands of related publications on animals and only hardly a handful about plants.

The most important comment about leaf colors and their potential defensive functions published before the year 2000 was a short paragraph by Harper (1977), who wrote in his seminal book on plant demography "botanists have been reluctant to accept precisions of adaptations that are commonplace to zoologists and often seem reluctant to see the animal as a powerful selective force in plant evolution except in the curiously acceptable realm of adaptation to pollination! It may be that much of the fantastic variation in leaf form, variegation, dissection and marking that is known in the plant kingdom is accounted for by the selective advantage to the plant of associating unpalatability with a visual symbol" (page 416). This, however, should be changed, and this book on defensive plant coloration is a continuation of the recent efforts to bridge this gap, reviewed in Ruxton et al. (2004), Lev-Yadun (2006a, 2009a, 2014a), Lev-Yadun and Gould (2007, 2009), Archetti et al. (2009a), Schaefer and Ruxton (2009, 2011) and Farmer (2014).

S. Lev-Yadun, *Defensive (anti-herbivory) Coloration in Land Plants*,
DOI 10.1007/978-3-319-42096-7_1

Almost every color imaginable has been observed in foliage of some plant taxa due to variations in concentrations and combinations of vacuolar pigments (e.g. anthocyanins and betalains), photosynthetic pigments (chlorophylls and carotenoids), cuticular waxes, and iridescent structures (e.g., iridisomes) (Lee 2007; Glover and Whitney 2010). Organs or tissues with colors other than green, may in many cases, but certainly not always, have an extra cost. The cost to the plant of producing colored organs has three aspects. First, it requires allocation of resources for the synthesis, application, degradation, and regulation of the expression patterns of the pigments. Second, any color of a non-woody aerial plant's organ, other than green, may in many cases, but certainly not all, since anthocyanins, betalains and carotenoids defend photosynthesis, be linked to lower photosynthesis (see Chalker-Scott 1999; Matile 2000; Hoch et al. 2001, 2003; Lee and Gould 2002; Gould et al. 2002a, 2002b; Close and Beadle 2003; Gould 2004; Ougham et al. 2005; Manetas 2006; Hatier and Gould 2008; Jain and Gould 2015). Third, conspicuousness may potentially attract herbivores. In general, the benefits in fitness from defensive coloration should be higher than its fitness costs for such characters to evolve.

Plant pigments and coloration caused by air spaces or other physical effects serve many physiological and communicative functions, such as photosynthesis, protection from UV, scavenging of oxygen radicals, pollination, seed dispersal, thermoregulation and defense from herbivores and pathogens (e.g., Gould et al. 2002a; Close and Beadle 2003; Lee 2007). Gould et al. (2002b), Lev-Yadun et al. (2002, 2004a), Gould (2004), Lev-Yadun (2006a, 2009a), Schaefer and Wilkinson (2004), Lev-Yadun and Gould (2007, 2009), and Archetti et al. (2009a), have already argued that the non-photosynthetic plant pigments have the potential to serve more than one function concurrently, including physiological ones. I stress that I fully agree with Endler (1981) who commented concerning animal coloration "we must be careful not to assume that because we have found one apparent function to a color pattern, it necessarily means that we have a complete explanation". Thus, various hypotheses concerning the function of coloration of leaves and other plant parts need not contrast with or exclude any other functional explanation of specific types of plant coloration, and traits such as coloration, that might have more than one type of benefit, may be selected for by several and even very different agents. Consistent with Grubb's (1992) view that defense systems are not simple, I usually consider that defensive plant coloration may defend by more than one method (including by signaling), reflect an adaptation to physiological pressures, and also serve non-defensive relations with other organisms.

Since some of the proposed defensive functions of leaf, flower and fruit coloration and shape (morphology) involve mimicry, I briefly describe the two major types (out of a longer list that includes various subtypes, see Wickler 1968; Pasteur 1982; Starrett 1993) of plant defensive mimicry. An innovative and elegant attempt to overcome the problematic, complicated and not fully satisfying definitions of mimicry by defining them as "adaptive resemblance" (Starrett 1993) did not manage to convince many and was not cited sufficiently to be used instead of the imperfect term mimicry.

Müllerian mimicry is a phenomenon in which two or more species with effective defenses share a similar appearance or signaling, and by this sharing reduce the cost of associative learning and evolution of refraining from attack by their enemies, a refraining that may sometimes have a genetic basis (Ruxton et al. 2004). Batesian mimicry is a phenomenon in which members of a palatable species or a group of such species, gain protection from predation by resembling an unpalatable or defended species (Bates 1862; Müller 1879; Wallace 1889; Cott 1940; Wickler 1968; Ruxton et al. 2004). There are, however, intermediate types (quasi-Batesian mimics) between these two extremes when defended and signaling species differ in their strength of defense or signaling (e.g., Rowland et al. 2010). Most of our knowledge about defensive mimicry has emerged from animal studies, and while even the better-studied animal mimicry systems such as butterfly aposematism and its Müllerian and Batesian mimicry, are still not fully understood (e.g., Forbes 2009), the operation of defensive mimicry in plants and its ecology and evolution is dramatically much less understood (e.g., Hinton 1973; Wiens 1978; Lev-Yadun and Inbar 2002; Lev-Yadun 2009a, b, c, 2014a; Schaefer and Ruxton 2009, 2011; Lev-Yadun and Ne'eman 2012; Farmer 2014).

In plants there is an additional recognized type of Müllerian and Batesian mimicry. Accordingly, flowers that attract pollinators with rewards are called Müllerian mimics and those without rewards are called Batesian (e.g., Dafni 1984; Roy and Widmer 1999; Schaefer and Ruxton 2011). Being non-defensive, they will not be discussed here. Only defensive Müllerian and Batesian plant mimicry will be discussed in this book. Moreover, I think that borrowing the terms Müllerian and Batesian mimicry for rewarding and rewardless flowers is inappropriate and has a logic discrepency because defensive Müllerian and Batesian mimicry is aimed to repel animals and in pollination they are aimed to attract. I suggest that concerning pollination, rewarding flowers that mimic other rewarding flowers are named Darwinian mimics because Darwin contributed critically to the understanding of pollination, and that for obvious reasons non-rewarding mimics are named Wallacian mimics.

The evolution of mimicry requires a model or models, a mimic, and a predator or predators (an operator). The model should be another species or a group of species, or their actions (e.g., release of chemicals or physical damage to other organisms) (Wickler 1968). Organisms may also mimic a biological or non-biological substrate on which they grow as a camouflage against enemies or to hide from potential prey (Cott 1940; Wickler 1968; Edmunds 1974; Ruxton et al. 2004; Caro 2005; Stevens and Merilaita 2011). Masquerade (close resemblance of inedible and often inanimate objects that will be discussed in more detail later) (Endler 1981; Allen and Cooper 1985; Skelhorn et al. 2010a, b; Lev-Yadun 2014b; Skelhorn 2015) is a related visual but not a crypsis-based defense, or at least, it is non-cryptic at close range. This character may sometimes initially evolve not for defense but for physiological reasons, e.g., the common light plant coloration that reflects excess light in sandy coastal or desert environments, resulting (probably as a secondary gain) in actual substrate mimicry that may potentially also reduce herbivory by camouflage or by just not being green (Wiens 1978; Lev-Yadun 2006b, 2014b, 2015a; Reeves

2011). Such multifunctional situations make it difficult to evaluate the relative role of the mimicry in various morphologies and types of coloration, but may explain the advantages during early stages of the evolution of such cases (Schaefer and Ruxton 2009).

Another significant theoretical problem that was highlighted recently is the difficulty in distinguishing between cases in which plants exploit perceptual biases of animals that may by convergent evolution arrive at a morphology that just looks like mimicry but is not a true mimicry (Schaefer and Ruxton 2009). For the convenience of discussion, and because of the very small number of experiments directly testing defensive plant mimicry, and because of our current very limited ability to distinguish between the situations of exploitation of perceptual biases of herbivores by plants and true mimicry by plants, I will usually refer only to mimicry. However, the exploitation of perceptual biases should be taken into consideration when "mimicry" is discussed or tested. I am sad to state that it is still not practical to do so in most cases and hope that this frustrating situation will change in the coming decades.

Chapter 2
Plants Are Not Sitting Ducks Waiting for Herbivores to Eat Them

Human patients with severe brain damage due to trauma or ischemia may never regain recognizable mental functions and they never speak because of absence of function in the cerebral cortex. Such patients were described in a classic medical paper as in a "persistent vegetative state" (Bryan and Plum 1972), and are referred to as "plants" or "vegetables" by laymen depending on language and culture. This attitude towards plants was recently manifested by Laundré (2014) in an essay about the hunting hardships of large carnivores. Laundré (2014) posited that "meeting daily energetic needs by large carnivores is not as easy as just going out and gathering plants that are waiting around to be found and eaten". I clearly and sharply posit that except for rewarding flowers towards legitimate pollinators (Faegri and van der Pijl 1979), ripe animal-dispersed fruits towards seed dispersers (van der Pijl 1982), or various food rewards to mutualistic ants (Jolivet 1998), plants do not wait around to be found and eaten. They do everything evolution has allowed them to do in order not to be eaten.

This common approach towards plants is a great underestimation of the huge and variable arsenal of defensive plant strategies. Moreover, many if not all land plants may employ several simultaneous constitutive and induced defensive strategies. Even plant parts that eventually wait to be found and eaten such as animal-dispersed ripe fleshy fruits that are intened to be eaten evantually by seed dispersers, defend themselves when they are unripe by means of chemistry, timing, camouflage, aposematism, and mechanical defenses (Herrera 2002; Schaefer et al. 2008; Lev-Yadun et al. 2009a; Lev-Yadun 2013a). Moreover, they, in the case of spiny *Opuntia* fruits may be still defended even when they are ripe. Plants employ bodyguards such as ants (Jolivet 1998), and call, using volatiles, to predators and parasitoids of their attacking invertebrate herbivores (Kessler and Baldwin 2001; Karban 2015). From time to time, they overwhelm and satiate herbivores by their sudden appearance in huge numbers in cases of mast fruiting or synchronous flowering (Janzen 1976; Kelly and Sork 2002), cause false satiation by various molecules (Lev-Yadun and Mirsky 2007), intimidate herbivores visually and chemically (Lev-Yadun 2009a), and possibly even by movement (Lev-Yadun 2013b). Plants also use camouflage

S. Lev-Yadun, *Defensive (anti-herbivory) Coloration in Land Plants*,
DOI 10.1007/978-3-319-42096-7_2

(e.g., Wiens 1978; Givnish 1990; Fadzly et al. 2009; Kloster et al. 2009; Burns 2010; Lev-Yadun and Ne'eman 2013; Niu et al. 2014), and pathogenic bacteria that they insert into the tissues of herbivores (Halpern et al. 2007a, b, 2011), they undermine the camouflage of herbivorous insects (Lev-Yadun et al. 2004a), and use strong wind-induced leaf movements to get rid of herbivores (Yamazaki 2011; Warren 2015), all these in order to defend themselves from herbivory. Plants have many types of mechanical defenses (Lucas et al. 2000) including even anisotropic arrangement of structural defenses that can lead invertebrate herbivores away from the plants (Vermeij 2015). Plants can even chemically identify (Karban 2015) and remember herbivore attacks for a long time (Gagliano et al. 2014; Karban 2015) and receive volatile cues from kin individuals that result in reduced herbivory damage compared to plants that receive such cues from non kin (Karban et al. 2013). This list is only a short partial summary of the many ways plants defend themselves from herbivores or manipulate them, and manipulate various herbivores' predaceous or parasitic enemies to the benefit of the plants.

Considering plants as defenseless and motionless victims of herbivores, because they are not fast and agile like large mammalian herbivores or carnivores, is a huge underestimation of plant defensive abilities, as many plants outsmart and even exploit many invertebrate and vertebrate herbivores and carnivores for pollination and for seed dispersal, and even carnivores and parasitoids, as was mentioned above, for defense. For instance, when thorny branches and spiny leaves move in the wind, they are actually "patrolling", and will hit anything positioned in their way and inflict wounds on large herbivores that enter their territory. Similarly, when such branches and leaves are bent by a large animal, they often return quickly and forcefully to their original position, stabbing their herbivores. I was painfully wounded by such plant spring actions numerous times during field work and fully appreciate their nasty reaction.

Chapter 3
The Many Defensive Mechanisms of Plants

Defensive plant coloration, morphology and movement, i.e., visual defenses, although variable as will be described and discussed in this book, are only a small part of the complicated and sophisticated anti-herbivory defense system employed by plants. In order to give the reader the right perspective, and to add to the previous chapter, below I describe the basics of their defensive system against other organisms. I stress that there is not a single species that has all these defenses together, but that this is the arsenal that evolution has allowed plants to possess. Since a full discussion of the defense system of plants will need a monograph larger than this book, I will present the data almost in the shape of an abstract.

Defenses are aimed towards enemies, and the biotic enemies of land plants include herbivores (vertebrates including mammals, reptiles and birds, and invertebrates), fungi, bacteria, viruses, parasitic plants and competitors (Harper 1977; Crawley 1983; Futuyma and Agrawal 2009; Gong and Zhang 2014; Karban 2015). Evidence from fossils indicates that herbivory on land plants began not later than the Early Devonian, some 400 million years ago (Labandeira 1998), and since earlier land plant fossils are not common and not well-preserved, it probably started at least dozens of millions years earlier.

Plants have ways to sense in many cases who attacked them and thus employ specific induced defenses through cues such as chitin (fungi, insects), volatiles, and probably by other cues and signals (Crawley 1983; Karban and Baldwin 1997; Karban and Agrawal 2002; Kant et al. 2015; Karban 2015). In spite of the large arsenal of anti-herbivory defenses that plants possess, herbivory in terrestrial ecosystems may consume up to about 75 % of the net primary production, however, the range of plant consumption is enormous, and can be as low as less than 0.1 % (Cebrian and Lartigue 2004), with a median of 18 % (Cyr and Pace 1993). I do not wish to elaborate on the question: why is the world green? The debates about this that began with Hairston et al. (1960) and developed into two contrasting views: top-down by predators, or bottom-up by plant defenses and low nutritional level (see review in Schmitz 2008). From my general experience and my involvement in several hot scientific debates, I think that both hypotheses are valid, one under

S. Lev-Yadun, *Defensive (anti-herbivory) Coloration in Land Plants*,
DOI 10.1007/978-3-319-42096-7_3

certain conditions, and the other under different conditions. For instance, mammalian herbivory in the Middle East, or even all over the Mediterranean basin, is a good example, although greatly influenced by human activity (hunting and habitat destruction) in the last several millennia. During the Pleistocene, many wild mammalian herbivores and their mammalian predators co-existed there and a certain level of balance prevailed between them and the vegetation. Especially during the second half of the Holocene, but starting in the very late Pleistocene, most of the many larger Near-Eastern mammalian taxa (larger than 30 kg), both herbivores and carnivores, became extinct (Tsahar et al. 2009) or are now found in very small numbers that cannot influence the regional ecology significantly. However, after the extinction, the landscape became dramatically less green over large areas because large-scale herding of domesticated mammals became common everywhere (Thirgood 1981). Classic top-down Pleistocenian ecology changed to bottom-up ecology because the vegetal productivity determined the amount of livestock, and because the enormous grazing pressure selected for resilient and defended plants (Zohary 1983; Perevolotsky and Seligman 1998). Similarly, diseases, migration, establishment of a toxic or spiny plant species, infertile soil or extremely low or high temperatures *versus* increased soil fertility following volcanism or influences of fire, hurricanes or El-Niño/La-Niña years, may all cause local changes in bottom-up or top-down processes to any direction (see Owen-Smith 2015).

Plant defenses can be classified as static (constitutive) and dynamic (induced) and many defenses are expressed as both constitutive and induced. The main constitutive ones are: the cuticle, cork, thorns, spines, prickles, cutting leaf edges, trichomes, resins, gums, latex, toxins, raphids and other calcium oxalate structures, silica bodies (phytoliths), protease inhibitors, lignification, stone cells (sclereids), aposematism (visual, olfactory, movement), camouflage, masquerade, mimicry, statistical (rarity, flowering once in many years, mast fruiting), collaboration with other organisms (ants, fungi, bacteria, birds, parasitoids and predaceous arthropods, nurse plants). The main induced ones are: wound cork, thorns, spines, prickles, trichomes, traumatic resin and gum ducts and their products, toxins, PR proteins, protease inhibitors, lignification, silica bodies, jasmonate (inducing plant defenses and attraction of predators and parasitoids) and hypersensitivity (Karban and Baldwin 1997; Lucas et al. 2000; Karban and Agrawal 2002; Barton 2016; Kant et al. 2015; Karban 2015; Schuman and Baldwin 2016).

Elaborating on any one of the defenses mentioned above may result in a massive book dedicated to only that one.

Chapter 4
No Defense Is Perfect and Defense Is Always Relative

Another basic defensive aspect I discuss is that no plant defense system is perfect, an issue already stated long ago (Rothschild 1972; Janzen 1979; Crawley 1983). Assuming that a perfect defense system could exist is naive. All types of defense probably have a cost, although the cost can be mitigated by multifunctionality of some of the defenses and following several types of non-defensive gains by these characters. An organism has to pass through an evolutionary filter and make physiological, developmental and ecological decisions on how much defense can be achieved with the existing resources at any point in time and life cycle (Herms and Mattson 1992). This implies that various defenses must be age, cell type, organ and ecology dependent (see Ochoa-López et al. 2015), the defense has to change with plant ontogeny as a whole, and according to organ age and condition (Boege and Marquis 2005; Barton and Koricheva 2010).

I illustrate the relativity of defense with the following anecdote: Two friends were hiking in an African savanna when a hungry lioness approached them with clear intentions. One of them took light running shoes from his bag and replaced his heavy hiking ones with them. His friend asked "Will you run faster than the lioness?" "No", he answered, "just faster than you". This principle stands behind many types of defense. The small edge gained over other organisms may be the evolutionary drive behind the general phenomenon of defensive coloration or any other type of defense. Indeed, the relative strength of plant's defense against herbivores as expressed in relation to the defense of his neighbors was experimentally demonstrated many times (e.g., Bergvall et al. 2006; Barbosa et al. 2009; Kim and Underwood 2015; Sato and Kudoh 2016). This relativity in plant defense was tested experimentally concerning defensive plant coloration only a few times, e.g., Archetti et al. (2009a) concerning yellow *versus* red autumn leaves. There is therefore a lot to do in this issue.

S. Lev-Yadun, *Defensive (anti-herbivory) Coloration in Land Plants*,
DOI 10.1007/978-3-319-42096-7_4

Chapter 5
Operating Under Stress and Fear in the Military as a Lesson Concerning Difficulties for Herbivory in Nature. Factors That Lower the Need for Perfect Defensive Mechanisms Including Mimicry

Because of the complexity of situations in nature, there is an understandable and even unavoidable need to isolate factors in experimental studies. This is part of the highly fruitful minimization approach that revolutionized many areas in science, including biology. However, the critical need to isolate and examine particular functions, molecules, or aspects, may lead to results that are significant for the experimental conditions, but may sometimes be of limited or of no ecological/evolutionary relevance, or may even be misleading concerning nature's complex reality.

Here I point to some relevant military considerations concerning defense mechanisms that illuminate the risks and hardships that herbivores have to face. The vast and ongoing documented human military experience provides an excellent opportunity to understand functionality under stress of war, which is in many ways comparable to the operation of herbivores under the fear of predation. As a military example of the problems facing a plant predator (herbivore), I discuss wartime submarine operation.

Before the advanced use of radar and other modern electronic aiming devices, submarines could practically view their potential targets at torpedo attack distance for only very short periods of time because of the risk of their periscope or silhouette being discovered by sight by defending submarine-hunting vessels or aircraft. This gave the attacking commander only a very short time to make decisions concerning the type of target vessel, its distance, speed and direction, and in that same short time he had to evaluate the risks of attack by the target vessel itself and also by the defending escort of vessels or aircraft. This, however, is only part of the multitude of difficulties that the underwater attackers had to face. The reality of war (and in this respect in nature *versus* many lab experiments), is nothing like Hollywood war movies with their usually accurate direct hits (see a relevant discussion on the CSI effects in the courtroom in Durnal 2010). The visual capability of submarine commanders was frequently compromised by natural agents such as waves, reflected sunlight, mosaics of cloud cover resulting in lighter and darker sea areas, fog, rain, and snow blizzards. In addition, smoke, fire and explosions further reduced visibility (e.g., Pope 2005). Not less important, extreme fatigue, shock due to close hits by

S. Lev-Yadun, *Defensive (anti-herbivory) Coloration in Land Plants*,
DOI 10.1007/978-3-319-42096-7_5

torpedoes, artillery shells and underwater explosions, malnutrition, general physical unease due to poor hygiene at sea and long exposure to stuffy air, fear of not performing well under fire, and the overall fear of death, being wounded and crippled or taken prisoner and possibly being tortured in captivity, are all factors that commonly cause most soldiers to act far below their highest theoretical potential.

Similarly to soldiers in battle, herbivores, which are usually not well defended against fierce carnivores at close range, operate under various fears (see Brown 1999; Fedriani and Boulay 2006; Preisser 2009; Sheriff et al. 2009; Hawlena and Schmitz 2010; Embar et al. 2011). Rothschild (1984) clearly described the fear factor: "birds are nervous predators and, due to their precarious lives and the need for constant vigilance, they are easily startled and are inclined to take wing hastily" (see also Janzen et al. 2010). Thus, part of the attention of herbivorous animals is always drawn away from their prey, the plants, and this helps the defending plants, which have evolved characters that add various visual difficulties that prolong the time needed for decision making, thus exploiting for defense the herbivore's limited attention. For instance, staying a long time in one place to graze, browse or to find and collect camouflaged seeds increases the risks of predation (Jones et al. 2006). If an herbivore has to spend a longer time in searching or decision making because of a plant's defensive coloration, morphology or other defenses, it may increase the risk of predation, sometimes making it safer to skip the problematic plants or seeds altogether. Accumulating evidence shows that the fear of predation is so significant that it influences herbivore activity in ways that may even significantly change vegetation structure (e.g., Pfister et al. 1990; Nolte et al. 1994; Terlouw et al. 1998; Brown and Kotler 2004; Ripple and Beschta 2004; Fendt 2006; Russell and Banks 2007; Valeix et al. 2009; Ford et al. 2014).

I propose that various types of plant coloration, and especially camouflage and masquerade, may operate in this way because they cause in certain cases a much longer foraging and decision-making time by the herbivores in comparison with plants that do not have these types of visual defenses. Looking for evidence of decreasing predation risks in nature by herbivores, one can indeed see such expected relevant differences in herbivorous caterpillar behavior. Poisonous and visually aposematic caterpillars are known in many cases to consume leaves at their near reach, leaving tattered edges, with no fear or with reduced fear of being detected by birds, whereas non-poisonous caterpillars try to eliminate evidence of foraging by eating the contours and keeping leaf shape intact, snipping partially-eaten leaves, or by moving away from them, and by feeding from the lower side or at night (Heinrich 1979, 1993; Heinrich and Collins 1983). A related aspect is that of limited attention. For instance, experiments with various animal taxa showed that animals are unable or at least have difficulties to simultaneously allocate sufficient attention both to vigilance towards predators and to searching cryptic prey (Dukas and Kamil 2000). Altogether, fear of predation combined with limited attention helps plants to visually defend themselves by crypsis, imperfect mimicry and other ways. I wonder if and suspect that the cruel reality in nature towards herbivores, and the complexity of situations, help plants to escape herbivory even more than what is found in partially or fully out of context laboratory experiments and simulations.

Chapter 6
Evaluating Risk: The Problematic and Even Erroneous Common View of "No Damage or No Attack Equals No Risk"

A theoretical issue related to both the functionality of defensive plant coloration as well as to any other anti-herbivory defense, and to the operation of herbivores under stress, which has been discussed above, is the question of risk evaluation in ecological and evolutionary plant/herbivore studies (e.g., Lev-Yadun 2006b, 2009a; Lev-Yadun and Gould 2007, 2009). There is a general difficulty in understanding this issue, which became clear to me first from reading the literature and later following personal discussions with other scientists and while attending international conferences. In order both to change this common and in many cases erroneous view, and to enhance relevant theoretical and experimental studies, I will elaborate on this issue. This is not just a mere theoretical issue, rather, the understanding of this principle may significantly influence the planning and interpreting of both experiments and actual herbivory data collected in nature. I stress that there are inherent theoretical difficulties in evaluating the quality of defense by experiments, by theoretical modeling, and especially in field observations when the issue of risk evaluation is not considered. A good plant defense may operate so well towards certain herbivores, or the defense of an animal towards predators, that there are almost no attacks on the defended organism, and thus, the experimental or observational noise may be bigger than the signals. A well-defended plant may have so few attacks that the statistical analysis of experiments may be highly problematic if not impossible.

Many scientists find it difficult to accept that "no damage does not automatically indicate that there is no risk". They usually say, no attacks equals no or reduced risk. However, in many cases, lower attack levels indicate just the opposite – that the defense is strong and well-known to potential enemies. A clear case of no attack in spite of a high risk from herbivory was demonstrated by Lev-Yadun and Ne'eman (2004) concerning various summer-green plants growing near Bedouin settlements with their huge flocks of sheep and goats, camels and donkeys in the Negev Desert (Israel). These plants grow in similar arid grazed habitats elsewhere in the Near East and North Africa. Several common alkaloid-rich poisonous or thorny plants form green islands in the dry summer when all surrounding plants in this desert have turned yellow or gray, and in many cases have been grazed down to their roots over

S. Lev-Yadun, *Defensive (anti-herbivory) Coloration in Land Plants*,
DOI 10.1007/978-3-319-42096-7_6

large areas. Even under such extreme grazing pressure, those green plants are ignored by the large herds of sheep, goats, donkeys and camels that pass them daily (Lev-Yadun and Ne'eman 2004). If each individual animal in these large flocks tasted a single leaf once a day, these green plants would disappear in a very short time. The absolute absence of attacks on these summer-green desert plants by the many mammalian herbivores that pass next to them in an otherwise plant-less environment, is a clear indication of their very good defensive and repelling qualities rather than of a low herbivory risk level. Therefore, understanding the principle according to which a low level of attacks indicates in many cases a good defense rather than low risk, is critical for studies of individual species (e.g., Soltau et al. 2009), or for comparative studies involving many taxa (e.g., Archetti 2000; Hamilton and Brown 2001).

The classic study by Soltau et al. (2009) is an excellent demonstration of the issue of the problematic statistics of studying well-defended organisms. Schaefer and Ruxton (2011:165–166) were absolutely correct from the common current statistical point of view when they were cautious with the field data presented in Soltau et al. (2009). However, the reason for a low level of attacks because of various defenses was not considered in depth by Schaefer and Ruxton (2011). This common approach among "statistically correct" ecologists (and there are very many of these) may allow for accepting results only when less-defended organisms are studied, and the role of a strong defense may be overlooked in many cases because of considering them as cases of no risk. For instance, there are fewer lion attacks on elephants, hippopotamus and rhinoceros than on gazelles, gnu and zebras, and even with many years of field observations, the statistics of lion attacks notwithstanding successful ones, on these well-defended herbivorous animals will be extremely low compared to those on the much less-defended herbivore taxa. I think that the statistically problematic, but in my understanding, real and very important results of Soltau et al. (2009) fall into the same category of the low level of lion attacks on elephants, hippopotamus and rhinoceros.

There is a strong need to develop (or borrow from other academic disciplines) suitable statistical tests and make them a standard for testing cases of low frequency events against a huge background noise in ecology. Such an approach was used recently by Menzies et al. (2016) and may help in understanding rare events. Otherwise, only large effects will be agreed on and published, while smaller effects, which are probably much more common and of considerable ecological and evolutionary importance, will not be recognized (see Martínez-Abraín 2008; Gotelli and Ulrich 2012; Mudge 2013). Alternatively, since statistics is not proving anything in any case, and cannot handle various other ecological/evolutionary questions, the actual numbers may be sufficient in such cases. Very small differences in survival rate, which are statistically insignificant, may over many generations result in strong evolutionary advantages (see von Helversen et al. 2013).

Chapter 7
Partial Descriptions of Color Patterns in Floras and Handbooks Has Consequences on the Study of Plant Coloration Biology

A major obstacle to studying defensive or functional plant coloration is the fragmentary and inconsistent descriptions of plant coloration, especially of vegetative organs and fruits. Taxonomists have usually refered only to flower colors and even this character has not always been described in full. Thorn, spine and prickle color, unripe and ripe fruit color, leaf colors (adaxial, abaxial, leaf margins, veins, petioles, when young and when old), bark color and color changes in all these organs were usually not systematically described if at all. The significant progress in understanding the defensive role of coloration in zoology and the basics of the involved genetic mechanisms in a few selected species took over a century (e.g., Majerus 1998; Ruxton et al. 2004; Caro 2005; Hoekstra 2006; Stevens and Merilaita 2011; Zhan et al. 2014), and the effort needed to reach the same level of progress in botany is certainly not smaller, and is probably even greater, because many plant pigments also serve various physiological functions and non-defensive signaling functions.

Lev-Yadun and Gould (2009) emphasized that in spite of all current difficulties to accept, understand and prove defensive plant coloration, there is no reason to continue with the long tradition of botanists of neglecting the study of defensive plant coloration, including aposematism. Moreover, even zoologists studying animal aposematism who studied plant-animal interactions related to herbivory, overlooked this issue. Because of the misunderstanding of the potential defensive functions of vegetative plant coloration (see Lev-Yadun et al. 2002), and only a partial understanding of various physiological functions of the same coloration (see Gould 2004; Archetti et al. 2009a), there is a great difficulty in using published handbooks and floras as a reliable systematic source of data about plant coloration. Several times reviewers of my submitted papers wondered why I didn't use floras to get the information about plant coloration. It was especially annoying in the beginning (e.g., Lev-Yadun 2001) because the reviewers were certain that the data exist, but it does not. Dry plants stored in herbaria lose most of their coloration or express modified colors and are therefore of very limited value for studying plant coloration (Rubino and McCarthy 2004). Grose (2007) in a discussion of the colors of Australian trees phrased the common superficial color description in an excellent

S. Lev-Yadun, *Defensive (anti-herbivory) Coloration in Land Plants*,
DOI 10.1007/978-3-319-42096-7_7

way "green above, paler below" and in a later contribution (Grose 2014) further elaborated on this, demonstrating the dramatic differences between trees from Europe with their bright green leaves and many Australian trees with their grayish-green leaves. The increasing number of electronic resources that include color pictures helps in compiling plant color data, but the gap is still huge. For instance, color polymorphism (an issue commonly neglected concerning wild plants) is known as defense from predation in animals (e.g., Karpestam et al. 2014), but has hardly been treated as such concerning herbivory, with only a few exceptions concerning leaves (e.g., Cooney et al. 2012; Gerchman et al. 2012; Lev-Yadun and Keasar 2012; Niu et al. 2014; Sinkkonen et al. 2012), seeds (Capon and Brecht 1970; Brayton and Capon 1980; Nystrand and Granström 1997; Saracino et al. 1997, 2004; Lev-Yadun and Ne'eman 2013; Porter 2013; Aviezer and Lev-Yadun 2015), spines (Rubino and McCarthy 2004; Lev-Yadun and Ne'eman 2006) and flowers (Irwin et al. 2003).

The direct consequences of the lack of reliable and comprehensive plant color descriptions is that we are in a situation of an intermediate stage of imperfect explanations, a stage that characterizes various parts of this book, which in any case are common in many areas of biology and other sciences. However, these imperfect explanations still allow progress on the issue of defensive plant coloration and may stimulate thinking by other scientists who may first document aspects of plant coloration that have not been documented, and second, develop even better theoretical or experimental ideas than the ones that exist today.

Chapter 8
Animal Color Vision

Herbivores use visual cues and signals as part of their complicated, multifactor way of selecting food plants (Arnold 1966; Krueger et al. 1974; Cahn and Harper 1976; Rausher 1978; Prokopy and Owens 1983; Reeves 2011). When discussing defensive plant coloration or any visual plant defense, it should be clear that the relevant herbivorous animals can see such visual cues, signals, deceptive coloration of various types, and distinguish (or not in crypsis) between these and backgrounds. It has been hotly debated whether mammalian or insect herbivores see colors or not, and if they do, which colors and to what degree. Reviews of this issue have provided conclusive proof that they do see colors (Jacobs 1993; Dafni et al. 1997; Kelber 2001; Kelber et al. 2003). While most large mammalian herbivores don't see colors the way trichromatic humans do, there are still good reasons to conclude that even with the dichromatic vision characterizing many large mammalian herbivores, colorful plant parts look different from regular green tissues because of their hue, saturation, contrast, or brightness (see Sumner and Mollon 2000a, b; Kelber et al. 2003). The herbivores may also use reflectance cues of the colorful plant parts (see Schaefer et al. 2006; Motoyoshi et al. 2007). This view is supported by the case of "color-blind" cuttlefish that successfully camouflage themselves in various backgrounds (Marshall and Messenger 1996) and by the importance of the luminance contrast component of aposematic and other types of defensive plant coloration even for color-blind animals (Prudic et al. 2007). In any case, colors can be seen by animals both as chromatic (wavelength related) signals and achromatic (intensity related) signals (e.g., Schaefer et al. 2006) and this issue should also be studied on a larger scale in the context of visual plant defenses. Herbivore's character of not seeing part of the spectrum may also be used by plants for defensive coloration. For instance, red appears black or very dark to certain insects. In such cases, "black" red coloration can be used to look dead, for mimicry, camouflage etc., while that red coloration can be used for signaling to animals that see red, or for physiological functions.

S. Lev-Yadun, *Defensive (anti-herbivory) Coloration in Land Plants*,
DOI 10.1007/978-3-319-42096-7_8

The UV Spectrum and Defense from Herbivores

The physiological role of trichomes and waxes in defending plants from excess sun irradiation in the UV spectrum is well known (e.g., Fahn and Cutler 1992; Grammatikopoulos et al. 1994; Ballaré 2003; Manetas 2003; Paul and Gwynn-Jones 2003). Many insects and birds, as well as some other vertebrates (Briscoe and Chittka 2001; Honkavaara et al. 2002; Shi and Yokoyama 2003) see in the UV-A (400–315 nm). Since birds are major insect predators (Heinrich 1993; Honkavaara et al. 2002; Van Bael et al. 2003; Lichter-Marck et al. 2015) an ecological role for the way insects and birds see plants in the UV band should be investigated. Reflection or absorption of UV-A from leaves should theoretically influence the behavior of insects that may choose to land and lay eggs on or consume the plants, or on the contrary, may decide not to land on visually unsuitable surfaces. When the issue of undermining herbivorous insect camouflage by plant coloration (Lev-Yadun et al. 2004a; Lev-Yadun 2006a, b, 2009a) is considered, the possibility that this principle also operates in the UV-A should be taken into account. Lev-Yadun (2006b) proposed just on a theoretical basis, with no actual data, that certain insects may choose to occupy or not to occupy light-colored coastal and dune plants on the basis of their appearance in the UV-A spectrum. The absorption and reflectance of plant surfaces in the UV-A, especially concerning herbivory and herbivore predation, remains largely unexplored (Paul and Gwynn-Jones 2003). Sinkkonen (2009) proposed that UV reflectance should be considered concerning the defensive coloration of autumn leaves because many insects see in the UV spectrum. Archetti et al. (2009b) agreed that how leaves look in UV should be explored, but explained that leaves should be studied as a whole and not as extracted molecules, because there is a great difference between extracted phytochemicals and whole leaves, especially since certain molecules found in leaves may reflect UV, but other may absorb it. Archetti et al. (2009b) showed that measurements of UV leaf reflectance in autumn leaves of more than 2400 species revealed that 99 % of them have a maximum UV reflectance of <8.6 %, a low value compared to general leaf reflectance, and compared to >30 % in flowers. Therefore, in addition to the probable UV-related physiological roles of the white or silvery coloration of many coastal and dune plants discussed in Lev-Yadun (2006b), an anti-herbivory role of this coloration in the UV-A spectrum should be considered and studied.

Chapter 9
The Nature of Signals

Visual plant defenses can be based on being cryptic, but can also be based on the opposite strategy, i.e., on being conspicuous and even by signaling. To make things more complicated, certain coloration patterns may serve for crypsis at long distances, but signal at close range, an issue that will be discussed in some detail in the chapter about aposematism. Maynard Smith and Harper (2003) defined signals "as any act or structure which alters the behaviour of other organisms, which evolved because of that effect, and which is effective because the receiver's response has also evolved". The recent increasing interests and debates on the signaling functions of red and yellow autumn leaves (e.g., Archetti 2000, 2007a, b, 2009a, b; Hamilton and Brown 2001; Holopainen and Peltonen 2002; Wilkinson et al. 2002; Hagen et al. 2003, 2004; Archetti and Brown 2004; Schaefer and Wilkinson 2004; Lev-Yadun et al. 2004a, 2012; Archetti and Leather 2005; Brown 2005; Schaefer and Rolshausen 2006a, b, 2007a, b; Sinkkonen 2006a, b, 2008; Chittka and Döring 2007; Lev-Yadun and Gould 2007; Schaefer and Gould 2007; Hatier and Gould 2008; Karageorgou et al. 2008; Ougham et al. 2008; Ramirez et al. 2008; Yamazaki 2008a, b; Archetti et al. 2009a; Lev-Yadun and Holopainen 2009, 2011; Lev-Yadun 2010; Sinkkonen et al. 2012) that will be discussed in detail in a later chapter are an excellent demonstration of the complicated experimental, physiological and interpretative issues that have to be dealt with when defensive visual plant signals are involved.

The situation of a single character having several functions is quite common in biology in general, and concerning plant coloration in particular. In spite of the strict definition of signals by Maynard Smith and Harper (2003), this is also true concerning visual plant signals. Therefore, there is no theoretical reason to dismiss simultaneous roles of pigments, pigment combinations, or morphology in both physiological functions and signaling (i.e., Endler 1981). There is probably a gradient between visual signals that serve solely or mostly as signals and those that mostly serve other functions with only some contribution to visual signaling. There is also no reason to ignore the possibility that signals that currently serve only for defense might have evolved primarily for other functions, or *vice versa*. The assumption that signals

S. Lev-Yadun, *Defensive (anti-herbivory) Coloration in Land Plants*,
DOI 10.1007/978-3-319-42096-7_9

serve only when they are costly (e.g., Zahavi 1975, 1977) and never contribute to other functions, implies that their evolution either ended or that such a trait evolved at once, in a single step. Such a scenario seems to be an over simplification of a much more complicated evolutionary and current functional situation. We are much too far from a full or even a good understanding of signal evolution (see Endler 1986; Karban 2015) according to any type of definition (and there are alternative ones to Maynard Smith and Harper 2003) to adopt such inflexible criteria (Lev-Yadun 2005). Augner (1994) proposed that any cue that an herbivore can use to discriminate between defended and non-defended plants is a signal (but see a different view by Schaefer and Rolshausen 2006a, b). I think that when all known facts about plant signaling are considered together, that there is no single definition that can fully cover all plant signals. I think that it is probable that efforts to reach such a definition are futile.

A very important if not critical issue in many cases of defensive plant signaling is the probable involvement of olfactory signaling along with the visual component. Out of the very many volatile molecules that plants omit, it is very difficult to identify the specific molecules that may deter specific herbivores, and then there is the possibility that the deterrence operates only when several molecules are sensed simultaneously or that they influence only certain herbivores but not others. A classic example of this is the repellence of various herbivores by odors emitted by members of the Brassicaceae that simultaneously attract specialized insect herbivores (Renwick 2002). The fact that not all animals sense and therefore respond similarly to any chemical signal or visual cue should thus also be considered. For instance, from studies of deceptive pollination we know that signaling to animals may involve both visual and olfactory components (Dafni 1984; Ayasse et al. 2000; Schiestl et al. 2000; Jersáková et al. 2006), although the olfactory component seems to be much more important, and there are good reasons to think that a differential response may be true in defense. I think that even with the current technological and theoretical rate of progress in these issues, but with the dramatic global decline in university botanical education, even many decades of intensive research will not be sufficient to allow a safe general definition of visual plant signaling.

Chapter 10
White as a Visual Signal

Concerning defensive signaling by white marking, two evolutionary issues should be specifically mentioned: first, the significance of white marking in contrast to colorful marking and second, the probable convergence of this character as expressed by geography and taxonomy. When a certain herbivore is color-blind to a certain sector of the spectrum or is color-blind altogether; or when illumination is not strong or it has an altered color temperature for some meteorological, ecological, or daytime related reason, then white marking increases the possibility that a signal will still be visible. As will be discussed below in several chapters, defensive white marking seems to be common in many plant species originating in various taxa, continents, ecologies, life forms and in many plant organs (see Lev-Yadun 2001, 2003a, 2014a, b, c, d).

White variegation has been proposed to have several functions that can compensate for the assumed or sometimes proven reduced photosynthetic ability of white tissues (Cahn and Harper 1976; Givnish 1990; Allen and Knill 1991; Lev-Yadun 2001, 2003a, 2006a, 2009a, b, c, 2013a, 2014a, b, c, d). However, in many species, white variegation is the result of air spaces formed by lack of adhesion between the epidermis and the photosynthetic parenchyma (Hara 1957; Scarchuk and Lent 1965; Tsukaya et al. 2004), a type of coloration probably not associated with a loss of photosynthetic capacity (Konoplyova et al. 2008). Several studies suggested or even showed in experiments and field work that white variegation or white pigmentation confirms some defense from herbivory (e.g., Cahn and Harper 1976; Wiens 1978; Shifriss 1981; Niemelä et al. 1984; Smith 1986; Givnish 1990; Agrawal and Spiller 2004; Lev-Yadun 2001, 2003a, 2006a, b, 2013a, 2014a, b, c, d; Lee 2007; Campitelli et al. 2008; Soltau et al. 2009; La Rocca et al. 2014), but the exact mechanisms by which it operates as defense from herbivory are only partly understood.

S. Lev-Yadun, *Defensive (anti-herbivory) Coloration in Land Plants*,
DOI 10.1007/978-3-319-42096-7_10

Chapter 11
Visual Signaling by Plants to Animals *via* Color

Colors play an important role in plant/animal communication. The best-known visual communication system is color *per se* and color patterns (including in the UV) serving as an advertisement to attract various animals, which serve as their pollinators for their flowers, hence as gene-dispersing agents as they transfer pollen among flowers (Darwin 1877; Faegri and van der Pijl 1979; Clegg and Durbin 2003; Schaefer and Ruxton 2011; Renoult et al. 2014). The next best known system is fruit colors, which alert seed-dispersing frugivores to the ripening stage of the fruits (van der Pijl 1982; Willson and Whelan 1990; Schaefer et al. 2007, 2014; Schaefer and Ruxton 2011; Stournaras et al. 2013; Renoult et al. 2014; Karban 2015). The third system is the phenomenon of fruit flags, namely, bright autumn leaves, which may signal to frugivores the existence of fruits (Stiles 1982; Facelli 1993), a debated hypothesis (e.g., Willson and Hoppes 1986; Burns and Dalen 2002). The fourth system is found in carnivorous plants whose colorful patterns, in both visible and UV light, attract insects to their deadly traps (Joel et al. 1985; Moran et al. 1999; Schaefer and Ruxton 2008). The fifth system is the conspicuous colors of autumn leaves, found in many autumn deciduous trees and shrubs, which have been proposed as a signal of a tree's vigor to its parasitic insects, especially aphids, and thus a case of Zahavi's handicap principle operating in plants (Archetti 2000; Hamilton and Brown 2001; Archetti and Brown 2004; Archetti et al. 2009a), or aposematism and other types of defensive signaling (Lev-Yadun 2006a; Lev-Yadun and Gould 2007; Archetti et al. 2009a; Lev-Yadun and Holopainen 2009). The sixth system is a variety of color patterns enabling plants to mimic insects so as to reduce herbivory: mimicry of butterfly eggs (Benson et al. 1975; Shapiro 1981a, b; Williams and Gilbert 1981), or of ants, aphids, beetles, spider webs, eyes, snakes, and aposematic caterpillars (Lev-Yadun and Inbar 2002; Lev-Yadun 2009d, 2015a; Yamazaki and Lev-Yadun 2014, 2015; Aviezer and Lev-Yadun 2015). The seventh system is red leaves mimicking dead or old ones or just looking different from young ones (Stone 1979; Lee et al. 1987; Juniper 1994; Richards 1996; Dominy et al. 2002). The eighth system is young colorful leaves that attract herbivores and divert them from the more costly older ones (Lüttge 1997). The ninth system is delayed greening of

S. Lev-Yadun, *Defensive (anti-herbivory) Coloration in Land Plants*,
DOI 10.1007/978-3-319-42096-7_11

young leaves in the tropics (Kursar and Coley 1992; Coley and Barone 1996; Numata et al. 2004). The tenth system is based on leaves that have different coloration on their adaxial and abaxial sides, or other kinds of non-variegated color patterns in other plant parts, which undermine herbivorous insect camouflage and was proposed to deter invertebrate herbivores (Lev-Yadun et al. 2004a; Lev-Yadun 2006a, b, 2009a; Lev-Yadun and Gould 2007, 2009). The eleventh system is of dazzle and trickery coloration (Lev-Yadun 2014c). Altogether there is a wealth of known coloration types by which plants signal to animals true or false information (e.g., Faegri and van der Pijl 1979; van der Pijl 1982; Lee 2007; Lev-Yadun 2009a; Schaefer and Ruxton 2011; Karban 2015) including some that are not listed here. However, it should be remembered that there are very many aspects of these complicated color systems that are not even well described, not well tested (if at all), (see Schaefer and Ruxton 2009, 2011), and that even after more than 150 years of systematic studies of parallel aspects in zoology, we are still very far from a comprehensive understanding of such aspects in zoology. Concerning plant coloration we are still at the tittering pain stage, very very far behind even the imperfect understanding in zoology.

Chapter 12
Müllerian and Batesian Mimics Are Extended Phenotypes

Aposematic signaling (by coloration, odor, texture, shape, movement, auditory) is a well-known defense across the animal and plant kingdoms, and the current hypothesis about its wide occurrence is that there is a selective advantage to this trait. Aposematic signaling evolved because predators and herbivores learn to associate it with bad taste, danger, poison, spines or unprofitable handling (Poulton 1890; Cott 1940; Edmunds 1974; Gittleman and Harvey 1980; Ruxton et al. 2004; Mappes et al. 2005; Lev-Yadun 2009a). Aposematism and mimicry of aposematic animals, especially by butterflies, were one of the first demonstrations of evolution already in Darwin's lifetime or shortly after (Bates 1862; Müller 1879; Wallace 1889). The detailed field data collected by Bates in South America showed that chemically defended (bitter) butterflies of different taxa have similar coloration patterns (later defined as Müllerian mimicry) and that chemically non-defended ones may look like defended ones and by this gain protection (later defined as Batesian mimicry), being defense parasites. Müller (1879) gave a mathematical explanation for the phenomenon, showing that use of the same signal by several taxa may lower the cost incurred on the defended and signaling prey by unexperienced predators that attack till they learn to avoid defended and signaling types. It was later shown that increasing the number of Müllerian mimics indeed helps to reduce the cost of learning by predators, even if the various species differ in their strength of defense (Rowland et al. 2007; Sherratt 2007). It has also been shown that association of non-defended types with defended ones defends the former, a situation that may in due time result in the evolution of Batesian mimicry (de Wert et al. 2012).

The concept of the extended phenotype was introduced by Dawkins (1982) concerning animals. According to this concept, an animal's behavior tends to maximize the survival of its genes whether or not those genes happen to be in the body of the particular animal performing it, or if the phenotype is expressed outside its body (Dawkins 1982; Schaedelin and Taborsky 2009). However, Dawkins (1982) did not mention aposematism based on signals and defense mechanisms derived from other organisms as a case of an extended phenotype, in spite of the fact that many aposematic animals acquire by feeding both the defensive chemicals and the signaling

S. Lev-Yadun, *Defensive (anti-herbivory) Coloration in Land Plants*,
DOI 10.1007/978-3-319-42096-7_12

components of their aposematism that together make them aposematic from plants or from their animal prey (e.g., Rothschild 1972). Similarly, in the best, very detailed update on aposematism (Ruxton et al. 2004), aposematism is not referred to as an extended phenotype, and similarly, in a recent review of extended phenotypes as animal signals (Schaedelin and Taborsky 2009), aposematism is not mentioned. Surprisingly, several recent monographs discussing visual and other defenses (Caro 2005; Lee 2007; Stevens and Merilaita 2011; Diamond and Bond 2013; Stevens 2013; Farmer 2014) also did not mention aposematism as an extended phenotype. Only colorful and chemically defended plant galls (abnormal plant growths induced by various organisms) were recently proposed to be the extended aposematic phenotypes of their inducing aphids (Inbar et al. 2010a). The aposematic signaling about the gall's defense has two components: (1) many galls are colorful, red, yellow, etc., colors produced by host's tissues, and the galls can be clearly distinguished by color and shape from the other parts of their host plant, which can serve in visual aposematism, and (2) some of the secondary metabolites found in galls are volatiles, also produced by the host's tissues, and may thus serve as olfactory aposematic signals about the chemical defense (Rostás et al. 2013; Rand et al. 2014). This plant-related extended phenotype, functions, however, to defend these induced abnormal plant parts (galls) from enemies of the gall-inducing organisms (Inbar et al. 2010a).

I conclude that since both Müllerian and Batesian mimics of all taxonomic groups use phenotypes of other individuals and taxa for defense, both types of mimicry are cases of extended phenotypes, an overlooked aspect that deserves much more research attention. The issue of extended phenotype will be mentioned several times in this book.

Part II
Gentle Defenses

Chapter 13
Camouflage

Camouflage is the most common visual defense in the animal kingdom (Stevens and Merilaita 2011) and potentially the best of all defenses. If an organism is well camouflaged its enemies do not see or sense it *via* other modalities and do not even consider an attack. Camouflage, when functioning well, is therefore the ultimate way to avoid attack.

Crypsis is usually a primary defense, common in many animal taxa (both vertebrates and invertebrates) of marine, and terrestrial, including tropical, subtropical, desert, temperate and arctic habitats (Cott 1940; Kettlewell 1973; Edmunds 1974; Endler 1978, 1984; Stamp and Wilkens 1993; Merilaita 1998; Lev-Yadun et al. 2004a; Ruxton et al. 2004; Caro 2005; Stevens and Merilaita 2011). Crypsis is commonly visual, but can also be chemical or immunological. Crypsis by coloration, surface pattern and morphology defends against visually-oriented predation or herbivory by making prey organisms indistinguishable from their background, by being at least less conspicuous, or by causing difficulties in recognizing their shape because of disruptive coloration. Since the natural environment is usually heterogeneous, there are several ways to be visually cryptic. It is possible to: (1) resemble the color of a specific microhabitat or substrates, (2) make a compromise of several colors of several microhabitats or substrates, or (3) have a disruptive coloration patterns that break up body outlines, (4) use counter shading, and (5) various combinations of the above (Thayer 1918; Cott 1940; Edmunds 1974; Merilaita 1998; Ruxton et al. 2004; Caro 2005; Stevens and Merilaita 2011; Penacchio et al. 2015a).

Certain animals have a flexible crypsis strategy to match various background colors or patterns. Chameleons may change their color according to their background within minutes, but flatfish can do it within seconds (Ramachandran et al. 1996). Various invertebrates cover themselves with substrate particles to become cryptic (e.g., Edmunds 1974; Eisner et al. 2005), and it has been proposed that certain plants do the same (Wiens 1978; Jürgens 1996; Lev-Yadun 2006b; LoPresti and Karban 2016). In certain lepidopterans, food content and quality are key factors determining larval color and morphology, which enhance crypsis. Green's (1989) classic study demonstrated the importance of intra-plant variation in caterpillar

S. Lev-Yadun, *Defensive (anti-herbivory) Coloration in Land Plants*,
DOI 10.1007/978-3-319-42096-7_13

coloration. On a given oak (*Quercus* sp.) host, caterpillars of *Nemoria arizonaria* may develop into a 'catkin' or 'twig' morph to maximize their camouflage. Some stick insects (Phasmatodea) may adjust their color according to changes in the color of their host-plant foliage (Bedford 1978).

Since many animals see colors and distinguish between shapes, they select their habitat accordingly for better visual protection, i.e., they avoid habitats non-suitable for crypsis. This issue will be discussed in detail in the chapter about plants that undermine herbivore's camouflage. There are both theoretical and experimental indications that all these strategies might operate (e.g., Cott 1940; Kettlewell 1973; Edmunds 1974; Endler 1978, 1984; Merilaita et al. 1999, 2001; Ruxton et al. 2004; Merilaita and Lind 2005; Schaefer and Stobbe 2006; Stevens et al. 2006; Stevens and Merilaita 2011). Masquerade, namely camouflage without crypsis, is also possible (e.g., Skelhorn et al. 2010a; Skelhorn 2015) and will be discussed separately.

Unlike in zoology, where camouflage has been studied regularly, and practiced by humans while hunting animals and in warfare and even by criminals, the possible defensive advantage of camouflage of whole plants, or even for obvious targets such as seeds lying on the ground was never an important issue in botany. The issue of camouflage as defense against herbivory in plants in general and of their canopy in particular has received very little attention (e.g., Cole 1970; Barlow and Wiens 1977; Wiens 1978; Benson 1982; Givnish 1990; Allen and Knill 1991; Cole and Cole 2005; Lev-Yadun 2006a, b; Klooster et al. 2009; Fadzly et al. 2009; Burns 2010; Fadzly and Burns 2010; Schaefer and Ruxton 2011; Lev-Yadun and Ne'eman 2013; La Rocca et al. 2014; Niu et al. 2014; Farmer 2014) and even this limited attention was sometimes anecdotal.

When background matching is discussed, Cole (1970) and Cole and Cole (2005) described how difficult it is to distinguish in the wild between pebbles and *Lithops* plants because of their excellent resemblance to the stone-covered ground. Wiens (1978) in an essay about mimicry in plants, which was usually cited concerning deceptive pollination in orchids, proposed that light coloration, achieved by sand attached to sticky glandular trichomes, may visually camouflage plants from herbivores, an issue discussed further concerning Namib desert plants (Jürgens 1996) and coastal and desert plants of the Near East by Lev-Yadun (2006b). Benson (1982) proposed that members of the cactus genus *Ariocarpus* (living rocks) are visually well hidden among the limestone desert pavement. A different method of camouflage was proposed by Givnish (1990), who focused on the common phenomenon of leaf variegation in understory herbs in the forests of New England, suggesting that it serves as camouflage since the understory light regime is characterized by sun-flecks (Figs. 13.1 and 13.2), and that the variegation may act as disruptive coloration.

In the last decade, much more attention has been given to plant camouflage (e.g., Lev-Yadun 2006a, b, 2013a, 2014a, b; Klooster et al. 2009; Fadzly et al. 2009; Burns 2010; Fadzly and Burns 2010; Lev-Yadun and Ne'eman 2013; Porter 2013; Strauss and Cacho 2013; Farmer 2014; La Rocca et al. 2014; Niu et al. 2014; Niu and Sun 2014; Aviezer and Lev-Yadun 2015). Klooster et al. (2009), Fadzly et al. (2009), Lee and Gould (2009), Fadzly and Burns (2010), Lev-Yadun and Ne'eman

Fig. 13.1 Light spots on leaves in a broad-leaf forest in southern Germany. The very bright light spots make the non-illuminated parts of leaves almost invisible

Fig. 13.2 Variegated leaves of *Hypoestes phyllostochya* (polka dot plant) growing in the botanical garden of Edmonton (Canada). It is difficult to figure where the edges of the leaves are

(2013), (Niu et al. 2014), and Strauss et al. (2015) also compared the spectral reflectance of the cryptic plants relative to their natural background, and the results supported the crypsis hypothesis well. Klooster et al. (2009), Strauss and Cacho (2013), and Niu et al. (2014) even showed that the camouflage increased the plants' fitness, because they produced more fruits and seeds than non-camouflaged ones that suffered more damage from herbivores.

Concerning camouflage as defense on whole shoots, special attention was given to mistletoes (Barlow and Wiens 1977; Pasteur 1982; Ehleringer et al. 1986; Canyon and Hill 1997; Blick et al. 2012; Pannell 2014). Barlow and Wiens (1977) proposed that species belonging to various Australian mistletoe genera (*Amyema*, *Lysiana*, *Muellerina*, *Diplatia*, and *Dendrophthoe*) of the Lorantaceae have a vegetative similarity to their usual host especially in the leaves, and especially of the host genera *Eucalyptus*, *Acacia*, and *Casuarina*. Barlow and Wiens (1977) proposed that out of 64 Australian mistletoe species, 36 are potential visual mimics. From the pictures given in their paper it is clear that the mistletoe *Amyema cambagei* on *Casuarina torulosa*, *Amyema quandang* on *Acacia cambagei*, *Lysiana murrayi* on *A. aneura*, *Amyema mackayense* on *Rhizophora* sp., *Amyema sanguineum* on *Eucalyptus crebra*, *Amyema pendulum* on *E. pauciflora*, and *Dendrophthoe homoplastica* on *Eucalyptus shirleyi* are indeed very good visual (morphological) mimics. Barlow and Wiens (1977) discussed but dismissed the possibility that simple convergence because of physiological issues was involved, and posited that defense from herbivores *via* crypsis (crypsis mimicry), especially against the common possum, is involved. Pasteur (1982) proposed that since the mistletoes are much more edible than their hosts, it is a typical case of Batesian mimicry. Ehleringer et al. (1986) tested the mistletoe mimicry hypothesis by comparing the levels of nitrogen in mistletoes and their hosts in 48 pairs of proposed mimetic and non-mimetic species. Their finds supported two hypotheses, the first, that on average, mimetic mistletoes have higher leaf nitrogen content than their hosts (not without exceptions), and the second, that non-mimetic mistletoes have on average a lower leaf nitrogen content than their hosts (also not without exceptions). Ehleringer et al. (1986) proposed that the nitrogen-rich mimics use visual crypsis for defense, while many of the nitrogen-poor non-mimics advertise their low quality by looking different from their hosts. Bannister (1989) tested Ehleringer et al.'s (1986) mimicry/nitrogen hypothesis in 41 pairs of mistletoe species from New Zealand and their respective hosts. He found that in New Zealand, cryptic (mimetic) mistletoe species have nitrogen concentrations that generally equal or exceed those of their hosts, whereas the non-cryptic mistletoe species, in particular those growing on small-leaved divaricate shrubs, have generally lower nitrogen concentrations than their hosts, supporting Ehleringer et al.'s (1986) mimicry/nitrogen hypothesis. I think that the low-nitrogen non-mimic (non-cryptic) mistletoes of both Australia and New Zealand should be considered aposematic. Canyon and Hill (1997) tested Ehleringer et al.'s (1986) mimicry/nitrogen hypothesis in two mistletoe/host pairs, one of mimic/host and the other non-mimic/host, but found no support for the Ehleringer et al.'s (1986) mimicry/nitrogen hypothesis in their much too small sample. Blick et al. (2012), who examined the mistletoes/host interactions by network topology of almost all Australian mistletoes, found only a partial support for the mimicry hypothesis. Since many of the mistletoes were never considered to be mimics but sampled anyway, the ratio of noise to signal in such an analysis seems to me to be too large. Pannell (2014) in his short review of the issue supported the mistletoes mimicry hypothesis, although with exceptions. Concerning the flora of Israel, the mistletoe *Viscum cruciatum* visually resembles its major host *Olea europaea*, but *Loranthus acaciae* does not

Fig. 13.3 A group of *Otanthus maritimus* plants in the sandy shore north of Tel Aviv, Israel. It is difficult to see these almost white plants from a distance. Being non-green they probably attract fewer visually-oriented herbivores

resemble any local member of the genus *Acacia* or other plant taxa it parasitizes. I conclude that this issue of visual crypsis in mistletoes is a classic case of a biological solution employed by certain relevant taxa but not by all.

Lev-Yadun (2006b) proposed that the white or whitish colors of many Near-Eastern coastal and desert plants, and plants that stick sand particles (Figs. 13.3, 13.4, 13.5 and 13.6), clay, or dust to their surface are probably camouflaged. Klooster et al. (2009) showed by removing the bracts that camouflage *Monotropsis odorata* plants in natural populations, that the non-cryptic treated plants suffered 20–27 % more herbivore damage, and 7–20 % lower mean fruit production. The camouflaging bracts strongly resemble ambient leaf litter both visually and under spectral analysis.

Fadzly et al. (2009), Burns (2010), and Fadzly and Burns (2010) studied leaf color as camouflage against the recently extinct herbivorous moa birds in New Zealand. In several species, i.e., *Pseudopanax crassifolius* and *Elaeocarpus hookerianus*, the leaves of young and low plants are cryptic, and when the plants attain the height of about three meters and their leaves are out of reach of moas, they cease being cryptic, and even have conspicuous coloration that advertises their spines (Fadzly et al. 2009). Burns (2010), in a short review of plant camouflage focusing on the New Zealand flora, further described ontogenetic changes from camouflaged leaves when the plants are young and low to conspicuous aposematic ones when the plants are much larger and older, as defense from the recently extinct moas. He also described camouflage in scree plants that have gray leaves matching the color of their stony habitat, an issue later studied in detail in scree plants in China (Niu et al. 2014; Niu and Sun 2014). Burns (2010) also discussed the difficulties involved in studying plant crypsis, including the problem of extinction of large herbivores that

Fig. 13.4 A close-up of a plant of *Otanthus maritimus*. In addition of being non-green, the leaves accumulate sand that can help in camouflaging the plants and also act as sand-paper on mouth parts of arthropods and teeth of vertebrates to reduce herbivory

Fig. 13.5 A group of small *Silene succulenta* plants growing in the sandy coast of Mount Carmel, Israel. These very sticky plants are covered by many sand particles that camouflage them to reduce herbivory. The sand may also act as sand-paper on mouth parts of arthropods and teeth of vertebrates if attacked

were the probable selection agents of certain defenses, a problem that may increase with the recent and current growing rate of full or even just near extinction of many large mammalian (Johnson 2009; Daskin et al. 2016) and avian herbivores (Burns 2010) in many parts of the world.

Strauss and Cacho (2013) also showed that *Streptanthus breweri* growing on bare serpentine habitats, where plant apparency is great, use leaf color that matches soil color as camouflage. They showed that when plants were transplanted to a soil that mismatches leaf color they suffered from increased herbivory. Similarly to

Fig. 13.6 Two small *Silene succulenta* plants growing in the sandy coast of Mount Carmel, Israel. These very sticky plants are covered by many sand particles that not only camouflage them to reduce herbivory but also act as sand-paper wearing moth parts of arthropods or teeth of vertebrates in case they are seen and eaten

sandy habitats (Lev-Yadun 2006b) and bare serpentine habitats (Strauss and Cacho 2013), alpine plants in China that grow on scree, and match the gray color of that habitat, suffer less herbivory than green types and produced more seeds, i.e., had a higher fitness (Niu et al. 2014; Niu and Sun 2014). La Rocca et al. (2014) described the variegated leaves of *Erythronium dens-canis*, an early-flowering understory lily growing on forest floors in southern Europe. The leaves are mottled with silvery flecks and brown patches. The brown patches, originating in vacuolar anthocyanins, gradually change into green. The authors suggested that the variegation as a whole serves both as camouflage from herbivores and also for attraction of pollinators. Aviezer and Lev-Yadun (2015) proposed that the pods of the climber legume *Pisum fulvum*, which are ornamented with red/brown/black blotches, are by this camouflaged within the canopy.

Patterns of coloration that may serve as camouflage may simultaneously serve other functions. Lev-Yadun (2013a) showed that the unripe fleshy fruit of many species originating in Asia (southeastern, eastern and central), the Middle East, Africa, America (South, Central and North), Australia, Polynesia and Micronesia that are larger than 1 cm, have white or light green spots (Figs. 13.7, 13.8 and 13.9). Lev-Yadun (2013a) proposed that the white or whitish variegation in these unripe fleshy fruits may serve at least five functions: (1) windows for photosynthesis, (2) camouflage, (3) signaling to frugivores that they are not ripe (possibly sometimes a type of mutualism with territorial frugivores), (4) signaling to frugivores that they are poisonous, or have a low nutritive value (aposematism), and (5) mimicking insect eggs to reduce egg laying. All these functions may be partly or fully simultaneous. The

Fig. 13.7 Immature toxic fleshy green fruits of *Bryonia* sp. typically marked with white spots. Hundreds of species with unripe fleshy green fruits from all vegetated continents express this character

Fig. 13.8 Immature toxic fleshy green fruits of *Ficus drupacea* typically marked with *white* spots. Riper fruits in *orange* and later *brown* do not show such *white* spots

role of variegation in plant camouflage *sensu* Givnish (1990) in leaves of understory plants seems to fit the possible role of white variegation in unripe green fleshy fruits growing among the leaves in dense canopies. While Givnish's (1990) hypothesis

Fig. 13.9 Immature toxic fleshy green fruits of *Juglans regia* typically marked with *white* spots. Ripe fruits are *black*

[and its later ramification (Lev-Yadun 2013a)] has to be tested carefully before anything significant can be said about it (Schaefer and Ruxton 2011), strong theoretical support for fruit or leaf mottling as camouflage within and under plant canopies emerges from zoology, since many mammals inhabiting forest habitats use coat mottling as camouflage (Ortolani 1999; Stoner et al. 2003a, b; Caro 2005, 2009; Allen et al. 2011; Bar-Oz and Lev-Yadun 2012).

Chapter 14
Seed Camouflage

Seeds must enjoy protection by visual camouflage when they are dispersed on the ground or on fallen leaves. According to my field experience, the seeds of many species show excellent background matching or disruptive coloration. However, even for seeds, only a few studies of their visual camouflage have been conducted.

During my field work on seed camouflage of several members of the genus *Lupinus* growing in Israel, especially *L. luteus* in habitats with red sandy Hamra soils in the coastal plain, and *L. micranthus* in habitats with gray Rendzina soils in the carbonate mountains of the upper Galilee, I had very great difficulty to find their highly cryptic, several mm in diameter roundish seeds, even when sitting on the ground, at a range that allowed me to spot and examine even minute soil particles. *L. luteus* has light-colored seeds ornamented with a dark crescent on their flat sides (Fig. 14.1), and *L. micranthus* seeds are brown and mottled with orange spots (Fig. 14.2). It usually took me some 10–20 min to find the first seed near the mature plants with their open pods that indicated that their seeds were dispersed, and I have looked dozens of times at the exact cm^2 where the seeds were lying before I could see them. Granivores such as rodents and birds cannot stay for a long time in the same spot without being exposed to severe risks of predation (e.g., Jones et al. 2006), as well as the energetic cost of a very long search *versus* the negative or small return.

The possibility of olfactory seed camouflage as the primary reason for the evolution of what is known as physical dormancy in seeds, was suggested by Paulsen et al. (2013, 2014, 2015), a hypothesis that met fierce opposition (Jayasuriya et al. 2015). The data and discussions presented in Paulsen et al. (2013, 2014, 2015) indeed show clearly that olfactory crypsis of seeds against rodent seed predators is important under various conditions and concerning various taxa, but how general is this phenomenon is still unknown. I think that there are indeed situations and taxa where olfactory seed camouflage does operate, and that Paulsen et al. (2013, 2014, 2015) provided sufficient data to demonstrate it. However, following the small number of studied cases, I am not certain that it is always so, and I am certain that the classic hypothesis about the value of plasticity in germination time is not an outdated

S. Lev-Yadun, *Defensive (anti-herbivory) Coloration in Land Plants*,
DOI 10.1007/978-3-319-42096-7_14

Fig. 14.1 *Lupinus luteus* seeds as seen on the soil. Ten seeds are seen (two out of focus) in this closeup, when the distance of inspection is not so close, and when plants or plant parts cover or shade the soil, the camouflage is simply perfect. After seed dispersal, it took me many minutes to find the first seed on the ground

Fig. 14.2 *Lupinus micranthus* has light-colored seeds ornamented with a *dark* crescent on their flat sides. It took me many minutes to find the first dispersed seed and not less time for the following ones. Whenever a seed dropped from my hand to the soil, I could not find it again

naive hypothesis. For me, the data in hand is not indicative enough to decide if olfactory crypsis as the primary, the secondary, or of equal importance for the evolution of this trait along with securing the genotype in arid zones as the classic hypothesis suggests (e.g., Rolston 1978; Baskin et al. 2000). Operating in the arid Near East, avoidance of dessication of seedlings because of a high germination rate following the first significant rainfall, followed by the common interrupted rain season, seems to me to be the primary reason for the evolution of this character, but the Near East is an arid zone with unpredictable patterns of winter rainfall and no summer rains, an exceptional climate. The fact that physical seed dormancy occurs in minute seeds such as those of *Cuscuta australis* (Jayasuriya et al. 2008) is in favor of the classic physiological rather than the crypsis hypothesis, at least in such cases. However, it is possible if not probable that defense from rodent seed predators may sometimes be the primary selective agent in humid regions, but interrupted rainfall and other germination-related issues seem to be the primary reason in arid regions. This complicated issue is however one out of many cases where visual and olfactory defenses operate simultaneously, that are discussed in this book.

The earliest mention of visual seed camouflage I am aware of is a note by Sherzer (1896), who proposed that certain seeds from the Philippines resemble beach pebbles. Although meager, there is clear experimental evidence of an important role of camouflage in seed survival (e.g., Capon and Brecht 1970; Brayton and Capon 1980; Nystrand and Granström 1997; Saracino et al. 1997, 2004; Lev-Yadun and Ne'eman 2013; Porter 2013; Aviezer and Lev-Yadun 2015). This, however, is very far from being satisfactory.

Reviewing the issue of seed dispersal in the genus *Pinus*, Lanner (1998) concluded that pine seed camouflage has not been studied even though it might be significant for seed survival. These true things were written just as the first study about seed camouflage in the genus *Pinus* were studied for the first time. Nystrand and Granström (1997), who studied *P. sylvestris* seed predation by the bird *Fringilla* under controlled experiments, showed that seeds with various colors were predated less when they matched soil color. They proposed that selection by visually searching seed predators helps maintain seed color polymorphism in *P. sylvestris* and the evolution of crypsis *via* seed color. *Pinus halepensis* seeds, like those of most pine species (about 100 winged species), have large light-brown wings that are detected in the post-dispersal, including post-fire environment. Seed camouflage becomes relevant only after the detachment of the very conspicuous wings. Camouflage by matching soil color by *P. halepensis* seeds of Italian origin, has been proposed as an important factor in their survival (Saracino et al. 1997, 2004). Saracino et al. (1997) found that following a forest fire, the color of *P. halepensis* seeds released changed with time: dark gray-brown seeds were released at first, whereas late-dispersed seeds showed a yellowish color, but no color differences were found between seeds extracted from cones of different ages or positions on the tree. They posited that this trend is an adaptation for better crypsis of seeds on the soil, which decreases seed predation. In later field experiments with post-fire bird predation of *P. halepensis* seeds, Saracino et al. (2004) found that light gray seeds were predated less when presented on light gray ash, suggesting crypsis as a protective strategy against bird predation. However, they found no differences in predation of black seeds on different colored substrates (Saracino et al. 2004). In both studies (Saracino et al. 1997, 2004) color variation between seeds was described, but no bimodal color pattern of each seed that will be discussed below. Porter (2013) studied seed camouflage of the Californian legume *Acmispon wrangelianus* on serpentine and non- serpentine soils of different colors. This small forb produces seeds 2–3.5 mm in size. Porter (2013) found genetic lineages that had seeds that matched more closely the color of their native soil than other soils. Moreover, even within the same soil type the seed colors matched better the local soil variety. These finds suggested that natural selection by visually oriented granivores at least partly shaped the evolution of seed coat coloration. Myczko et al. (2015) found that in the vicinity of Poznań (Poland), empty seeds of *P. sylvestris* are light colored and that light-colored but full seeds seem to mimic them because they are predated less by the bird *Fringilla* that preferred dark seeds. They also found that matching ground color did not improve seed survival.

The importance and functionality of seed camouflage were tested for agricultural purposes. Sown seeds, especially of cereals such as wheat, rice, and corn, attract various animals that consume them and cause significant crop losses. Therefore, sown seeds are commonly treated with toxic chemicals and colored red to alert humans to avoid consumption. Various granivorous bird species are attracted to sown fields, where they consume large amounts of such treated seeds and are poisoned and die. In order to reduce their poisoning and death, seeds were camouflaged by painting them to match soil color and this decreased seed consumption and bird death (Almeida et al. 2010a, b).

Bimodal Color Pattern – A New Pine Seed Camouflage Strategy

A very special type of visual seed camouflage is found in the eastern Mediterranean type of *Pinus halepensis* (Aleppo pine) trees (Lev-Yadun and Ne'eman 2013). In order to allow good understanding of the value of this special type of seed camouflage, the ecology of this pine is given in some detail. Aleppo pine simultaneously produces two types of seed cones, regular mature brown cones (Fig. 14.3) that open during hot and dry days from the day of their maturation, and serotinous cones that turn gray with time (Fig. 14.4) and open almost only after exposure to fire temperatures (Shaw 1914; Nathan et al. 1999, 2000; Goubitz et al. 2003, 2004). Post-fire regeneration of *P. halepensis* and its ability to colonize such newly opened areas depend on the combined effects of the canopy-stored seed bank, short- and long-distance seed dispersal, post-dispersal predation, seed germination, seedling and sapling competition, establishment, and maturation (Schiller 1978; Acherar et al. 1984; Saracino et al. 1997; Nathan et al. 2000; Trabaud 2000; Nathan and Ne'eman 2000, 2004). The rate of post-dispersal seed predation on the soil in un burnt stands of *P. halepensis* in Israel was found to be 98.5% after 2 months (Schiller 1978; Ne'eman and Izhaki 1999; Nathan and Ne'eman 2000). After severe forest fires, mature *P. halepensis* trees disperse in a relatively short time large quantities of seeds that land on the ground and are exposed to predation, although probably a much lower predation than in non-burnt areas. Rodent populations in Aleppo pine forests suffer considerable losses during large and intensive forest fires (Haim and Izhaki

Fig. 14.3 Mature, ripe, brown seed cones of *Pinus halepensis* (Aleppo pine) from Israel, a short time after their maturation. At this stage they are *brown*. Some of the cones of this stage will open on hot dry days and disperse their seeds. Others will remain closed till they will be exposed to high temperatures during forest fires (serotinous cones). Serotinous cones will gradually change their color and become *gray-silvery*

Fig. 14.4 A *gray-silvery* several-year-old serotinous *Pinus halepensis* cone. The function or significance of color change in serotinous cones is unknown

1994; Haim et al. 1996; Saracino et al. 2004) and thus pose almost no predation threat for the pine seeds. Many ant nests survive forest fires and sustain their diversity, mainly by opportunistic species (Izhaki et al. 2009), but only some of them are granivores. Thus, after forest fires, visually oriented, diurnal granivorous birds are probably the major pine seed predators (Nathan and Ne'eman 2000). Under such conditions, seed crypsis may be of critical survival value.

Serotinous Aleppo pine cones differ in their anatomical structure from non-serotinous cones growing on the same tree (Salvatore et al. 2010; Moya et al. 2008) and their seeds vary also in factors controlling their germination (Goubitz et al. 2003). Therefore, these cone and seed traits are probably regulated not only by common genetic differences, and possibly represent a component of phenotypic plasticity with epigenetic regulation (e.g., Jaenisch and Bird 2003). An important question is whether we have evidence that these traits of *P. halepensis* are fire adaptations that evolved with fire. The answer is yes, because serotinous cones that open only after fire are common in *P. halepensis* (Shaw 1914; Nathan et al. 1999, 2000; Goubitz et al. 2004; Ne'eman et al. 2004; Moya et al. 2007), especially in young trees that germinated and were established in burnt areas after a forest fire (Goubitz et al. 2004). One fire event was sufficient to increase the degree of serotiny in *P. halepensis* from 45 to 90 % (Goubitz et al. 2004), which is strong evidence for fire-related selection and evolution.

Lev-Yadun and Ne'eman (2013) examined the potential of post-dispersal seed camouflage in this fire-adapted species and described for the first time for any plant species the bimodal color pattern of eastern Mediterranean *P. halepensis* seeds, and proposed an adaptive explanation for this type of seed coat coloration in light of *P. halepensis* life history. As shown by Lev-Yadun and Ne'eman (2013), eastern Mediterranean *P. halepensis* seeds (Mount Carmel, and elsewhere in Israel) display a bimodal seed coat color pattern with dark (black) and light (brown or gray) colors (Figs. 14.5 and 14.6), which present different degrees of camouflage when viewed against various backgrounds. The adaxial (upper and flat) side of the seeds from all non-serotinous cones was light, and their abaxial (lower and convex) side was black. When seeds from serotinous cones were examined, seeds from 83 % of these cones had a similar bimodal black and light seed coat color pattern, while seeds from 17 % of the serotinous cones were light (either brown or gray) colored on both sides. The examined seeds of several western Mediterranean *P. halepensis* trees (Mt. Luberon, Avignon, France) demonstrated contrasting dark/light color patterns; two trees had all light and two other trees had all dark seeds (Lev-Yadun and Ne'eman 2013).

The dual life strategy of *P. halepensis* as a pioneer-invasive species expanding into disturbed areas (Acherar et al. 1984) and as a post-fire obligate seeder (Nathan et al. 1999, 2000; Goubitz et al. 2004; Ne'eman et al. 2004) is connected with alternating phases of seed dispersal. After a forest fire, seeds are dispersed mainly on black or dark ground or on lighter ash patches, while without fire they are dispersed on light-brown needle-bed or brown to gray soil. As fires and soil disturbances are unpredictable in time and space, there seems to be no unidirectional selection for light- or dark-colored seeds, and under such conditions the bimodal color pattern of individual seeds is preserved, and supplemented by some variability of unimodal

Fig. 14.5 A close up of a mixture of seeds of *Pinus halepensis* from Israel. Some are with their *dark* side up and some with their *light-colored* side up. Each seed has a *dark* side and a *light-colored* side

Fig. 14.6 Winged seeds of *Pinus halepensis* from Israel. In one side of the picture the *dark* side is shown and on the other side the *light-colored* side. In the field, when many of the seeds were already dispered away from the mother tree the wings are detached, and the seeds may be camouflaged according to the combination of soil color and seed side

seed coat coloration. This is different from the previously proposed only unimodal seed color adaptation for *P. halepensis* (Saracino et al. 1997) and *P. sylvestris* (Nystrand and Granström 1997) and possibly represents regional adaptations.

Thus, within the same pine species there are two different levels of seed coat color variability (intertree and intraseed). In the eastern Mediterranean *P. halepensis*

Fig. 14.7 Hundreds of *Pinus halepensis* seeds spread on a dark soil. Most of the seeds but especially those with their *darker* side up blend into the background

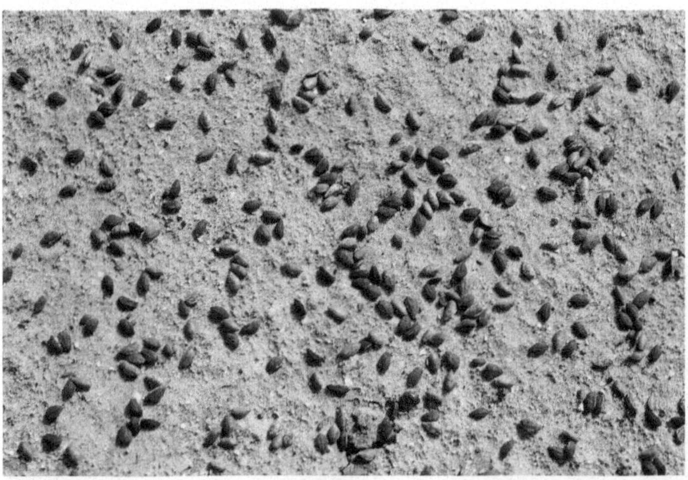

Fig. 14.8 Many *Pinus halepensis* seeds spread on a *light-colored* soil. Most of the seeds, especially those with their *darker* side up are very conspicuous. Those with their *light-colored* side up blend better into the background

type there is an intra-seed-coat variability, in which one seed coat side is light-colored, and the other is dark-colored. Since the seed coat is a maternal tissue, the mother tree imposes its interests and the gain is for the mother because of the large number of dispersed seeds. There is a lottery of what side will be facing the light, a bet that may potentially and practically be lethal for many individual genets because of the expected higher risk of seed predation for half of the seeds (Figs. 14.7 and 14.8). By contrast, in the western Mediterranean *P. halepensis* type, the variability is at the tree level, with trees that have lighter seed coat color and others with a dark one. There, the parent-offspring conflict seems to be smaller since the manipulation of seed camouflage by the mother tree is not so drastic.

In addition to camouflage *via* background matching *per se*, the bimodal color pattern may also make defensive use of the tendency of many animals to use a search image while looking for cryptic food (e.g., Dawkins 1971; Dukas and Ellner 1993). For instance, if a bird finds either several dark or light seeds at the beginning of its search, it might look for this rewarding search image while skipping the other morph.

The use of alternating dark and light colors within the same seed, especially regarding camouflage in burnt and un-burnt habitats, is a more "sophisticated" version of the related classic phenomenon of industrial melanism of various arthropods and its recent reverse process. It illustrates the great importance of plant-based camouflage for insects in that specific case, but they should also apply with other taxa such as pine seeds. Air pollution following the widespread use of coal in England and the USA in the nineteenth century resulted in the overall darkening of the environment (like some of the soil in pine forests after fire), and a consequent increase in the proportion of dark morphs in many insect populations. This change was brought about by selective predation by birds of the lighter morphs, which did not match the new, darker background of soil, walls, tree trunks, branches and foliage (Kettlewell 1973; Majerus 1998; Cook 2000, 2003). The marked reduction in air pollution in recent decades and the consequent lightening of the environment were followed by a decrease in the proportion of the dark morphs in both England and the USA (Cook 2000, 2003). This large-scale natural experiment, as well as the preference of dark *Sceloporus occidentalis* lizards for dark branches after fires in southern Californian chaparral (Lillywhite et al. 1977), other similar cases of selection of animals to soil color (Hocking 1964; Hoekstra 2006; Karlsson et al. 2008), and the predation experiments with *P. sylvestris* seeds by Nystrand and Granström (1997), all clearly demonstrated the adaptive significance of the dark morphs in darker environments and of the lighter morphs in lighter environments.

To conclude, Lev-Yadun and Ne'eman (2013) proposed that the bimodal color pattern of *P. halepensis* seeds is a special case of a more general phenomenon, and I know that this phenomenon occurs in some other pine species. The dual life strategy of *P. halepensis* as a post-fire obligate seeder and as an invader into disturbed habitats in a constantly changing environment of various fire intensities and frequencies, as well as other disturbances, may have caused the stabilization of the bimodal seed color pattern as an efficient camouflage strategy in the arid eastern Mediterranean region. The possibility that this phenomenon is common in other pine species and other fire-dependent serotinous taxa should be addressed in future studies.

Chapter 15
Pod and Seed Camouflage in the Genus *Pisum*

The genus *Pisum* includes according to different views either two well-defined species (*P. fulvum* and *P. sativum*), or three species (*P. fulvum*, *P. elatius*, *P. humile*) and a domesticated derivative (*P. sativum*) of its wild progenitor *P. humile* (Ben-Ze'ev and Zohary 1973). Even for those that think that only two species comprise the genus, the two common wild types of the domesticated *Pisum sativum* are known as *P. elatius* and *P. humile*. In Israel, these two closely related *Pisum* taxa seem in the field to be two very different biological entities, i.e., seem to be valid species and the genus *Pisum* seems to include three species.

Pisum fulvum typically climbs on various shrubs, low trees and even on tall annuals. *Pisum elatius* is an omni-Mediterranean tall climber of mesic habitats that climbs on shrubs and on both low and tall trees, and *P. humile* is a short, more xeric steppe type that climbs on various low annual or perennial plants. All *Pisum* taxa are predominantly selfers (Ben-Ze'ev and Zohary 1973), a character that helps to express unique phenotypes. The young pods and both the unripe and mature seeds of all three wild *Pisum* taxa are edible. The pods of the three *Pisum* taxa contain seeds that are not only highly nutritious and non-toxic, but are also practically the only offspring of those annual individuals that have only a single chance to carry their genes on, especially because they are mostly selfers and their gene dispersal *via* pollen in very low. Protecting such a viable organ is thus critical. Aviezer and Lev-Yadun (2015) supposed that this is the reason for the evolution of what seems to be several types of defensive coloration and probable animal mimicry in the pods and seeds in these three wild *Pisum* taxa.

Two types of seed defense not related to coloration were already known in the genus *Pisum*. All *Pisum* taxa have strong ballistic seed dispersal originating in their pod anatomy (Fahn and Zohary 1955). The distance of ballistic seed dispersal in *P. elatius* may reach more than three meters, but the average distance is about 80 cm (Ambrose and Ellis 2008). From my field experience, it is very difficult to identify the dispersed seeds under natural conditions. In addition, various *P. fulvum* genotypes have not only their regular aerial pods, but also some subterranean pods or

S. Lev-Yadun, *Defensive (anti-herbivory) Coloration in Land Plants*,
DOI 10.1007/978-3-319-42096-7_15

pods formed very close to the soil surface, where they are better defended from grazers and various granivores (Mattatia 1977).

The major differences between the wild-type pea and the domesticated one are a strong seed dormancy and seed dispersal by pod shuttering, two characters that are not found in the domesticated types (Zohary et al. 2012), although seed dormancy seems to be much more important concerning the early and critical stages of pea domestication (Abbo et al. 2011). Interestingly, the pods and seeds of the most common varieties of domesticated *Pisum sativum* are plain green, and they do not express the types of pod and seed coloration found in the wild type and proposed by Aviezer and Lev-Yadun (2015) to be defensive, and thus they suggested that these coloration patterns were selected against at an unknown time and for unknown reasons after pea domestication. The types of defensive coloration described and discussed by Aviezer and Lev-Yadun (2015) were not considered part of the various genetic changes related to pea domestication (Weeden 2007). This is not surprising, since the progress in understanding defensive plant coloration is mostly a post-2000 development (see Archetti 2000; Lev-Yadun et al. 2002, 2004a; Ruxton et al. 2004; Archetti et al. 2009a; Lev-Yadun 2001, 2006a, 2009a; Schaefer and Ruxton 2011; Farmer 2014), and was not yet integrated into many other areas of research in plant biology and ecology, including plant domestication. Contrary to domesticated *Pisum*, the trend in many domesticated animals was to express various types of coloration not found in nature (Cieslak et al. 2011), but the loss of a probable defensive coloration of red autumn leaves was also found in domesticated apple (Archetti 2009c).

Following the initial finds that the pods of various wild Near Eastern legume species may mimic aposematic caterpillars (Lev-Yadun and Inbar 2002), and field work that clearly showed that wild *Pisum* pods and seeds have variable types of coloration, Aviezer and Lev-Yadun decided to study the color patterns of all three wild *Pisum* taxa (*P. fulvum, P. elatius* and *P. humile*) in order to understand their potential defensive coloration (camouflage and Batesian mimicry). A careful and detailed examination of several wild populations of each species and several common garden experiments revealed that all three taxa are polymorphic concerning pod and seed coloration. All the very young pods and some mature ones of all *Pisum* taxa are green (Fig. 15.1), a color that provides a basic camouflage within green canopies, but as the pods (and seeds) grow in size and in age, various types of coloration start to decorate many of them.

The pattern of coloration found in the fully grown pods of *P. fulvum* was either plain green, or green with blotches in a dark color (black/dark purple) (Figs. 15.2–15.5), a character found in various other legume taxa. When comparing pod coloration in the different sites (two common-garden experiments and a large natural wild population at Yokne'am forest), there were conspicuous differences in the frequency of color types. At a partly shaded common-garden, of the 476 examined pods, mature green ones dominated and accounted for 95 % of the total. In contrast, in a much more sun-exposed common-garden population, of the 288 examined pods, only 11 % of the mature ones were green, while the majority had a blotched

Fig. 15.1 Green immature pods of *Pisum fulvum*. All young pods in the genus *Pisum* and in most other land plant taxa practically almost every young cone or fruit are *green*, a color that serves as basic primary camouflage withing green canopies and allows for photosynthesis

color pattern. Of 290 examined pods of a wild *P. fulvum* population in the Yokne'am forest site, 45 % were green and 55 % had a blotched color pattern (Aviezer and Lev-Yadun 2015).

While the two closely related taxa (*P. elatius* and *P. humile*) seem to employ various patterns of pod coloration to create defensive animal (Batesian) mimicry, which will be discussed in detail in the chapter about defensive visual animal mimicry by plants, the defensive coloration of *P. fulvum* pods seems to rely on a very different strategy, e.g., visual camouflage. The blotches found on its pods (Aviezer and Lev-Yadun 2015) look like camouflage coloration found in many animals (see Cott 1940; Edmunds 1974; Merilaita 1998; Ruxton et al. 2004; Stevens and Merilaita 2011) and in the military (Stanley 1998; Newark and Miller 2007).

While at least ten various seed-coat colors exist in the genus *Pisum* (e.g., Hoey et al. 1966), they were neither described by them nor was their biological significance discussed. In other legume and non-legume taxa, seed-coat colors may be associated with differences in germination (Gutterman and Evenari 1972; Gutterman 1978), but while Aviezer and Lev-Yadun (2015) did not examine seed germination in relation to seed-coat coloration, and discussed seed-coat coloration patterns only concerning a visual anti-predatory defensive mechanism, they did not see any theoretical difficulty to assume that simultaneous physiological functions could also be involved in the evolution of seed-color polymorphism in the genus *Pisum*. All three *Pisum* taxa growing naturally in Israel have several seed-coat color morphs, and some morphs were taxon specific. The seeds of all three species are camouflaged by either: (1) color matching that of soil (especially in *P. fulvum* and *P. elatius*)

Figs. 15.2–15.5 Variability in camouflage coloration in fully developped pods of *Pisum fulvum*. The pods differ in the strength of coloration and in the number of *dark* blotches

Fig. 15.6 Typical seed colors of *Pisum fulvum* that may serve as background matching when dispered on the soil

Fig. 15.7 Typical seed colors of *Pisum elatius* that may serve as background matching when dispered on the soil. They differ from the seed coat colors of *P. fulvum* but serve the same purpose

(Figs. 15.6 and 15.7), (2) disruptive coloration (especially *P. humile*) (Fig. 15.8), (3) variability in seed-coat color that may partly undermine the use of search images by visually oriented granivores that look for cryptic seeds (all taxa) (Aviezer and Lev-Yadun 2015). The polymorphism in seed crypsis types and coloration in the three studied *Pisum* taxa may thus help directly in lowering the efficiency of their detection by visually oriented herbivores (e.g., Jones et al. 2006) because it is harder to search simultaneously for two or more different cryptic items that differ in their cryptic type, than to search for only a single type (Bond 2007). *Pisum fulvum* had

Fig. 15.8 Variability in seed coat coloration in *Pisum humile* including seeds with classic disruptive coloration

the least seed-coat color morphs of the three taxa. The other two taxa (*P. elatius* and *P. humile*) express much more seed-coat color polymorphism even though they differ from each other. The morphs found in *P. fulvum* were: (1) smooth black seeds, (2) smooth brown seeds and (3) mottled brown, although the mottling was not very conspicuous. The dominant pattern was the smooth one, brown or black (Fig. 15.6).

In *P. elatius* the dominant seed-coat color pattern was sometimes population dependent, although in all three sampled populations the color pattern was widely diverse. While *P. elatius* has the greatest variation of seed-coat color morphs, all these morphs, like those of *P. fulvum*, are of the background matching color type, looking in many cases like a soil particle because of their size, color, and texture. In the Ikrit and Sursuk populations of *P. elatius*, the dominant pattern was dotted gray, but in the Kziv population, dotted brown dominated (Fig. 15.7).

The seeds of *P. humile* had many color patterns, including gold-yellow color with black spots, brown stripes on different background colors, brown stripes and black spots on different background colors (Fig. 15.8) and also seed coats with a smooth black color (Fig. 15.9). The seed coat coloration of *P. humile* is much more conspicuous than in the other *Pisum* species in many of the seeds, and since camouflage is considered, it seems to have a double strategy. The first is that from a greater distance the seed-coat color of any type can serve as background matching. At a closer range, except for plain black seeds, many *P. humile* seeds employ disruptive coloration, i.e., large and contrasting colored areas that make it difficult to recognize the shape, or the contours of the seed (Fig. 15.8), a well-known animal crypsis strategy (e.g., Cott 1940; Merilaita 1998; Ruxton et al. 2004) and a common military camouflage (e.g., Stanley 1998; Blechman 2004).

Fig. 15.9 Black seeds of *Pisum humile*. Such *black* seeds may mimic the toxic seeds of *Lathyrus ochrus*

Fig. 15.10 Toxic black seeds of *Lathyrus ochrus* mixed with the light *gray* morph

While black seeds of both *P. humile* and *P. fulvum* are highly edible, there are various black toxic seeds of legumes and of other taxa that grow in the same region. The black seeds of both *P. humile* and *P. fulvum* may thus not only be cryptic, but also mimic at close range the black seeds of various toxic legumes, e.g., black seeds are formed in about half of the plants of the toxic legume *Lathyrus ochrus* (Figs. 15.10, 15.11–15.13), or even non-legume toxic seeds. Black, non-toxic *P. humile* and *P. fulvum* seeds may thus be the Batesian mimics of such toxic seeds.

Figs. 15.11–15.13 The interplant seed coat color variability of *Lathyrus ochrus* seems to allow better or lesser crypsis on soils of different levels of darkness

Chapter 16
Defensive Functions of White Coloration in Coastal and Dune Plants

This detailed part of the discussion about plant camouflage focuses on light-colored plant species growing in sandy or desert habitats in the Near East and on the several other simultaneous defensive and physiological functions of this coloration.

Lev-Yadun (2006b) showed that under eastern Mediterranean conditions many sand-dune and sandy shore plants growing in Israel and in adjacent countries are white, whitish, or silver/gray. Light-colored plants may be almost white (*Otanthus maritimus*) (Figs. 13.3, 13.4, 13.5 and 13.6), or have a very light green or silvery color (*Lotus creticus*, *Convolvulus secundus*). There are several ways in which plants can be light-colored without losing photosynthetic tissues by not expressing chlorophylls: (1) by trichome cover, (2) by air spaces under the epidermis, (3) by producing light-colored cork, (4) by secreting light-colored waxes, and (5) by sticking sand, clay, or dust.

One group of 11 species of light-colored plants includes plants covered with white trichomes (*Echinops philistaeus*, *Otanthus maritimus*, Asteraceae; *Lotus creticus*, *L. lanuginosus*, *Medicago marina*, Fabaceae; *Convolvulus secundus*, Convolvulaceae; *Neurada procumbens*, Neuradaceae; *Plantago albicans*, *P. ciliata*, Plantaginaceae; *Crucianella maritima*, Rubiaceae; *Zygophyllum album*, Zygophyllaceae), and composes 25 % out of the 44 common Israeli coastal plant species described in Waisel et al. (1975). This is a much higher proportion of plants covered with dense white or silvery trichomes than in the whole Israeli flora. Another group of coastal and inland Near Eastern species includes plants that have glandular trichomes that secrete sticky substances that stick sand grains and clay particles to their surface. Most of the species are described in Danin (1996) and include *Aegialophila pumilio*, *Ifloga spicata*, Asteraceae; *Eremobium aegyptiacum*, Brassicaceae; *Fagonia arabica*, *F. glutinosa*, *F. mollis*, Zygophyllaceae; *Ononis serrata*, *O. natrix* subsp. *stenophylla*, *O. variegata*, Fabaceae; *Gypsophylla viscosa*, *Savignia parviflora* Brassicaceae; *Silene arabica*, *S. sedoides*, *S. succulenta* (Figs. 13.5 and 13.6), *S. villosa*, Caryophyllaceae. Such plants that stick sand on their

S. Lev-Yadun, *Defensive (anti-herbivory) Coloration in Land Plants*,
DOI 10.1007/978-3-319-42096-7_16

surfaces are common the the southern African flora (Jürgens 1996; Neinhuis et al. 1996) but are also found in other world regions (LoPresti and Karban 2016).

Trichomes, the unicellular and multicellular appendages of the epidermis (Fahn 1990), are well known for their multiple functions in plants. Trichomes may serve in protecting plants from excess sun irradiation of various wavelengths, including UV (Johnson 1975; Ehleringer et al. 1976; Fahn and Cutler 1992; Manetas 2003), secrete toxic ions, especially in saline habitats (Waisel 1972; Fahn 1988), function in water absorption (Fahn and Cutler 1992; Benzing 2000), reduce transpiration (Fahn and Cutler 1992; Werker 2000), defend from insect or other herbivorous invertebrates by physically reducing accessibility of plant tissues, by actual trapping of their legs, or even by various chemical means (Levin 1973; Fahn 1979, 1988; Werker 2000) and defend from large herbivores when they sting as in *Urtica* (Thurston and Lersten 1969; Levin 1973; Fahn 1990; Wheeler 2004; Fu et al. 2006; Taylor 2009). In addition, in certain carnivorous plants like *Drosera* and *Dionea*, they may take part in the attraction, capture or digestion of insects (Juniper et al. 1989; Fahn 1990).

To avoid misunderstanding, I emphasize that both the physiological functions and the defensive functions of plant trichomes may operate simultaneously. Thus, various hypotheses concerning the functions of trichomes need not conflict with or exclude any other functional explanation. Traits such as trichomes, that commonly have more than one type of benefit, may have been selected for by several agents simultaneously or independently.

The common non-herbivore-related explanation for light coastal and desert plant coloration is that it protects from irradiation, and that in addition, the glued sand defends them from abrasion by moving sand (Waisel 1972; Fahn and Cutler 1992; Danin 1996; Jürgens 1996; Neinhuis et al. 1996). Wiens (1978) was probably the first to propose that plants stick sand particles for camouflage, a hypothesis also briefly discussed by Jürgens (1996) and by Neinhuis et al. (1996). Lev-Yadun (2006b) proposed several additional novel anti-herbivory functions for white coloration and for sticky glandular trichomes in sandy or light-colored environments: (1) since dust is a strong insect repellent and even lethal for insects, its attachment to sticky glandular trichomes should defend plants from insect herbivory, (2) sand attached to plant surfaces will like sandpaper wear down teeth of large herbivores (and possibly also mouth parts of arthropods), (3) such white plant surfaces undermine the camouflage of herbivorous insects that have other colors and expose them to predation, and (4) white coloration of leaves and branches (Figs. 13.3 and 13.4) may mimic fungal infestation (Fig. 16.1) to reduce herbivory. Lev-Yadun (2006b) posited that while direct experimental data for the functionality of these defensive mechanisms are missing for many of the previous and for all new hypotheses (but see recent experimentation by Strauss and Cacho 2013; Strauss et al. 2015; LoPresti and Karban 2016), there are many indirect supporting indications that will be discussed below.

Fig. 16.1 Leaves of the annual legume *Onobrychis* sp. from northern Israel. Young green leaves contrast with whitish more mature leaves, the outcome of a fungal attack

Camouflage by Sand Particles Attached to Sticky Glandular Trichomes and Light-Colored Trichomes

Wiens (1978) proposed that light coloration, achieved by sand stuck to sticky glandular trichomes, may camouflage plants from herbivores. The examples that attracted his attention and convinced him that such a strategy is feasible were the five species of the genus *Psammophora* (Aizoaceae) and *Crassula alpestris* (Crassulaceae) from South Africa, and several other species from the deserts of North America (Wiens 1978). In addition to Wiens' (1978) hypothesis, Lev-Yadun (2006b) proposed that light coloration, achieved by non-sticky light-colored trichomes or any other light-colored surface such as cork (*Caralluma* sp.), may also camouflage plants from herbivores in light-colored rocky habitats or in sandy ones. Similar coloration in animals that live in white, snow-covered habitats or habitats dominated by light-colored sand is commonly referred to as camouflage (Cott 1940; Hamilton 1973; Edmunds 1974; Endler 1978, 1984; Cloudsley-Thompson 1979, 1999; Owen 1982; Majerus 1998; Ruxton et al. 2004; Hoekstra 2006; Rosenblum 2006; Linnen et al. 2013), and the same logic should also apply to plants.

Various invertebrates attach substrate or plant (sand, gravel, algae, lichens, mosses, dry branches) particles from the environment for better background matching as defense (Gressitt et al. 1965; Edmunds 1974; Slocum and Lawrey 1976; Owen 1982; Shears 1988; Brandt and Mahsberg 2002; Domínguez and Jiménez 2005; Ruxton and Stevens 2015) – the parallel of attaching clay or sand particles in plants by sticky glandular trichomes. For instance, the green lacewing (*Chrysopa slossonae*) attaches packets of sycamore leaf trichomes or lichen pieces. The tortoise beetle (*Gratiana pallidula*) and other insects carry feces to form a defensive packet. Caterpillars of the family Geometridae, e.g., *Synchlora* sp. cover themselves with pieces of petals of flowers they harbor (Eisner et al. 2005). As for the defensive potential of dust- and sand-covered plants, Lev-Yadun (2006b) posited that the basic principle is simple, the plants just have to not look green, and this character will grant them a certain level of defense from various types of visually oriented herbivores. Since plants and herbivores have co-evolved, detection of plants *via* visual cues is a basic aspect of herbivore life (e.g., Prokopy and Owens 1983; Reeves 2011). The search image for plants in visually oriented herbivores should be green in general, or if the plants are eaten when dry, shades of yellow, gray or brown. Other colors are not immediately considered by the visual sensory apparatus of many herbivores as evidence of forage availability. It is probable that for specialized monophagic herbivores such defensive coloration does not operate well, especially if odors play a part in their foraging strategy, but this coloration may defend from generalist herbivores that look for whatever green plants they can reach. A similar principle led Stone (1979) to propose that young red leaves of certain palms mimic dead or old ones or just look unlike young edible ones, a hypothesis that has been accepted by several other authors (Juniper 1994; Dominy et al. 2002; Gould 2004; Karageorgou and Manetas 2006; Manetas 2006).

Looking like the soil or gravel serves as excellent camouflage in members of the genus *Lithops* that grow in the desert in South Africa (Cole 1970; Cole and Cole 2005), members of the genus *Pseudolithos* (Lavranos 1974) or certain cacti (Benson 1982; Owen 1982). Lev-Yadun (2006b) proposed that when plants are light colored for whatever reason, having silvery-white trichomes or *via* sand or clay cover that adheres to the sticky trichomes, they resemble the sandy soil and are thus camouflaged. This hypothesis was recently strongly supported by the resemblance of scree plants to their gray substrate (Burns 2010; Niu et al. 2014; Niu and Sun 2014; Strauss et al. 2015).

Dust as a Strong Insect Repellent

Another defensive mechanism proposed in Lev-Yadun (2006b) is that since dust is a strong insect repellent and may even be lethal to insects (Ebeling 1971; Cook et al. 1980; Showler 2002, 2003), attached soil particles, especially clay, may defend plants with sticky glandular trichomes from insect herbivory. Mammals and birds take "dust baths" and by doing so receive considerable defense against ectoparasites (Ebeling 1971). Plants with sticky glandular trichomes that are covered by soil clay particles (see Fahn and Shimony 1996) should enjoy a similar benefit. For instance, the ash from the eruption of Mount St. Helens on May 18, 1980, considerably damaged the insect fauna of the region because of this effect (Cook et al. 1980). Several field experiments clearly demonstrated the defensive benefits of covering plant surfaces with clay particles (Ebeling 1971; Bar-Joseph and Frenkel 1983; Lapointe 2000; Showler 2002, 2003). This mechanism is not restricted to sandy habitats and there are good reasons to assume that it acts in many dusty habitats, i.e., in many hot and arid regions. Moreover, if whitish/grayish plants appear to insects to be covered by dust even when they are not, this is a case of dust mimicry that may defend the plants from certain visually oriented herbivorous insects.

Sand May Wear Down the Teeth of Large Herbivores

Plant camouflage, undermining herbivorous insect camouflage and insect repellence seem to be only a part of the antiherbivory effect of sand and clay attachment to sticky glandular trichomes. Sand particles are commonly made of SiO_2, a very hard material. Many plants, especially grasses, produce a large number of silica bodies, known as phytoliths (Rapp and Mulholland 1992; Meunier and Colin 2001; Prychild et al. 2004). Along with other biological functions of silica, plants use phytoliths to defend themselves from herbivory (Baker et al. 1959; Walker et al. 1978; McNaughton and Tarrants 1983; McNaughton et al. 1985; Massey and Hartley 2006; Katz 2015). Sand attached to plant surfaces probably wears down the teeth of large herbivores possibly even better than phytoliths. After all, we use sand paper and sand spraying to polish hard materials including concrete and steel. There are indeed some indications that soil particles may wear down teeth of mammalian herbivores more than phytoliths (Ungar et al. 1995; Mainland 2003; Sanson et al. 2007), but this issue is still very far from a good understanding. Lev-Yadun (2006b)

Fig. 16.2 The only white coastal dune species *Elymus arenarius* (= *Leymus arenarius*) I found in the Baltic coast of Estonia

proposed that especially for the Near Eastern dicotyledonous plant species (*Aegialophila pumilio, Eremobium aegyptiacum, Ifloga spicata, Fagonia glutinosa, F. mollis, Ononis serrata, O. natrix* subsp. *stenophylla, O. variegata, Gypsophylla viscosa, Savignia parviflora, Silene arabica, S. sedoides, S. succulenta* and *S. villosa*) that have sticky glandular trichomes that glue sand and clay but produce only small numbers of phytoliths compared to grasses, the attached sand is an external shield compensating for the low level of phytoliths they produce, and that the attached sand may defend these plants from vertebrate herbivory. It is possible that similar damages may be caused to the hard mouth parts of insects when they eat sand-covered plants, but this issue has not yet been studied in detail.

Lev-Yadun (2006b) suggested that the fact that there are so many light-colored plants in the coastal and desert vegetation of Israel and adjacent countries, an arid zone, should be studied first in similar habitats but with a taxonomically very different flora, such as the Australian flora where many trees are much less green than trees of the temperate zone (Grose 2007, 2014). It should be studied also in much colder climates and in ecosystems where light is not in excess and in coastal dunes of seas or large freshwater lakes that are not salty or are exposed to lower levels of herbivory and sun irradiation, such as the Baltic Sea. A field examination of the coastal and lake shore vegetation in Estonia I conducted in the summer of the year 2006 in order to examine this issue turned up with only one coastal dune species (*Elymus arenarius=Leymus arenarius*) that is whitish (Fig. 16.2). Only a comparative picture that will emerge from such studies in many ecosystems will enable the evaluation of the generality of the hypotheses discussed in Lev-Yadun (2006b).

Chapter 17
Gloger's Rule in Plants: The Species and Ecosystem Levels

Gloger's rule, one of several zoological rules that identify patterns of adaptations at a global scale (rules that do have exceptions), posits that darker birds are found more often in humid environments than in arid ones, especially in the tropics (Gloger 1833). In many animal taxa, desert animals tend to be paler compared to other members of the same taxon that inhabit wetter environments. Being paler in desert habitats is considered to reflect the selection by both physiological factors such as reflecting sun irradiation, and by visual defense from predators or prey (e.g., Cloudsley-Thompson 1999; Stoner et al. 2003b). This rule was later demonstrated to also be true for certain mammalian groups including humans (Stoner et al. 2003b; Hancock et al. 2011). Gloger's rule in animals is manifested at two levels, (1) at the species level (different populations of the same species have different patterns of pigmentation at different latitudes or habitats), and (2) at the species assembly level (different groups of taxa have different pigmentation than other taxa groups found in different habitats or latitudes). It is not easy to pinpoint all the reasons for the operation of Gloger's rule in animals. It seems, however, that three factors, (1) crypsis *via* background matching, (2) resistance to keratin-degrading micro-organisms in hair or feathers rich in eumelanin, and (3) thermoregulation, are involved (Burtt and Ichida 2004; Kamilar and Bradley 2011).

Concerning plants, Gloger's rule was first proposed to operate in many plant species growing in sand dunes, sandy shores and in Near Eastern deserts (Lev-Yadun 2006b). As demonstrated and discussed above, many species found there are white, whitish, or silver colored, based on white trichomes, because of sand grains and clay particles glued to sticky glandular trichomes, or because of light-colored waxes (Lev-Yadun 2006b). For instance, as discussed in the previous chapter, about 25 % of the Israeli coastal plant species are light-colored. These light-colored species indicate that Gloger's rule may also apply to plants. Like in animals, similar factors seem to have selected for lighter coloration in plants: (1) physiological factors such as reduction of the load of sun irradiation, and (2) defense from visually oriented herbivores (Lev-Yadun 2006b). Since such "rules" are not without exceptions in the

S. Lev-Yadun, *Defensive (anti-herbivory) Coloration in Land Plants*,
DOI 10.1007/978-3-319-42096-7_17

animal kingdom, they should also have the same limitations to their generality in the plant kingdom, as will be discussed below.

For decades, the common classic explanations for light coloration of such coastal or desert plants were that it protects them from sun irradiation (including UV) (Fahn and Cutler 1992), that the glued sand defends them from abrasion by sand particles moving in strong winds (Danin 1996), and serves as camouflage from herbivores (e.g., Wiens 1978; Jürgens 1996). Lev-Yadun (2006b) proposed that since many desert animals tend to be paler than other members of the same taxa that inhabit wetter environments, according to Gloger's rule (e.g., Cloudsley-Thompson 1979), the above-mentioned light-colored plant species are a good indication that Gloger's rule applies to plants at the level of the whole canopy.

Recently it was elegantly demonstrated that Gloger's rule also applies at the intraspecific level and at the organ level concerning UVB absorption *via* dark areas in flowers to defend their anthers and pollen from exposure to UV light reflected from petals (Koski and Ashman 2015), this being the second system where Gloger's rule operates in plants. In both cases, flowers (Koski and Ashman 2015) and coastal/sand-dune/desert plants (Lev-Yadun 2006b, 2015b), the characters involved in the operation of Gloger's rule in plants belong to complex functions, the probable outcome of several simultaneous selective agents that together resulted in the proposed Gloger's rule syndrome.

In addition to the numerous, complicated and not well-defined and sometimes unknown factors involved in the evolution of types of light or dark coloration of both animals and plants, which commonly complicate the study of functional coloration, there are various salient exceptions to Gloger's rule. A common exception is of dark (melanic) animal morphs in certain temperate animals selected to be dark because this color allows for better warming (Majerus 1998; Caro 2005; Trullas et al. 2007; Hegna et al. 2013). Being darker at higher latitudes for warming was also proposed to be important for some plant taxa. For instance, in the species *Collinsia parviflora* and *Mimulus guttatus*, growing in the Flat Top Islands, British Columbia, plants with upper-epidermis anthocyanin coloration are found more frequently in cold, exposed habitats (Ganders et al. 1980). Similarly, the immature female (seed) cones of European larch (*Larix decidua*) and Norway spruce (*Picea abies*) growing in Austria were found to be red in high altitudes, habitats with low temperatures, while green cones dominated lower and warmer habitats (Geburek et al. 2007). On the other hand, there are plant taxa in which light-colored morphs are found in more northern and humid habitats that are not snow-covered. A good example of this is the Mediterranean species *Anemone coronaria*, in which the flower's color is only red in the drier habitats of central and southern Israel, while in the much more humid and colder habitats of the north, white, light bluish and pink morphs grow alongside the red morph (Horovitz et al. 1975) (Fig. 17.1). This "anti-Gloger" pattern was proposed to reflect edaphic issues (Horovitz 1976), but a close inspection of *A. coronaria* populations in the field indicated to me that there are many cases where the edaphic factor is not exclusive. The complexity of patterns of selection and evolution in flower color was further demonstrated when the selection and evolution of flower color polymorphism was studied in the wild radish

Fig. 17.1 An "anti-Gloger" pattern of the Mediterranean species *Anemone coronaria* with *light-colored morphs* (*white, light bluish* and *pink*), along with the *red* one. The light-colored morphs are found in more northern and humid habitats, while the *darker red morph* is the only one found in the drier habitats of central and southern Israel. The picture with the *red* and *white morphs* was taked in a mixed population in northern Israel

(*Raphanus sativus*). In wild radish, pollinators select for yellow and white flowers, while herbivores select for pink and bronze flowers (Irwin et al. 2003). A strong "anti-Gloger" pattern was demonstrated by Lev-Yadun and Ne'eman (2013), who found that in *Pinus halepensis*, which disperses its seeds either on hot dry days or after fires, and which suffers very high seed predation rates by birds, ants, and rodents, many seeds have a bi-modal color pattern. One of the seed's sides is light-brown or light gray and the other side is black. The seeds are dispersed by wind, fall to the ground, and are commonly exposed to secondary dispersal by wind before losing their conspicuous wings. The seeds expose only one color when lying on the ground and one side of such seeds with a bimodal color pattern provides better camouflage than the other on any light or dark background. Thus, in this case, being dark- or light-colored was independent of latitude, humidity or annual rainfall.

Since very little attention has been given to Gloger's rule in plants, we still do not know the extent of its significance in plant biology.

Chapter 18
Defensive Masquerade by Plants

Another defensive method in the "family of crypsis" is the defensive strategy known as masquerade, or camouflage without crypsis, which has received very little scientific attention even in animals (e.g., Allen and Cooper 1985; Ruxton et al. 2004; Skelhorn et al. 2010a, b, c; Skelhorn and Ruxton 2011, 2013; Skelhorn 2015), and concerning plants even less (Ruxton et al. 2004: 24; Skelhorn et al. 2010b; Lev-Yadun 2014b; Skelhorn 2015).

Masquerade by animals is a situation of a prey resembling inedible objects such as leaves, twigs, stones or bird droppings. Thus, unlike typical crypsis, a masquerading organism is not only fully visible, but even identified by its enemies, but with a false identity. The issue of masquerade is not easy to deal with because many cases of masquerade fall in between crypsis and mimicry (see Endler 1981). For instance, Ruxton et al. (2004) brought up the question, when herbivores pass over a stone-mimicking plant, is it a case of crypsis *via* background matching or of masquerade? In this case I suppose that the answer can and should be both. Accordingly, crypsis *via* background matching is more important when the stone-mimicking plants are viewed from a distance, and masquerade of these plants is more important at close range. Masquerade is usually considered to operate visually, although there are cases of chemical masquerade by caterpillars as defense against predaceous ants (Ruxton 2009), and by plants that mimic odors of feces and carrion (Lev-Yadun et al. 2009b; Lev-Yadun 2014e). In recent years, studies in animal behavior have revealed that successful masquerade depends not only on the quality of mimicry, but also on the context, i.e., they are either found in a context in which predators have never before experienced the model, or in a context in which the predators have experienced an inedible model (Skelhorn and Ruxton 2011), or else it depends on the size and density of the defending animal (Skelhorn et al. 2011; Skelhorn and Ruxton 2013).

Till recently (Lev-Yadun 2014b), the question of the operation of defensive masquerade by plants has been addressed only indirectly (the several lines on page 24 of Ruxton et al. 2004 concerning stone-mimicking plants mentioned above, and even a shorter mention in Skelhorn et al. 2010b were the exceptions) although both

S. Lev-Yadun, *Defensive (anti-herbivory) Coloration in Land Plants*,
DOI 10.1007/978-3-319-42096-7_18

Table 18.1 Plant- and non-plant-mimicry types of masquerade

Plant-mimicry	Non-plant-mimicry
The plants don't look appealing for herbivores since they look like dead leaves or branches, don't look green, look as if they are infested or as if they were already attacked, or look like leaves poor in nitrogen	The plants look (or smell) like uninteresting objects to herbivores (look like a stone, an animal, like droppings, or smell like droppings or carrion)

visual (e.g., Wiens 1978; Brown and Lawton 1991) and chemical (e.g., Lev-Yadun et al. 2009b) defensive masquerade by plants were described but without considering them as defensive masquerade. Lev-Yadun (2014b) reviewed the issue of masquerade in plants and described several cases and types of defensive masquerade, and also proposed that defensive visual plant masquerade belongs to two different sub-types: (1) plants that don't look like plants (non-plant-mimicking defensive masquerade), in which they look (or smell) like uninteresting objects to herbivores (look like a stone or an animal, or smell like droppings or carrion, etc.), and are therefore ignored or not attacked, and (2) plants that look like plants (plant-mimicking defensive masquerade), in which plants or plant parts do not look appealing for herbivores, by not being green, by looking dead or old, or as if they are harboring insects, already attacked, less nutritious, toxic, etc., and are ignored because of this (Table 18.1). Lev-Yadun (2014b) also proposed that for plants it is probably easier to evolve to form the second masquerade type because of the flexible developmental system of plants and because such changes may simultaneously serve other functions, including physiological. For instance, changing one leaf shape or type for another is genetically and developmentally simple (e.g., Nicotra et al. 2011; Gianoli and Carrasco-Urra 2014). Lev-Yadun (2014b) posited that defensive masquerade by plants may in many cases be non-exclusive, and serve additional physiological and defensive functions.

Classic Non-Plant-Mimicking Defensive Visual Masquerade

In his classic review on plant mimicry, Wiens (1978) described various cases of mimicry that should be considered as cases of defensive masquerade by plants. Wiens (1978) described various succulent species belonging to the families Asclepidaceae, Cactaceae, Crassulaceae, Euphorbiaceae, Liliaceae, Mesembryanthemaceae and Portulacaceae, especially those growing in arid regions, which are cryptic. In the arid habitats where these plants grow, the pressure of herbivory and selection by visually oriented herbivores, mostly mammals, but also by large ground-dwelling birds, e.g., ostriches, and by certain Lepidoptera may be higher than in more humid and vegetated habitats. Wiens (1978) also discussed what he thought to be the most common type of crypsis in arid regions of southern Africa – stone mimicry. According to Wiens (1978), stone mimicry is very common in arid regions of southern and northeastern Africa, and also occurrs in some cacti species from North America (i.e., Lavranos 1974; Wiens 1978; Benson 1982; Owen 1982; Sajeva and Costanzo 1994; Cole and Cole 2005). Wiens (1978) estimated in the hundreds the number of species that mimic stones, belonging in Africa to at least four families, Crassulaceae, Euphorbiaceae, Liliaceae and Mesembryanthemaceae, all resembling various types of rocks or stones. Both Wiens (1978) and Cole and Cole (2005) described how difficult it is in the wild to distinguish stone-mimicking plants from real stones. It seems that stone mimicry by many alpine scree species that have gray leaves (Burns 2010; Niu et al. 2014; Niu and Sun 2014; Strauss et al. 2015) is a parallel phenomenon. Wiens (1978) also described visual feces mimicry, but said that this is a doubtful category. Examples of possible visual feces mimicry listed in Wiens (1978) are *Anacampseros papyracea* (Portulacaceae), which has some resemblance to the fecal droppings of large birds, and other, smaller species, e.g., *Anacampseros crinita* and *Crassula alstonii*, which look like ungulate feces. Olfactory defensive feces and carrion mimicry as defense from mammalian herbivores was proposed by Lev-Yadun et al. (2009b). However, the monograph by Wiens (1978) was published before masquerade as visual defense was better defined, e.g., Endler (1981) and therefore he referred just to mimicry and did not regard the strategy known as masquerade.

Another aspect of defensive visual masquerade by plants is arthropod mimicry. I do not posit here that the proposed types of such mimicry do not function as direct defense by arthropod mimicry, but rather that certain herbivores may consider the mimics not to be appealing plant organs. Lev-Yadun (2014b) proposed that there are good prospects that defensive arthropod mimicry by plants may not only cause certain herbivores to be repelled by the mimicked organisms if they are dangerous or inedible ones, but may also cause various herbivores to consider the plant or plant organ to be uninteresting, i.e., classic masquerade. A well-known, classic case of defensive masquerade by plants that was not considered to be masquerade is that of butterfly egg mimicry expressed by various plant taxa that causes female butterflies to consider these plants as occupied and therefore look for another, unoccupied egg-laying site (Gilbert 1980, 1982; Shapiro 1981a, b; Williams and Gilbert 1981; Lev-Yadun 2013a). Some of the recent proposals of insect mimicry by plants that also seem to fit into this category are ant, aphid, beetle, caterpillar, and spider web mimicry (Lev-Yadun and Inbar 2002; Lev-Yadun 2009d, 2015a; Yamazaki and Lev-Yadun 2014, 2015; Aviezer and Lev-Yadun 2015), visual bee or wasp mimicry by orchid flowers for the sake of defense (Lev-Yadun and Ne'eman 2012), which operates simultaneously with olfactory bee mimicry for the sake of pollination (Schiestl 2005; Jürgens and Shuttleworth 2016), and the enigmatic fast leaflet rotation in *Desmodium motorium* proposed to mimic butterfly or other winged arthropods (Lev-Yadun 2013b) that will be discussed in later chapters.

The best studied example of plant mimicry as defensive masquerade by plants is probably that of insect tunneling damage mimicry. Smith (1986) was the first to discuss in detail the defensive role of white leaf variegation in the species (*Byttneria aculeata*) he studied, and suggested that mimicry of insect tunneling is the probable function of the variegation. Mimicry of tunneling insect damage was also one of the several types of proposed defensive functions of white leaf variegation of *Silybum*

Fig. 18.1 Several leaves of *Sonchus oleraceus* infested by tunneling insects. The variegation of *Silybum marianum* leaves (Fig. 1) looks very similar and in the beginning I considered the variegation in *S. marianum* to serve only as tunneling mimicry

marianum (Lev-Yadun 2003a) (Fig. 1), where it looks as if it were already infested to prevent further insect attacks (Fig. 18.1). Lee (2007) also proposed that certain types of white variegation may serve as a defense by mimicking leaf infestation. Soltau et al. (2009) experimentally manipulated the visual appearance of *Caladium steudneriifolium* (Araceae), an understory plant from the *Podocarpus* National Park in southeast Ecuador, by painting artificial white variegation on non-variegated green leaves. The leaves of *C. steudneriifolium* are either plain green or patterned with whitish variegation. In nature, about a third of the leaves are variegated and both morphs are frequently attacked by mining moth caterpillars. The variegated zones of the leaves strongly resemble recent mining damage, and so it was hypothesized to mimic recent mining attacks. Infestation was 4–12 times higher in plain green leaves than in variegated leaves. Painting white variegation on plain green leaves reduced the attacks from 7.88 % to 0.41 %. Therefore, Soltau et al. (2009) proposed that the variegation is probably the mimicry of mining damage to deter ovipositing moths. Mimicry of insect tunneling damage in leaves seems to be an obvious case of looking uninteresting to herbivores while not really being so. The fact that other mechanisms are probably involved in defense by white leaf variegation (e.g., Lev-Yadun 2003a, 2014a, b, c, d) just adds to the selective advantage of such a character. This type (mimicry of already attacked leaves) also includes leaves that look partly consumed (Niemelä and Tuomi 1987; Brown and Lawton 1991).

Chemical Non-Plant-Mimicking Masquerade

Lev-Yadun et al. (2009b) proposed that the carrion and dung odors of various flowers that have traditionally been considered only as an adaptation for attracting the flies and beetles that pollinate them (Faegri and van der Pijl 1979) may also have another, overlooked, anti-herbivore defensive function. Lev-Yadun et al. (2009b) suggested that such odors may also deter mammalian herbivores, especially during the critical period of flowering, because carrion odor is a good predictor for two potential dangers to mammalian herbivores: (1) pathogenic microbes, and (2) proximity of carnivores. Similarly, dung odor predicts feces-contaminated habitats that present high risks of parasitism. Long-term measurements of the impact of cattle grazing on plant biomass accumulation in Israel (Lev-Yadun and Gutman 2013) showed clearly that cattle repeatedly refrained from grazing in plots used to dump carcasses, a very strong support for the proposed role of such plant odors in reducing mammalian herbivores. Herbivore fecal odor mimicry was recently found in seeds of *Ceratocaryum argenteum* (Restionaceae) and shown to be used for the attraction of dung beetles that mistake them for real dung, and roll and bury them and by this disperse them (Midgley et al. 2015). I propose that by the same token these seeds may be defended by these odors from granivores. In cases of plants emitting carrion and dung odors, chemical defensive masquerade seems to operate.

Plant-Mimicking by Other Plants as Defensive Visual Masquerade

Lev-Yadun (2014b) defined plant-mimicking defensive masquerade by plants as a masquerade type in which masquerading plants still look like plants but with characters that cause them to be uninteresting to herbivores for various reasons.

Shimshi (1979/1980) proposed that the hill ecotype of *Iris atrofusca* from the heavily grazed desert/steppe transition zone in Israel with its erect (ensiform) leaves that differ from the regular curved (falcate) leaves of the common *Iris* type in other ecosystems is a mimicry of the poisonous common geophyte *Asphodelus ramosus* (= *A. microcarpus*) of the Liliaceae that grows in the same habitat and is not grazed. Brown and Lawton (1991) postulated that the two non-spiny species *Celmisia lyalli* and *C. petriei* (Asteraceae) growing in New Zealand look rather like spiny members of the genus *Aciphylla* (Apiaceae). These and similar cases that will be described in more detail later can also be considered as Batesian mimics and not only as masquerade.

Wiens (1978) described dead grass mimicry by cacti (e.g., *Pediocactus papyracanthus*) and explained that the similarity to dead grass should have a considerable survival value during the dry season when green, palatable succulents might be heavily foraged by mammals because of both nutrition and being a source of water in their arid habitats. Similarly, dead leaf mimicry by insects and fish that was considered for many years as simple visual mimicry (Cott 1940; Edmunds 1974; Purser 2003; Suzuki et al. 2014) is now a known type of defensive animal masquerade (Ruxton et al. 2004; Skelhorn et al. 2010a; Skelhorn 2015). Wiens (1978) also described dead stick mimicry, especially known in members of six plant families (Asclepidaceae, Asteraceae, Crassulaceae, Euphorbiaceae, Liliaceae, Portulacaceae) from southern and eastern Africa that should be considered as masquerade. It involves leafless succulents with wrinkled, cylindrical, grayish stems that typically grow among weathered, gray branches of dead shrubs. Stone (1979) proposed that the dull brown coloration of young leaves in certain Malaysian palms causes them to look dead and therefore uninteresting to herbivores. This hypothesis was further discussed and supported by others (Juniper 1994; Dominy et al. 2002; Manetas 2006). Lev-Yadun (2006b) proposed that the dozens of white or light-colored plant species common in Mediterranean sandy or arid habitats may, among other physiological and defensive functions, be defended from visually oriented herbivores that look for green plants and skip those with sand or soil color. This hypothesis (don't look green, look old or dead) was also supported by other studies and discussions (e.g., Brown and Lawton 1991; Lucas et al. 1998; Dominy et al. 2002; Fadzly et al. 2009; Lee and Gould 2009; Klooster et al. 2009; Burns 2010; Strauss and Cacho 2013; Queenborough et al. 2013; Niu et al. 2014; Niu and Sun 2014; Strauss et al. 2015). Again, regular background-matching-based crypsis has very good prospects to be simultaneously involved in some of the above-mentioned cases. It is possible that some of the defense provided to young leaves in the tropics by the delayed greening associated with low nutritive value (Kursar and Coley 1992; Coley and Barone 1996; Numata et al. 2004) should be attributed to the same principle (don't look green), and not only to signaling aposematism (Lev-Yadun 2009a) or undermining herbivore's camouflage (Lev-Yadun et al. 2004a).

Leaf polymorphism that causes female butterflies to consider these plants unsuitable for egg laying because they look for a specific leaf shape (Rausher 1978) is another defensive plant masquerade subtype. Indeed, leaf polymorphism of the woody vine *Boquila trifoliolata* mimics the leaves of its supporting trees that belong to different taxa. When a single individual vine climbs on different tree species, it mimics the leaf shape of each supporting tree, thus having a sequence of leaf shapes. The unsupported vine parts and the parts that climb on leafless trunks have a different leaf shape from the shapes within canopies and suffer more damage from herbivory (Gianoli and Carrasco-Urra 2014).

As was discussed in the past concerning masquerade by animals, there is a functional and theoretical overlap between crypsis, mimicry and masquerade (Endler 1981; Allen and Cooper 1985; Ruxton et al. 2004; Skelhorn 2015). Such overlaps are also true for plants. Lev-Yadun (2014b) proposed that with the recent advances in understanding animal masquerade, its role in plant-animal interactions has to be considered more seriously. There is a strong need to study defensive plant masquerade systematically from the descriptive, theoretical and experimental points of view, something that has not been done yet at the necessary level. Conducting experiments in order to distinguish between crypsis and masquerade is not always an easy task because of the simultaneous functions of morphologies and coloration for crypsis and/or in physiology. A similar complicated situation exists in the phenomenon of colorful autumn leaves that will be discussed later; only after nearly a decade of heated debates, a more balanced understanding has emerged (e.g., Archetti et al. 2009a). The same seems to be true for plant masquerade, although the debates and the progress in understanding have just began after a detailed defensive plant masquerade hypothesis was formulated in Lev-Yadun (2014b). For instance, a recent study in crypsis of the bimodal-colored Aleppo pine (*Pinus halepensis*) seeds (one side dark and the other light-colored) dispersed after fires from serotinous cones on the mosaic of dark and light ash-covered soil (Lev-Yadun and Ne'eman 2013) may also be partly due to masquerade. The crypsis hypothesis *via* background matching of the seeds is obvious; however, it is also possible to consider that when the seeds lie on the ground of a burnt forest with the darker side up they may masquerade as a charcoal fragment, and when with the light-colored side up they may masquerade as a small carbonate stone or light colored ash. This visual resemblance caused me considerable difficulties when I tried to count *P. halepensis* seed density in the field after a forest fire. I actually saw many seeds very soon, but for long minutes considered many of them as "non-seeds" and I was sure that I missed some real seeds. Lev-Yadun (2014b) proposed that thousands of plant species inhabiting diverse habitats in many parts of the world probably employ various types of visual or chemical masquerade as part of their anti-herbivory arsenal.

The best option to test the masquerade hypothesis seems to depend on the existence of genotypes that differ only by the studied character and to examine the responses of relevant herbivores to the two types. There are such genotypes, e.g., green *versus* non-green types (Burns 2010; Niu et al. 2014; Niu and Sun 2014) and their study indeed indicated that the non-green types had a higher fitness of because of reduced herbivory (Strauss and Cacho 2013; Niu et al. 2014). Optimally, such experiments should be conducted under natural conditions, but it is not always possible or practical. However, it is possible that different leaf colors may also involve different olfactory cues or signals because of a different composition of secondary metabolites (Fineblum and Rausher 1997; Schaefer and Rolshausen 2006a), or even aposematism (e.g., Lev-Yadun 2009a; Maskato et al. 2014) and this may further complicate the experiments and their interpretation.

Chapter 19
Potential Defense From Herbivory by Dazzle Effects and Trickery Coloration of Variegated Leaves

Two types of defensive coloration theoretically and practically related to and even quite overlapping with masquerade are various types of military dazzle and trickery coloration. Dazzle and trickery coloration were developed and applied to defend navies when the range of artillery and torpedoes increased to several kilometers, and when aiming by visual means ruled naval warfare. The very conspicuous dazzle coloration invented for naval defense during World War I was used in pre-radar days to mislead attackers of naval units concerning vessel size, type, speed and direction. This is why I consider them to be related to or even to be a branch/type of masquerade. In spite of the large scale use of dazzle and trickery coloration during World Wars I and II (Wilkinson 1969; Stanley 1998; Williams 2001; Forbes 2009), and the apparent visual parallels of these patterns of coloration in animals and plants, very little attention has been given to these potential types of defense in biology in general, and concerning plants in particular. Defensive dazzle plant coloration received for some years only very brief and limited attention, without discussing the theoretical aspects (Lev-Yadun 2003a, 2006a, 2009a) and only recently (Lev-Yadun 2014c) the theoretical and botanical aspects were discussed in some depth. The definition of dazzle coloration in the scientific literature was also partly misleading because of the common confusion of dazzle coloration with crypsis *via* disruptive coloration (see Forbes 2009). In both the military and nature, the borders between these two types of defensive coloration are indeed not always clear, especially since certain color patterns may serve as camouflage when distant and for dazzling or trickery at a closer range or under different lighted conditions.

Four different major types of visual deception were involved in trickery naval painting (painted trickery *sensu* Stanley 1998). The first type was simple, painting the fore and aft sectors of naval units lighter. By this, the vessels looked shorter, or appeared to be sailing at a greater distance than the actual distance. Moreover, it may also cause difficulties in vessel type identification. The second type of painted trickery was painting false bow waves. It could be done in two ways: (1) painted on the bow side, giving the impression of a higher speed than the actual, or (2) when painted also or only on the stern side, this may have caused confusion concerning

S. Lev-Yadun, *Defensive (anti-herbivory) Coloration in Land Plants*,
DOI 10.1007/978-3-319-42096-7_19

the vessel's direction (Stanley 1998; Williams 2001). A similar deceptive visual defense tactic in animals is known as a "false head" or "deflection marks" and these are assumed to divert predator attacks (or prey attention) towards less critical body parts (Cott 1940; Wickler 1968; Edmunds 1974; Robbins 1980, 1981; Purser 2003; Ruxton et al. 2004). The third type was painting shapes of other vessels as if positioned alongside the hull, creating the illusion that either another vessel was blocking the way to the target for artillery shells, torpedoes and aerial bombs, or that there was an escort vessel that might attack (Stanley 1998; Williams 2001). The fourth type of trickery naval painting (dazzle painting) was visually and theoretically the most complicated. It was proposed to the British Admiralty as a visual defensive measure during the critical German submarine attack crisis in the year 1917 by the professional marine painter Norman Wilkinson, who served as an officer in the Royal Navy during World War I, and realized during his vessel's patrols at sea that under many types of both day and night illumination, regular naval camouflage not only failed to conceal vessels, but sometimes actually made them more conspicuous (Wilkinson 1969). Norman Wilkinson therefore proposed to the British Admiralty that instead of using the non-efficient camouflaging of navy units, they should be conspicuously painted with strong colors applied in a way that would make it hard to identify the vessel type and its distance, speed and direction. Dazzle coloration consists of high contrast patterns in either black and white, or in various colors that make the target very conspicuous, but result in difficulties in estimating the speed and trajectory and in identifying the type of target. Indeed, after this technique was tested by experienced, high ranking naval officers that agreed that it seems to function, in a short time the seas and oceans carried many thousands of vessels of all types, painted with conspicuous coloration that looked in a way like modern art. This continued to be so until the years 1942/1943 when RADAR targeting made visual targeting and aiming secondary, and after WWII, following further advances in electronic targeting, this type of visual naval defense disappeared.

Dazzle Coloration in Animals

The best commonly known suggestion for the operation of dazzle as an anti-predator defense is for the zebra's stripes by causing difficulties in distinguishing an individual zebra from the herd (e.g., Kruuk 1972; Kingdon 1984; Morris 1990). However, this hypothesis, which was not supported by carefully examined predation statistics, was basically rejected (Ruxton 2002). The recent detailed experimental study by painting life size plastic horses with zebra stripes (Egri et al. 2012) of the role of zebra stripes in horsefly (tabanids) repellence seems to give very strong indications that other, earlier suggestions for insect repellence by such patterns (Waage 1981; Brady and Shereni 1988; Doku and Brady 1989; Gibson 1992) indeed function, even if the exact mechanism is unknown, and that defense from blood-sucking and disease-transmitting insects rather than from large carnivores is involved. In any case, after decades of some theoretical discussions (e.g., Stevens

2007; Stevens and Merilaita 2009, and citations therein), only recently some experimentation of the potential for defensive dazzle effects by animals was conducted. Still, except for one study with captive birds (wild great tit, *Parus major*) on artificial food items (Hämäläinen et al. 2015), all other studies were conducted with human subjects as hunters of images on computer screens rather than in nature (Stevens et al. 2008a, 2011; Scott-Samuel et al. 2011; von Helversen et al. 2013; Hughes et al. 2014, 2015). A related study of the neuronal response in locust indicated that dazzle effects by high-contrast patterns indeed weaken movement detection (Santer 2013). The various results of all these experiments illuminate both the possibilities and limitations of the defensive potential of dazzle coloration.

Actual and Relative Plant Movement and Its Possible Role in Dazzle Effects

The operation of naval dazzle painting is related to giving the wrong impression of target identity, size, speed and direction. When this type of visual defensive tactic is discussed concerning plants, it immediately raises the question of the movement component in such a defense. The role of plant movement as defense against herbivory has been almost overlooked. The classic examples include the escape from herbivory by folding of *Mimosa pudica* leaves (Jensen et al. 2011; Amador-Vargas et al. 2014) or the exposure of thorns by leaflet folding in *Schrankia microphylla* (Eisner 1981). Lev-Yadun and Inbar (2002) proposed that black anthers of *Paspalum paspalodes* that move with a light wind may look like swivelling aphids. Later, Lev-Yadun (2009a) proposed that the swaying of leaves, stems or branches in the wind may help in ant mimicry by plants by giving the illusion that the coloration patterns proposed to mimic "ants" (e.g., Lev-Yadun and Inbar 2002; Lev-Yadun 2009d) move. A special case that will be elaborated later, is that of the enigmatic leaflet movements in elliptical circles every few minutes of the Indian telegraph (semaphore) plant *Desmodium motorium* (= *D. gyrans* = *Hedysarum gyrans* = *Codariocalyx motorius*), which has intrigued scientists, including Charles Darwin, for centuries. Lev-Yadun (2013b) proposed that it serves as butterfly or general winged arthropod mimicry for defense from herbivory. Yamazaki (2011) in the first broad, innovative and intriguing theoretical treatment of defensive plant movement, advocated the hypothesis of considering the possible roles of passive wind-induced leaf movement as a common anti-herbivory defense, although not in the connotation of visual dazzle effects, but rather on a mechanical basis. Recently, experiments by Warren (2015) strongly supported Yamazaki's (2011) suggestions.

The visual aspect of plant movement as a way of actual communication with (and deception of) animals has already been studied in two cases of pollinator attraction. In the first, some orchid species belonging to the genus *Oncidium* induce attacks by territorial male *Centris* bees when the inflorescences move in the wind, because the male bees mistake them for rival males. During the attacks of the pseudo rivals, the male bees transfer the pollen (Dodson and Frymire 1961; Wiens 1978). In the other case, Warren and James (2008) showed that movements ("waving") of the inflorescences of *Silene maritima* increase pollination success. Thus, following such functional examples from pollination biology, there is no theoretical reason to dismiss the possibility that actual and relative plant movements take part in visual defense from herbivory (see Lev-Yadun and Ne'eman 2012).

There are two ubiquitous possibilities for plant movement. The first is the actual swaying of leaves, stems, branches or flower parts in the wind (e.g., Lev-Yadun and Inbar 2002; Warren and James 2008; Lev-Yadun 2009a; Yamazaki 2011; Warren 2015). When the plants move, their morphological and color patterns move with them. From many attempts to photograph various types of defensive coloration under windy conditions in the field I realized that the movement, especially of leaves and thin branches, may be significant. The other type of movement is relative, the result of the movement of the head of large herbivores as they approach plants, or that of the entire body of an insect flying towards the plants. This is not relevant for caterpillars, aphids or other small herbivores that already inhabit leaves or walk slowly on them. Therefore, while plants are commonly considered to be motionless targets for foraging herbivores, in many instances this is not so.

Potential Defense by Dazzle Effects and Other Tricky Types of Leaf Variegation and Coloration

Several types of leaf variegation and coloration point to the possibility that variegated plants indeed use variegation and coloration as visual defense that may trick the sensory system of the herbivores concerning leaf size and location, as does naval trickery coloration *sensu* Stanley (1998). Lev-Yadun (2014c) posited that this by itself is not aposematism, although a simultaneous aposematic function for the same conspicuous markings may operate in certain cases when the variegated leaves are spiny or toxic. Lev-Yadun (2014c) proposed that zebra-like white leaf variegation (Fig. 19.1) may defend leaves and other plant organs from herbivory by dazzle effects. Two different dazzle effects may take part in defending plants from herbivory: (1) it may make it hard for herbivores to decide where in a three-dimensional space to bite the leaves (large herbivores), and (2) it may cause difficulties to land on them (insects). In addition, related types of leaf coloration described in Lev-Yadun (2014c) (Figs. 19.2–19.3) parallel trickery naval defensive paintings, which may also deceive herbivores about the actual shape, location and identity of leaves. Lev-Yadun (2014c) followed the naval terminology when considering the variegation types, trickery and dazzle coloration. The basic principle of such defense by trickery and dazzle coloration is its bearing on the tradeoff between speed of decision/operation and accuracy when herbivores attempt to bite the plants. A similar tradeoff seems to operate in imperfect-mimicry-based defense systems (Chittka and Osorio 2007; Janzen et al. 2010). The longer it takes a herbivore to decide if and

Fig. 19.1 *Silybum marianum* leaves with strong dazzel effect that makes it difficult to distinguish between different leaves

Figs. 19.2–19.3 Markings on leaves of *Trifolium* sp. from Mount Hermon, Israel. Similar markings were painted for instance on the World War II German warship Bismark

where to bite or land, the higher are the risks of attacks by its own predators (e.g., Lima and Dill 1990; Brown 1999; Ripple and Beschta 2004; Nersesian et al. 2011), along with the decrease in feeding efficiency. Some of these dazzle and trickery visual defenses may operate simultaneously with other visual defenses, such as aposematism, or serve various physiological functions.

Zebra-like white leaf variegation was first proposed, although very briefly, to serve as defensive dazzle coloration (*sensu* Wilkinson 1969) only for the three spiny annual species of open, well-illuminated Mediterranean habitats *Silybum marianum* (Figs. 1 and 19.1), *Notobasis syriaca* (Fig. 19.4) and *Scolymus maculatus* (Lev-Yadun 2003a; 2006a, 2009a, b; 2014a, c). Lev-Yadun (2014c) proposed that this initial hypothesis can be extended to various species belonging to the genera *Haworthia*, *Gasteria* and *Sansevieria*, which also have zebra-like white leaf variegation, and to any other plant taxa with such coloration. Interestingly, all three

Fig. 19.4 A variegated young leaf of *Notobasis syriaca*. This variegation is similar to that of *Silybum marianum* (Fig. 1). In many *N. syriaca* plants the white variegation of leaves that develop later in plant ontogeny turnes into white areas. Since the variegation is the outcome of air spaces, a greenhouse effect that allows better photosynthesis in cold winter days is expected

genera are common in open habitats in arid and semi-arid districts of South Africa, habitats resembling those occupied by the three Mediterranean spiny species listed above. This coincidence may indicate a similar function/s and similar evolutionary pressures. Such plants actually have classic dazzle coloration very similar to what was applied to naval vessels (compare photographs of the French cruiser Gloire in Williams 2001 or on the internet, and other photographs in Wilkinson 1969, and in Stanley 1998, with those in Lev-Yadun 2003a, and Figs. 1, 19.1, 19.2–19.3 and 19.4 here). The swaying of the leaves in the wind and the relative movements of plants in respect to flying insects or the head movements of large herbivores approaching the plants, increase the potential to form the desired dazzle effects on the visual nerve systems of the herbivores that make it difficult for them to land on or bite the leaf because of problems in locating its actual position in space.

The putative dazzle effects of such variegation are probably if not certainly not the sole visual defensive function in some of these plants. It has been proposed that the conspicuous zebra-like leaf variegation of the very spiny species *S. marianum*, *N. syriaca* and *S. maculatus* serves mainly as visual aposematism (Lev-Yadun 2003a; 2006a, 2009a, b, 2014a). Lev-Yadun (2003a) proposed that mimicking of tunneling by flies belonging to the Agromyzidae may deter female Agromyzidae flies from laying eggs, and that such conspicuous stripes may also reduce insect landing on the leaves in general, as was proposed for the evolution of zebra stripes.

Certain plant species have white or colorful zones on their leaves formed parallel to their contours, which create a visual illusion of several leaf layers instead of one. This pattern is typical of several wild taxa common in Israel: *Amaranthus* sp., *Cyclamen persicum*, and *Lamium moschatum*, all with white markings, *Polygonum*

Figs. 19.5–19.9 Markings on leaves of *Trifolium* sp. from Israel. In the genera *Trifolium* and *Medicago* there is a wealth of types of such markings. When the leaves are also toxic, visual aposematism may also be involved

lapathifolium with dark markings, and *Trifolium* spp. and *Medicago* spp. with either white or various similar but colorful markings (Figs. 19.5–19.9). The genus *Cyclamen* in general is especially characterized by such white variegated leaves (Grey-Wilson 1988) (Fig. 19.10). In a much more limited field search of the flora of Switzerland, I found the same phenomenon in *Lamium galeobdolon*, *Trifolium repens* and *T. pratense* (with white markings), and *Medicago arabica* and *Polygonum lapathifolium* (with dark markings). Except for *Cyclamen* species that occupy both open and forested habitats, all other taxa occupy open and well-illuminated habitats. Lev-Yadun (2014c) proposed that in certain plant leaves, coloration patterns formed parallel to their contours may cause a visual illusion of a fake leaf shape, size, distance and perspective, blurring actual leaf shapes, sizes, and distances, making it difficult for herbivores to estimate the distance to the leaf and aim their bites or to

Fig. 19.10 Typical markings on leaves of *Cyclamen persicum* from Israel. Each leaf looks as if it is composed of at least two leaves, making it difficult for a mammalian herbivore to aim a bite. Since the leaves are also toxic, aposematism may also be involved. All members of the genus *Cyclamen* express similar types of markings. *Cyclamen persicum* plants from Israel show a great variability of the markings

land on them. When there are several such parallel markings, the leaf may look as if it is made of several leaf layers that occupy different planes. Lev-Yadun (2014c) proposed that leaves of such taxa use the vegetal parallel of the third type of defensive naval painting, i.e., painting shapes of smaller vessels as if positioned alongside the hull (e.g., Stanley 1998; Williams 2001). As with several other types of defensive plant coloration, additional physiological or defensive functions of such coloration patterns should also be considered and this should be done specifically for each species and ecology. A dramatic leaf coloration of this type is that of the ornamental *Pelargonium x hortorum 'Mr. Henry Cox*' (Fig. 19.11). Moreover, as some of the plants with leaves variegated in white (or with other colors) (e.g., *Cyclamen* spp. and certain *Trifolium* spp. and *Medicago* spp.) may be poisonous, because there are cyanogenic types in *Trifolium* spp. and *Medicago* spp. (Crawford-Sidebotham 1972; Dirzo and Harper 1982) and *Cyclamen* species are known to be rich with alkaloids and saponins (Hornell 1941; Reznicek et al. 1989), the visual illusion may thus not be the sole visual defensive mode, and these patterns may also serve as aposematic coloration. Müllerian and Batesian mimicry rings are expected to exist in such cases. The issue of plant aposematism will be discussed in great detail in later chapters. Theoretically, coloration that produces an illusion of a different leaf shape may also cause identification problems for insects that search for specific leaf types or leaf developmental stages (e.g., Rausher 1978; Prokopy and Owens 1983; Brown and Lawton 1991; Reeves 2011).

Lev-Yadun (2014c) concluded that following the data from the recent experiments that indicated that dazzle effects may indeed defend animals from predation (Stevens et al. 2008a, 2011; Scott-Samuel et al. 2011; von Helversen et al. 2013;

Fig. 19.11 The ornamental *Pelargonium x hortorum 'Mr. Henry Cox'* variety demonstrates an extreme pattern of trickery coloration

Hughes et al. 2014, 2015; Hämäläinen et al. 2015), and from the morphological descriptions concerning potential dazzle patterns in leaves of various taxa (e.g., Lev-Yadun 2003a, 2009b, 2014c), which were proposed to defend from herbivory by dazzle and trickery coloration, that defense from herbivory by dazzle and trickery coloration should be considered as a working hypothesis concerning the hundreds if not thousands of plant species that express such types of leaf coloration.

Chapter 20
Plants Undermine Herbirorous Insect Camouflage

The everlasting war between plants and herbivores is not only a direct one, and other organisms are commonly involved in plant-herbivore interactions. Plants use microorganisms, fungi, invertebrates, vertebrates and even other plants for their defense, and their morphology and coloration, i.e., visual aspects are among the many factors by which plants help organisms that belong to other kingdoms, other plant taxa, and other trophic levels to attack or repel herbivores. This chapter is devoted to one of these visual systems.

Plants provide habitat and food for many animals, so it is logical to assume that the visual perception of animals and their visual crypsis (of both herbivores and predators) co-evolved with plants, as is well known from mutualistic plant/animal interactions such as pollination. Intuitively, the common optimal camouflage for herbivorous insects on photosynthetic organs should be green, and indeed, many herbivorous insect species, e.g., aphids, caterpillars and grasshoppers, have evolved green coloration (Fig. 20.1). This is also true for many predators such as mantids, frogs, lizards and snakes (Cott 1940; Purser 2003). The effectiveness of green camouflage that matches that of leaves and green branches, or gray/brown colors that match the colors of many barks, is impaired by diverse non-green, or even by a variety of green shades of plant backgrounds, as was evident with industrial melanism (Kettlewell 1973; Majerus 1998; Cook 2000, 2003; Grant and Wiseman 2002).

My interest in the function of the common different coloration of two leaf sides started because of a daily visual provocation. The municipal authority of my hometown Ramat-HaSharon (several km north-east of Tel Aviv), planted along the street that enters the town a type of an ornamental *Ficus* tree species with huge leaves that are green on the upper side and whitish on the lower, and *Vitex* shrubs with leaves that are green on the upper side and purple-blue on the lower. For years, whenever I drove home, I could see those leaves while waiting in the much too slow traffic, and I would wonder not only when would the mayor and the National Road Authority do something about the traffic jams, but also what the function of this type of coloration is. I knew hundreds of other species (annuals, geophytes, shrubs, climbers and trees) that had leaves with different colors on the upper (adaxial) and lower (abaxial)

S. Lev-Yadun, *Defensive (anti-herbivory) Coloration in Land Plants*,
DOI 10.1007/978-3-319-42096-7_20

Fig. 20.1 A green insect on a green young spike of *Hordeum spontaneum*, east of Mount Tabor, Israel. Green is the best camouflage for animals occupying green foliage, stems, flowers and fruits

sides. These plants originated in hot deserts, in tropical, in sub-tropical, in temperate and even in arctic regions. They were evergreens (*Ficus* spp., *Magnolia* spp.), deciduous (*Populus* spp.), monoecious (*Rubus* spp.) and dioecious (*Ficus* spp.). They belonged to mountain floras, lowland floras, wetland plants, coastal plants, winter plants and summer plants. Some expressed such coloration only at the cotyledon stage (*Matricaria* sp.) and others were tall trees that expressed it all over their large canopy (*Ficus* spp., *Magnolia* spp., *Populus* spp.). Some plants had it throughout the life span of a leaf (*Ficus* spp., *Populus* spp.), but others, e.g., *Cyclamen persicum* expressed it only or mostly when the leaves were young (Figs. 20.2, 20.3, 20.4, 20.5, 20.6, 20.7 and 20.8). While the color of the upper side was always green, the lower side could be black, blue, brown, gray, light green, orange, pink, purple, almost black, red, tan, white or yellow. The color of the underside was the outcome of epidermal or sub-epidermal cell color or of trichomes of various colors. Not even the slightest cue for any potential rule emerged when I tabulated the color combinations and plant characters. Light reflection to lower temperature or photoinhibition, or on the contrary to allow better illumination to lower and shaded leaves, heating, camouflage or counter-shedding against herbivory, none of these various potential functions could explain or even hint about the function of these leaf color types by their co-occurrence with ecological patterns and life forms. However, I was sure that there must be some simple principle that explains this very common and color-vise variable phenomenon that was expressed irrespective of the ecological type and growth conditions, morphological aspects and phylogenetic origin. The more I understood various aspects of defensive plant coloration, the more it intrigued me to find a solution for this world-wide leaf coloration pattern.

The solution emerged by surprise while I was traveling in a roofless car to an experimental forest plot in North Carolina. I visited the laboratory of Professor Ron

Fig. 20.2 *Acer* sp. from North Carolina. The three colors of the system: green upper leaf side, white lower leaf side and reddish veins and branches, will cause difficulties for camouflage for insects of any color that will consume the foliage

Fig. 20.3 Purplish undersides of young *Cyclamen persicum* leaves from Israel. This conspicuous color fades to *whitish* as the leaves mature

Sederoff in the Forest Biotechnology Group at the State University of North Carolina at Raleigh for discussions and collaboration on issues related to wood and fiber formation and tree biotechnology (e.g., Lev-Yadun and Sederoff 2000, 2001). North Carolina has many trees and shrubs with a different color on each leaf side, including several *Magnolia* species in which this character is very conspicuous. As usual, the double coloration of the leaves intrigued me and I took many pictures of such leaves. In one of the visits (August 2002), we drove to the forest in his convertible

Fig. 20.4 *Red* and *purple* young leaves of *Ficus elastica* grown as an ornamental in Tivon, Israel. Many tropical trees have such coloration in their young leaves. An insect camouflaged on typical mature green leaves will not be camouglaged on young ones, and *vice-versa*

Fig. 20.5 *Upper* (adaxial) *green* and *lower* (abaxial) *brown* leaf side in *Magnolia* sp. from North Carolina

car in order that I see a large field experiment with loblolly pine (*Pinus taeda*) trees that was part of his important, American size project of cloning and characterizing the genes involved in loblolly pine's wood formation. After some minutes of traveling and looking upwards on the tree canopies with their various underside leaf colors, but still at a point between the University and the forest, it occurred to me, in a process that took only several seconds to mature into a clear idea, that if an insect is camouflaged on one side of a leaf expressing different colors on each side, it cannot be camouflaged on the other, making half of the tree's leaf area unsuitable for it. Remembering from my M.Sc. days that poisonous and aposematic caterpillars are known to consume leaves at their near reach, leaving tattered edges, with no fear of being detected by birds (Heinrich 1979; Heinrich and Collins 1983), whereas non-poisonous caterpillars try to eliminate evidence of foraging by snipping

Fig. 20.6 A *Magnolia* sp. tree from North Carolina. The differences in color of the adaxial and abaxial leaf sides is very conspicuous

Fig. 20.7 A red underside of a large cotyledon of the annual winter herb *Matricaria* sp. from northern Israel

partially-eaten leaves, or by moving away from them, by keeping the leaf shape intact by eating the contours, or by feeding from the lower side or at night (Heinrich 1979; Heinrich and Collins 1983), I understood that I could define a potentially new type of defensive plant coloration – undermining herbivore camouflage. Two years of discussions with relevant colleagues and careful writing helped me refine this hypothesis (Lev-Yadun et al. 2004a), which will be described below.

This proposed visual defensive strategy to reduce herbivorous invertebrate attacks by undermining their camouflage is most strongly manifested by the

Fig. 20.8 The contrast between the *white underside* and *green upperside* of *Rubus* sp. leaves from Athens Ohio

common differences in the color of upper (adaxial) and lower (abaxial) leaf sides. However, it also operates by non-green colors of petioles, major leaf veins, stems, flowers, fruits, or colorful spring or otherwise young leaves and of autumn or otherwise senescing leaves.

The ecological and evolutionary importance of coloration and camouflage for insect survival has received much attention, leading to several hypotheses and their experimental testing (Théry and Gomez 2010). Some insects adapt to various types of plant coloration as a consequence of the evolutionary arms race between plants, herbivorous insects and predators of these insects. Many populations of insect herbivores show a remarkable polymorphism and optimize local crypsis to fit the heterogeneity of the microhabitats of their host plants. Such intraspecific color polymorphism has been found, for instance in stick insects (Bedford 1978), grasshoppers (Pener 1991) and butterflies (Stamp and Wilkens 1993; Fink 1995). Plant organs, including leaves and stems, often change their color with age or season, and there is evidence that certain arthropods can trace these changes and modify their coloration to optimize their crypsis, e.g., the coloration of hawkmoth caterpillars is determined by the reflection of the background that they perceive soon after hatching (Grayson and Edmunds 1989). In other lepidopterans, food content and quality are key factors in determining larval color and morphology, which enhance crypsis (Fink 1995; Green 1989). Green's (1989) classic study demonstrated the importance of intra-plant variation in caterpillar coloration. On a given oak (*Quercus* sp.) host, caterpillars of *Nemoria arizonaria* may develop into either a 'catkin' or 'twig' morph to maximize their camouflage. Some stick insects (Phasmatodea) may adjust their color with changes in the color of the host plant's foliage (Bedford 1978). Many insects see colors, distinguish between shapes, and select their habitat accordingly (Prokopy and Owens 1983; Stavenga and Hardie 1989; Dafni et al. 1997;

Fig. 20.9 *Red* leaf margins of *Aeonium* sp., a character found in thousands of plant species all over the world

Rausher 1978; Rivero-Lynch et al. 1996; Kelber 2001; Penacchio et al. 2015b). Therefore, producing organs of unsuitable color may cause insects not to occupy such plants (see Ruxton et al. 2004:87).

While flowers and fruits exhibit enormous variations in coloration, which was studied extensively concerning plant-animal signaling (e.g., Faegri and van der Pijl 1979; van der Pijl 1982; Schaefer and Ruxton 2011), leaves, petioles and stems can also be quite colorful, but were studied much less even at the descriptive level, notwithstanding concerning plant-animal signaling or anti-herbivory defense. Many higher plants inhabiting diverse terrestrial ecosystems worldwide exhibit remarkable inter- and intra- individual and organ color variation. For instance, color differences between the upper and lower sides of leaves, and between the veins or petioles and the leaf blade (Figs. 20.2, 20.3, 20.4, 20.5, 20.6, 20.7, 20.8, 20.9 and 20.10), are common across diverse plant forms, from short annuals or perennial herbaceous plants to tall trees, and in various habitats, from deserts to rain forests and from the tropics to the temperate region. Furthermore, leaf color frequently changes with age, season, or physiological condition. Young leaves of many tropical trees and shrubs (Richards 1996), as well as many non-tropical (Lev-Yadun et al. 2012), are red and later on become green, or are just not green when young (Kursar and Coley 1991, 1992, 2003; Coley and Barone 1996), whereas in the temperate zones leaves of many woody species change to bright or conspicuous yellow and red colors in autumn (Archetti 2000, 2009a; Archetti et al. 2009a; Lev-Yadun and Holopainen 2009). Although it is generally agreed that the bright or contrasting flower and fruit colors facilitate communication between plants and their pollinators and seed-dispersers (e.g., Ridley 1930; Faegri and van der Pijl 1979; Willson and Whelan 1990; Weiss 1995; Clegg and Durbin 2003; Schaefer et al. 2004; Schaefer and Ruxton 2011), there is no *a priori* reason to assume that their colors cannot also

Fig. 20.10 Spiny leaf margins of ananas (*Ananas comosus*)

serve as defense from herbivory. Thus, it was worthwhile to formulate a general unifying hypothesis that explains the adaptive significance of the many common color patterns found in plants.

Lev-Yadun et al. (2004a) proposed such a new unifying concept, a general explanation for many of the vegetal and some reproductive coloration types found in nature, proposing that all these can help in plant defense against herbivory. The essence of the hypothesis was based on a very simple, new principle, that many patterns of plant coloration undermine the visual camouflage of invertebrate herbivores, especially insects and other small arthropods, thus exposing them to predation, and in addition causing them to avoid plant organs with unsuitable coloration, to the benefit of the plants. This is a special case of "the enemy of my enemy is my friend", and in a way a visual parallel of the chemical signals that plants emit to call predaceous wasps when they are attacked by caterpillars (Kessler and Baldwin 2001; Kappers et al. 2005; Karban 2015). This hypothesis (undermining invertebrate herbivores' camouflage), however, does not contrast or exclude other previous or future explanations of specific types of plant coloration. Traits such as coloration that have more than one type of benefit may be selected for by several agents and evolve more rapidly than ones with a single type of advantage.

Lev-Yadun et al. (2004a) proposed that when a given leaf has two different colors – green on its upper (adaxial) side and blue, brown, pink, red, white, yellow or just a different shade of green on its lower (abaxial) side, a green insect (or one of any color) that is camouflaged on one of the leaf's sides will not be camouflaged on the other. The same is true for colorful major leaf veins, colorful leaf margins, petiole, branch, stem, flower, or fruit coloration. Colorful veins, for instance, make it easier for birds to detect damaged leaves, a hunting clue to search these leaves for insects. Light-colored coastal and desert plants (e.g., Lev-Yadun 2006b) may also undermine the camouflage of many herbivorous invertebrates. It has therefore been suggested (Lev-Yadun et al. 2004a) that green or otherwise colored herbivores that move, feed, or rest during the day on plant parts that have a different coloration from their own immediately become more conspicuous to their predators. The same is true for insect egg color, which should match the background color for defense. Thus, many plants are simply too colorful to enable a universal camouflage of herbivorous insects and other invertebrates to operate successfully, and so they force small herbivores to cross visual "killing zones" with colors that do not match their camouflage, for instance when they have to move from one leaf to another or from one branch to another. Since the variable coloration is usually either ephemeral (red young leaves or red or yellow autumn leaves) or occupies only a small part of the canopy (young leaves, petioles, leaf veins, flowers, and fruits), the gains for insects that have evolved to match such ephemeral or less common coloration are low (Lev-Yadun et al. 2004a), and with low gains it is difficult to overcome this type of plant defense by evolution. The excellent color vision of many predators of insects, in particular insectivorous birds (the most common and significant predators of herbivorous invertebrates) (Van Bael et al. 2003; Lichter-Marck et al. 2015), probably makes undermining herbivores' camouflage highly rewarding for plants (Lev-Yadun et al. 2004a). These patterns of coloration thus both expose such herbivores to their predators and cause them to avoid plant parts that do not match their color, to the benefit of the plants.

Lev-Yadun et al. (2004a) discussed several lines of evidence (industrial melanism, caterpillar and bird behavior, insect habitat selection, insect and bird vision) that supported their hypothesis. The multiple potential benefits from the two sides of the leaf having different colors include the following: improved photosynthesis, improved water balance, protection from UV irradiation, cooling and heating; and in addition reduced herbivory, which together might increase the adaptive value of such variable coloration and enhance the rate of evolution of variable coloration patterns within taxa and result in common convergence. Moreover, this is a common natural parallel to the well-known principle basic to the phenomenon of industrial melanism (e.g., Kettlewell 1973; Majerus 1998; Cook 2000, 2003; Grant and Wiseman 2002), and the reverse process (Majerus 1998; Cook 2000, 2003; Grant and Wiseman 2002), which illustrates the great importance of plant-surface color-based camouflage for herbivorous insect survival and can serve as an independent test for the herbivore camouflage undermining defensive plant-coloration hypothesis.

Air pollution following the widespread use of coal in England and the USA in the nineteenth century resulted in the overall darkening of the environment, and the consequent increase in the proportion of dark morphs in many insect populations (industrial melanism). This change was brought about by selective predation by birds of the lighter morphs, which did not match the new, darker background of tree trunks, branches and foliage (Kettlewell 1973). The marked reduction in air pollution over the last decades and the consequent lightening of the environment was followed by the subsequent decrease in proportion of the dark morphs and an increase in the proportion of light-colored morphs in both England and the USA (Majerus 1998; Cook 2000, 2003; Grant and Wiseman 2002). This large-scale natural experiment clearly demonstrates the adaptive significance of color matching of insects with their vegetal background (Majerus et al. 2000). Experiments of selective predation of various color morphs of caterpillars, not related to industrial melanism, provided similar results (Endler 1984; Edmunds and Grayson 1991).

In heterogeneous habitats, optimal camouflage coloration should maximize the degree of crypsis in the microhabitats used by the prey, and herbivores may enjoy better crypsis in heterogeneous habitats (Endler 1984; Edmunds and Grayson 1991; Merilaita et al. 1999; Merilaita 2003). Therefore, a plant with many colors may under certain conditions that will be discussed below provide better crypsis than a unicolored one. However, what is critical for both the plant and the herbivorous insects is the ratio between the size of the herbivore and the size of color patches of plants, and this ratio determines whether a certain coloration pattern will promote or undermine the crypsis of the invertebrate herbivores. Since insects are in general smaller than many of the color patches of leaves, flowers, fruits or branches, they will be more exposed to predators and parasites on most plant surfaces because of the variable plant coloration. By contrast, those types of variegation that form small-scale mosaics and regions of contact zones between two coloration types (Fig. 1) are unlikely to undermine insect camouflage, a point that has been addressed by Schaefer and Rolshausen (2006a). I emphasize that many types of variegation that form small-scale mosaics are not considered undermining insect camouflage and were never considered to be so in Lev-Yadun et al. (2004a), and that this point was not understood by Schaefer and Rolshausen (2006a) and by Schaefer and Ruxton (2011). These authors considered together variegation in general with large unicolored areas that are non-variegated concerning undermining insect camouflage. This important issue should be documented and analyzed under natural conditions in many taxa combinations of plants and insects to allow better understanding.

One aspect of the special importance of the hypothesis on herbivores' camouflage-undermining (Lev-Yadun et al. 2004a) is that it imparts a defensive function not only to the orange, pink, red, brown, blue, and black colors that stem from anthocyanins, which are correlated with various defense chemicals (see Fineblum and Rausher 1997; Schaefer and Rolshausen 2006a), but also to light green and white coloration, which lack such an inherited chemically-based explanatory connection.

Of the many experiments conducted on feeding strategies of herbivores in relation to predation risks, those focusing on timing, selection of background, mimicry of plants, position of the herbivore on the leaf, modes of feeding and the concealing of damage to plant tissues pertain to the hypothesis of undermining invertebrate herbivore camouflage. In such studies, in which birds (mostly tits *Parus* spp.) and caterpillars or grasshoppers were used in large aviaries. That hypothesis can be tested in such experimental setups where the colors of plant organs as well as those of caterpillar bodies (or artificial insect models) can be manipulated. Consequently, in such experiments the fitness of the herbivore and the efficacy of its predator can be evaluated (e.g., Heinrich 1993; Joern 2002). A different approach would be to manipulate organ and insect coloration on photographs and monitor the behavior of potential visually oriented predators (e.g., Bond and Kamil 1998). It is possible to alter plant coloration by genetic engineering, or by using color mutants, and to test both herbivore and predator behavior. Genetic engineering seems to be better as it might lower variability of other possible signals such as odor, unless the new genetic combinations provide metabolites to various biosynthetic pathways that will result in new odors. Plants, in which the color of one side of the leaf, or that of a petiole or vein, or of the bark was changed, can be used to examine if herbivore choices concerning landing and feeding sites have changed. Similarly, birds' ability to inspect prey, and herbivore survival with changing plant colors, can also be studied. There is also the question of the ability of insect predators that use colors for crypsis from both their prey and their own enemies, e.g., spiders, mantis, chameleons, snakes and frogs, to prey upon insects when plant colors change. Since many birds and insects see UV (Maddocks et al. 2001), testing the camouflage of many insect species in the UV, as was done for the peppered moth (Majerus et al. 2000), is essential. Moreover, we have only a very fragmentary view of how plants look in the UV, and this understanding should be established for both issues of herbivory and protection from UV radiation. Comparison of the level of insect herbivory of red or green morphs of young leaves of *Quercus coccifera* showed less herbivory of the red leaves. This has been interpreted as either the outcome of lower visibility of red leaves to insects (a type of camouflage) or as the operation of the undermining of the herbivorous insect camouflage (Karageorgou and Manetas 2006).

Chapter 21
Delayed Greening

Delayed greening of young leaves is a phenomenon in which instead of being green, these young leaves express various conspicuous colors (white, pink, red, very light green) (Fig. 20.4), is a common phenomenon in the tropics (Richards 1996). The hypothesis that delayed greening is associated with low nutritive value in young leaves of tropical plants, and that this property defends them from herbivory (Kursar and Coley 1991, 1992, 2003; Coley and Barone 1996) is a special case of a more general hypothesis about low nutritive value as defense (Feeny 1976; Moran and Hamilton 1980; Augner 1995). A similar principle based on a very different mechanism is known to operate well as defense in many leaves, stems, and young fruit that produce high levels of tannins and other protease inhibitors that decrease protein availability during digestion (Robbins et al. 1987; Bernays et al. 1989; Ryan 1990) or in flowers with diluted nectar as defense against nectar robbers (Irwin et al. 2004).

The defensive function of delayed greening was tested in the forest by Numata et al. (2004), who showed that seedlings of various species of the genus *Shorea* (Dipterocarpaceae) that express delayed greening indeed suffered less damage from insect herbivory than species with regular greening. A second test of the possibility of a defensive potential of delayed greening and red young leaves conducted in Ecuador and Panama also showed clearly that this character is associated with reduced herbivory (Queenborough et al. 2013). A similar or related phenomenon occurs in *Conocarpus erectus* (buttonwood), which has polymorphic leaf color. In buttonwood, some individual plants are silvery and some are green, but some change from green when young to silvery later. Silvery leaves in buttonwood suffer less insect herbivory (Schoener 1987, 1988; Agrawal and Spiller 2004). Yet despite the high likelihood that delayed greening is effective, and probably also operates outside the tropics, this hypothesis has not received the attention it merits.

Lev-Yadun (2009a) proposed that the associations of being unpalatable with conspicuous colors (delayed greening *per se*, and all other colors expressed by such young leaves) may signal herbivores about the lower nutritive value, a typical aposematism. At the same time, such coloration may also undermine herbivorous insect camouflage (e.g., Lev-Yadun et al. 2004a; Lev-Yadun 2006a, 2009a).

S. Lev-Yadun, *Defensive (anti-herbivory) Coloration in Land Plants*,
DOI 10.1007/978-3-319-42096-7_21

Chapter 22
Red/Purple Leaf Margin Coloration: Potential Defensive Functions

Another multifunctional type of leaf coloration is that of red leaf margins. The leaf margins of many plant species of various plant families and life history types, belonging to the floras of several continents, produce a conspicuous but sometimes very thin band of red/purple color around their periphery (Fig. 20.9). Common explanations for defensive leaf coloration such as plant camouflage or undermining herbivorous insect camouflage do not seem to explain the repeated evolution of this color pattern. Despite the widespread distribution of this leaf trait, very few studies have proposed, discussed, or tested hypotheses to explain its probable defensive significance. Hughes and Lev-Yadun (2015) reviewed previous studies and further explored this topic.

In some species, reddening/purpling of the leaf margin can be induced by environmental stress, such as phosphorus deficiency (Raghothama 1999; Lloyd et al. 2001) or freezing stress (Nicotra et al. 2003). In many other species, red/purple leaf margins is a constitutive trait, exhibited only during the early stages of leaf development or persisting throughout the entire lifespan of the leaf (e.g. Bayly and Kellow 2006; Hughes et al. 2010). In most described cases of red/purple leaf margin coloration, the pigment responsible for this coloration was either shown or presumed to be anthocyanin, although there is no reason to dismiss the possibility that carotenoids or betalains may be involved in other cases. Could red/purple leaf margins play a role in anti-herbivory defense? The answer, in spite of the meager data, is probably positive. Hughes and Lev-Yadun (2015) outlined three possible mechanisms by which this could be possible.

Many insects do not simply land at random on leaves to consume them, but rather utilize chemical and visual cues, see colors, distinguish between shapes, and choose suitable leaves and avoid unsuitable ones (Rausher 1978; Shapiro 1981a, b; Prokopy and Owens 1983; Rivero-Lynch et al. 1996; Dafni et al. 1997; Finch and Collier 2000; Kelber 2001; Reeves 2011). Because most insects lack a red photoreceptor, red leaves should appear less bright than green leaves to insects (Döring et al. 2009). Whether this deters insects by giving the appearance of dead/necrotic leaf tissues as proposed by Stone (1979), or is perceived as its own, distinct signal,

S. Lev-Yadun, *Defensive (anti-herbivory) Coloration in Land Plants*,
DOI 10.1007/978-3-319-42096-7_22

is still unclear. For instance, the old leaves of various taxa begin their death from the margins. Some studies have demonstrated that aphid species find red and brown similarly unattractive compared to green and yellow surfaces (Döring et al. 2009), and several studies have demonstrated that insects prefer landing (Kostal and Finch 1996; Ramírez et al. 2008; Archetti 2009b; Döring et al. 2009), feeding (Numata et al. 2004; Karageorgou and Manetas 2006; Markwick et al. 2013; Maskato et al. 2014), and ovipositing (Maskato et al. 2014) on green rather than on red leaves. However, results are not always consistent across all insect taxa (Irwin et al. 2003). Support for red leaf coloration as an honest signal of low leaf quality and/or high chemical defenses include a positive correlation between increased reddening and low foliar nitrogen (Schaberg et al. 2003; Kytridis et al. 2008; Nikiforou et al. 2011; Carpenter et al. 2014), low phosphorus (Dell et al. 1987; Bernier and Brazeau 1988; Raese 2002; Tng et al. 2014), increased phenolics (Karageorgou and Manetas 2006; Karageorgou et al. 2008) and other defense compounds, e.g., polygodial in Cooney et al. (2012). Some studies have further suggested that anthocyanins themselves may impair herbivore growth and vigor. A recent study on genetically-modified *Malus x domestica* engineered to express high levels of anthocyanin pigments in leaves demonstrated that the light brown apple moth, *Epiphyas postvittana,* not only prefers to feed on greener leaves, but exhibits significant reduction in development and growth when fed red leaves compared to green ones (Markwick et al. 2013). Similar results were found for *Pieris brassica* larvae reared on red compared to green cabbage leaves (Maskato et al. 2014), and slugs reared on either Anthocyanin + or Anthocaynin– phenotypes of wild radish (Irwin et al. 2003). However, aphids that oviposited on either green or red *Nothofagus alessandrii* leaves during the autumn did not differ in performance in the following spring (Ramírez et al. 2008).

Within the context of leaf margin reddening, initiation of herbivore feeding at the leaf edge has been reported for many insect groups including caterpillars, adult beetles and beetle larvae, grasshoppers, and weevils (Cranshaw 1998). Accordingly, enhanced defense of leaf margins has evolved many times in many plant taxa. Hughes and Lev-Yadun (2015) briefly described some of the more common ones. The leaves of *Desmostachya bipinnata* (Halfa grass), a perennial grass of the Poaceae, have very smooth, but sharp, edges that cut the tissues of soft-bodied animals like a razor when they are pulled (my personal bloody field experience). The same is true for the perennial aquatic plant *Butomus umbellatus* of the Butomaceae, a character that gave the genus its scientific name meaning cutting cow (Stant 1967). *Pennisetum purpureum* (elephant grass) is also known to cut mouth tissues of grazing animals (Katewa et al. 2001). Various plant species have leaves with a cutting edge that is not sharp and smooth, but rather armed with small spines that turn it into a saw. Famous examples are the two Pampas grass species *Cortaderia selloana* and *C. jubata* (Meyers et al. 2008), the saw grass *Cladium jamaicense,* and *Ananas comosus* (pineapple) (Fig. 20.10). Larger margin leaf spines have also been shown to deter herbivory, especially of larger mammalian browsers (Cooper and Owen-Smith 1986). Kangaroos presented with foliage exhibiting either intact or removed spines tend to prefer feeding on the leaves from which spines have been removed

(Rafferty 1999), and ungulates preferentially graze on less-spiny individuals of European holly (*Ilex aquifolium*). As might be expected, new growth of browsed *I. aquifolium* shoots exhibited increased spinescence compared to un-browsed shoots (Obeso 1997). Leaf-margin spines are also more common in low branches of tall individuals or in permanently grazed evergreen Near Eastern trees of *Quercus calliprinos* and *Phillyrea latifolia* (Perevolotsky and Haimov 1991).

Within the context of defense against invertebrate herbivores, increased chemical defenses, sclerophylly, trichomes, and fiber content are all believed to be more effective deterrents than larger leaf spines. However, several studies have focused on these characteristics at the leaf margin specifically. One study on American holly (*Ilex opaca*) demonstrated that fibrous veins and a thickened epidermis at the leaf margin appeared to be more effective than leaf spines at deterring herbivory by the generalist caterpillar *Hyphantria cunea* (Potter and Kimmerer 1988). In several New Zealand *Veronica* species (Hughes et al. 2010), some Australian eucalypts (Neish et al. 1995), and *Aloe* species (Gutterman and Chauser-Volfson 2000; Chauser-Volfson et al. 2002), tissues at the leaf margin have significantly higher defensive phenolic content than more interior tissues. This can be an adaptation to specifically deter herbivores from leaf margins.

Hughes and Lev-Yadun (2015) discussed the possibility that reddening at the leaf margins may serve to deter herbivory, perhaps by warning potential leaf-chewing herbivores of reduced nutritional quality, and/or enhanced mechanical or chemical defenses. Some support for this hypothesis already existed in the literature. Hansen et al. (2003) proposed that red leaf margins may mimic browsing damage and by this indicate that it is not worth spending time consuming them. Hughes et al. (2010) were the first to hypothesize that reddening of leaf margins could be a signal to potential herbivores of increased defense compounds (aposematism), either within the leaf margin or in the entire leaf. They compared marginal and laminar leaf phenolic concentrations, as well as the levels of herbivory at the leaf margins of five New Zealand *Veronica* species displaying anthocyanic leaf margins *versus* six species which lacked red marginal coloration, grown in a common garden. Although margins of all species contained higher levels of phenolics than more interior tissues, results showed no significant difference between levels of phenolic compounds (either in the margin or the centers of leaves) or in herbivory incurred at the leaf margin in species with red *versus* non-red margins. However, this study was rather crude in its approach, as some phenolics involved in anti-herbivory defense (e.g. formyl phloroglucinol) comprise only a small part of the total phenolic pool (<1 %), and did not correlate with total phenolic levels (Lawler et al. 1998, 1999). Hughes et al. (2010) recommended that future studies examine relative concentrations of specific defense compounds with anthocyanins, rather than entire metabolic pools. Accordingly, soon after, Cooney et al. (2012) demonstrated that the width of red margins in leaves of *Pseudowintera colorata* correlated positively with concentrations of the defense compound polygodial, and negatively with the extent of leaf herbivory observed in the field. Furthermore, when the authors conducted feeding trials using larvae of the generalist herbivore *Ctenopseustis obliquana* (brown-headed leafroller moth), which naturally feeds on *P. colorata* leaves, significant preference for leaves lacking red margins was observed under white light, but not under red light or darkness, which cancelled the visual red signal, consistent with a visual, herbivory-deterrent function of a red leaf margin (Cooney et al. 2012). It is obvious that many more species with red leaf margins should be studied. Moreover, Müllerian and Batesian mimics are expected to occur.

A second and very different type of visual defense is also potentially provided by red leaf margins. It belongs to the group of defenses (undermining herbivorous insect camouflage, mimicking caterpillar damage by lobed leaves, and call for help by predators with specific volatiles) employing the principle of "the enemy of my enemy is my friend". Plants are thus not always passive members in the tri-trophic system of plants-herbivores-predators, but on an evolutionary scale are rather active and tricky ones (e.g., Lev-Yadun et al. 2004a). "The enemy of my enemy is my friend" concept can be applied to colorful leaf margins as a potential explanation for the occurrence of the trait, although not an exclusive one. The visual strategy within this principle of defense on leaves was first proposed by Niemelä and Tuomi (1987), accordingly, the lobed leaf morphology of some plants belonging to the Moraceae mimic caterpillar damage and that this morphology attracts predaceous birds. Similarly, Dirzo (2002) proposed that the leaf ends in certain palms mimic chewing damage and this has been posited to repel herbivores. Thus, Hughes and Lev-Yadun (2015) proposed that disruption of a continuous colored margin by feeding can serve as a very good cue for foraging insectivorous birds and for predaceous or parasitic insects about the existence of herbivores, and so attract them. Accordingly, damaged red leaf margins show (advertise) previous damage as a way to avoid future attacks on leaves by herbivores (Fig. 22.1). The ecological and evolutionary importance of outline (contour) coloration has been described previously concerning butterflies. The wings of many butterflies are outlined with a narrow frame of

Fig. 22.1 *Red* leaf margins of *Crassula* sp. In the leaf that was chewed the red margin is disrupted, a potential cue for herbivore activity for predators

color. This pattern enables an easy spotting of damaged wings, a cue for females that males are either developmentally defective, clumsy, or old, i.e., less preferred types to mate with. The same principle was proposed to operate in birds and concerning ornaments of other animal groups, especially during sexual selection (Zahavi and Zahavi 1997). A damaged red leaf margin is also a visual parallel to the volatiles that damaged plants emit that attract predators of herbivorous insects (Kessler and Baldwin 2001). Support for this hypothesis emerges from the fact that birds prefer to forage on willow leaves with higher visible insect damage to the leaves (Sipura 1999) and that birds are attracted to defoliated canopies (Mäntylä et al. 2008). It is thus possible that damaged colorful leaf margins may serve as a clear indication of previous herbivore activity and by this deter various other herbivores.

A third potential visual defense type is that leaves with interrupted red margins because of previous feeding may be viewed less favorably by browsing mammals or by insect herbivores, which may refrain from attacking or occupying already damaged leaves. A case similar to colorful leaf margins that may signal damage when consumed is described concerning the leaves of the silvery morph of *Conocarpus erectus* (buttonwood), where the silvery morph is less damaged by caterpillars than the green morph (Agrawal and Spiller 2004). From their Fig. 1C it is obvious that their major coleopteran herbivore (*Biopsyche thoracia*) causes brown scars on the silvery leaves that are very conspicuous. These scars may be used by other insects as an indication of previous damage and that the leaves of the branch or plant are already occupied by competitors, thereby reducing future herbivory. This phenomenon is known from frugivores, avoiding eating damaged fruit (especially large fruit) (Janzen 1977; Herrera 1982; Manzur and Courtney 1984; Borowicz 1988; Buchholz and Levey 1990; Cipollini and Stiles 1993; Marchand and McNeil 2004), pollinators that prefer plants with undamaged leaves over plants with damaged leaves (Lehtilä and Strauss 1997), and herbivores that avoid damaged or occupied leaves (Benson et al. 1975; Gilbert 1980; Shapiro 1981a, b; Williams and Gilbert 1981; Hartley and Lawton 1987; Ralphs et al. 1997; Papaj and Newsom 2005; Soltau et al. 2009). Furthermore, damaged tissues often up-regulate production of chemical defenses (e.g., Green and Ryan 1972; Schultz and Baldwin 1982; Ryan 1990; Karban and Baldwin 1997; Kessler and Baldwin 2001; Karban 2015) and/or may be harboring poisonous/dangerous microorganisms that gained entry into the tissue upon damage (Janzen 1977; Ruxton et al. 2015); hence, consuming damaged plants may harm insects (Kaplan et al. 2007) or vertebrate herbivores. Caterpillars seem to partly consume leaves and then, move to other, undamaged ones whenever they can, a behavior that Roslin et al. (2008) attributed to the avoidance of induced chemical defenses. However, it is also possible that the caterpillars moved to undamaged leaves to distance themselves from visual evidence of their presence.

Part III
Aggressive Defenses

Chapter 23
Aposematism

This central part of the book focuses on intimidation of herbivores by signaling/advertisement, in many ways the opposite strategy of camouflage and some other related visual defenses discussed in previous chapters. In spite of the contrasting strategies (crypsis *versus* conspicuousness or looking non-tempting/masquerade *versus* looking dangerous) there are cases where the same visual characters can serve as camouflage or related defense at long distances, and for aposematism at a closer range.

Visual advertisement, including various colors and even signaling in the UV, is an important aspect of plant-animal communication because it attracts pollinators to flowers (Darwin 1877; Faegri and van der Pijl 1979; Willson 1983; Weiss 1995), and seed-dispersing frugivores to ripe fruit (Ridley 1930; van der Pijl 1982; Snow and Snow 1988; Willson and Whelan 1990; Schaefer et al. 2007; Schaefer and Ruxton 2011), although other plant parts (branches of the raceme, panicles or leaves) may also contribute by colors to those visual signaling systems (Ridley 1930; Faegri and van der Pijl 1979; van der Pijl 1982; Stiles 1982). Flower and fruit odors are also involved in signaling to attract animals (Faegri and van der Pijl 1979; van der Pijl 1982; Schaefer and Ruxton 2011). The vast majority of studies about plant-animal signaling (advertisement) were related to pollination. A significant number dealt with seed dispersal, and only a minute number with aposematism or other types of visual defensive signaling.

Aposematic (warning) coloration is a biological phenomenon in which poisonous, dangerous or otherwise unpalatable or unprofitable organisms visually advertise these qualities to other organisms, usually to animals (Poulton 1890; Cott 1940; Edmunds 1974; Gittleman and Harvey 1980; Ruxton et al. 2004) as defense from predation. Aposematic signaling may also be conveyed by movement, morphology, odor, taste and even by acoustics (Ruxton et al. 2004). Aposematism signals about defense, unpalatability or unprofitability from a lower trophic level to a higher one. Therefore, I proposed that it sometimes operates also between host plants towards parasitic plants (Lev-Yadun 2013c). The evolution of aposematic coloration or other types of aposematic signaling is based on the ability of targeted enemies to associate

S. Lev-Yadun, *Defensive (anti-herbivory) Coloration in Land Plants*,
DOI 10.1007/978-3-319-42096-7_23

the visual, chemical or acoustic signal with the risk, damage, or non-profitable handling, and later to avoid such organisms as prey (Cott 1940; Edmunds 1974; Gittleman and Harvey 1980; Ruxton et al. 2004), and in certain cases there is even an innate tendency to avoid objects with certain colors or color patterns (Smith 1975, 1977; Lindström et al. 1999; Ruxton et al. 2004). Typical colors of aposematic animals are yellow, orange, red, purple, black, white, brown and combinations of these (Cott 1940; Edmunds 1974; Wickler 1968; Savage and Slowinski 1992; Ruxton et al. 2004). The common defense achieved by aposematic coloration has resulted in the evolution of many mimicking animals. The mimics belong to two general categories, Müllerian mimicry and Batesian mimicry (discussed above), although there are intermediate situations known as quasi-Batesian mimicry. I wish to point to the fact that for the vast majority of species of all animal and plant taxa posited to be aposematic, aposematism has never been proved by showing avoidance learning or genetically-based avoidance. Still, proposed aposematism is an excellent research tool that explains various interactions among organisms and the existence of many characters in many taxa. Moreover, it stimulates further research.

Aposematic (warning) coloration is a common defense in plants, although it received very limited and mostly sporadic attention till the year 2001. Based on my knowledge of the flora of Israel, and the many visual and olfactory cases of aposematism by plant species that potentially exist there, I estimate the global number of aposematic plant species to be in the range of 100,000, or about 25 % of the vascular land plants. The fact that many aposematic animals, especially insects, use both plant-based pigments and sequestered poisonous molecules to become aposematic (Rothschild 1972, 1980; Nishida 2002; Ruxton et al. 2004; Heath et al. 2013; Keasar et al. 2013) emphasizes the absurdity of neglecting the aposematic character of so many toxic plants. As will be discussed later, as in animals, aposematic coloration in plants is commonly yellow, orange, red, brown, black, white, or combinations of these colors. Aposematic coloration is expressed in many thorny, spiny, prickly, poisonous, and in plants unpalatable or of low nutritive value for various other reasons as well as by plants that are unsuitable habitats for small herbivores because of their color or texture. Moreover, plants can be aposematic as the outcome of defenses acquired from other organisms including plants, fungi, bacteria, insects and possibly also various vertebrates. Plants that mimic aposematic plants or aposematic animals are also known. Many types of plant aposematic coloration simultaneously serve other functions such as physiological, communicative and even other defensive functions. It is therefore difficult in many cases to evaluate the relative functional share of visual aposematism in various plant color patterns and the specific selective agents involved in their evolution.

In this long chapter I will describe and discuss the facts, hypotheses, and questions related to aposematic coloration in plants, in an attempt to outline this enormously broad, diverse and complicated phenomenon and draw parallels to it from the broad knowledge about visual aposematism in animals. I will refer to aposematic coloration in the broadest sense, considering any visual warning phenomenon associated with defense, unpalatability, or unsuiability that may deter herbivores.

Distance of Action of Aposematic Coloration (Crypsis *Versus* Aposematism)

The difference between a smart person and a wise one is that the wise one will not get into difficulty from which the smart can find his way out. Ruxton et al. (2004) named their book about defensive mechanisms "Avoiding Attack". Camouflage of various types is the best way to evoid attack, but when it fails, in various animals the same coloration that matches a background or acts as disruptive coloration from a distance, may serve at close range as aposematic, for instance in coral snakes (e.g., Edmunds 1974; Roze 1996; Ruxton et al. 2004; Mappes et al. 2005; Tullberg et al. 2005; Barnett and Cuthill 2014). The different visual and cognitive abilities of various animal taxa add to the significance of the variability of aposematic signaling (Endler and Mappes 2004). This is probably also true when plants have several colors in their spine system (e.g., Lev-Yadun 2001) or in their leaves. While this double strategy (camouflage *versus* aposematism) has not yet been studied in depth in plants, there are indications not only that it may operate, but also that it may be frequent. For instance, certain cacti use their spines for camouflage from a distance (Benson 1982), probably by concealing their green color, while they may look colorful and aposematic at close range (e.g., Lev-Yadun 2001). This issue of plants being camouflaged from a distance and visually aposematic at close range deserves many descriptive, theoretical and experimental studies in order to be better understood.

Chapter 24
Olfactory Aposematism

While this book is mainly dedicated to visual plant defenses, I will briefly discuss olfactory aposematism, whereby poisonous plants deter mammalian or insect herbivores by volatiles (Eisner 1964; Rothschild 1972, 1973, 1986; Levin 1973; Atsatt and O'Dowd 1976; Wiens 1978; Eisner and Grant 1981; Harborne 1982; Rothschild et al. 1984; Guilford et al. 1987; Rothschild and Moore 1987; Kaye et al. 1989; Moore et al. 1990; Woolfson and Rothschild 1990; Launchbaugh and Provenza 1993; Provenza et al. 2000; Massei et al. 2007; Lev-Yadun et al. 2009b; Lev-Yadun 2014d, e; Karban 2015). I discuss it briefly because it is probable that like in pollination (Faegri and van der Pijl 1979; Dafni 1984; Jersáková et al. 2006) and seed dispersal (van der Pijl 1982; Schaefer and Ruxton 2011), where many plants simultaneously use visual and olfactory signals for animal attraction, such double signaling is also true for various cases of plant aposematism. For instance, concerning *Silybum marianum* it has been proposed (separately) that olfactory aposematism *via* pyrazine (Rothschild and Moore 1987) and visual aposematism *via* spine color and zebra-like conspicuous white variegation (Lev-Yadun 2003a, 2009a, 2014a) operate simultaneously. Certain toxic, nuterient-poor or otherwise dangerous and colorful autumn leaves (Lev-Yadun and Gould 2007) probably simultaneously use olfactory aposematism (Holopainen 2008; Blande et al. 2010; Holopainen et al. 2010). It is very likely that the few such proposed cases, where olfactory and visual aposematism are known to operate simultaneously in plants, are just examples of a much broader phenomenon. However, it is currently unknown if the aposematic signaling *via* color and odor are aimed towards the same type of herbivores, or to different receivers. The possibility that thorny, spiny, prickly and toxic plants simultaneously use visual and olfactory aposematism should be studied systematically. I stress that as has been shown for many fungi (Sherratt et al. 2005), olfactory aposematism may be especially important as a defense against nocturnal fungivores, and this may also be so concerning nocturnal or subterranean herbivores or in habitats such as the forest floor where light levels are very low.

As early as the year 1972, Rothschild proposed that various plants may gain certain olfactory aposematism not because of their own odors, but rather from odors

S. Lev-Yadun, *Defensive (anti-herbivory) Coloration in Land Plants*,
DOI 10.1007/978-3-319-42096-7_24

of aposematic insect herbivores that occupy toxic plants, something that is also an extended phenotype although not considered so in those years. Lev-Yadun et al. (2009b) proposed that carrion and dung odors of various flowers that have traditionally been considered an adaptation for attracting the flies and beetles that pollinate them, may also have another, overlooked, anti-herbivore defensive function. They suggested that such odors may also deter mammalian herbivores, especially during the critical period of flowering (see relevant field data demonstrating the fear of cattle from carrion in Lev-Yadun and Gutman 2013). Lev-Yadun et al. (2009b) proposed that carrion odor is a good predictor for two potential dangers to mammalian herbivores: (1) pathogenic microbes, (2) proximity of carnivores. Similarly, dung odor predicts feces-contaminated habitats that present high risks of parasitism. These are two new types of repulsive olfactory aposematic mimicry by plants: (1) olfactory feigning of carcass (thanatosis), a well-known behavioral defensive strategy in animals (Edmunds 1974; Ruxton et al. 2004), (2) olfactory mimicry of feces, which also has a defensive visual parallel in animals (Eisner et al. 2005), and probably also in plants (Wiens 1978; Lev-Yadun 2014b). A derived olfactory aposematism as an extended phenotype was proposed by Lev-Yadun (2014e). According to that hypothesis, insect carrion and trapped live insects are attached to the surfaces of many plant species that have sticky or hooked trichomes or are sticky following the exudation of latex or resin when wounded. Such attached dead or trapped living insects may serve as billboards, cueing and signaling by rotting carrion or stress volatiles emitted from the trapped insects that such plants are dangerous and even deadly.

Chapter 25
The Anecdotal History of Discussing Plant Aposematic Coloration

After it became clear to me in February 1996, following compelling evidence in the field that aposematic coloration characterizes many thorny, spiny and prickly plants, I conducted a thorough BIOSIS database search of "aposematism in plants" that yielded nothing. In those pre-Google and Google Scholar days it appeared to be a totally new issue. There were, however, several papers published since the 1960s mentioning toxic plant aposematism *via* visual and olfactory signals, but they were data-base-wise transparent and I have found them with time, following my very extensive library work. Concerning visual aposematism of thorny, spiny and prickly plants is was indeed a new hypothesis. Twenty years of thorough library study resulted in a very short pre-year-2000 list of authors who discussed visual aposematism (usually very briefly) and only concerning poisonous plants (Cook et al. 1971; Hinton 1973; Harper 1977; Wiens 1978; Rothschild 1972, 1980, 1986; Harborne 1982; Williamson 1982; Knight and Siegfried 1983; Smith 1986; Lee et al. 1987; Givnish 1990; Tuomi and Augner 1993). Moreover, several of these references (Knight and Siegfried 1983; Smith 1986; Lee et al. 1987), while discussing the theoretical possibility of aposematic signaling by toxic plants, dismissed the existence/operation of aposematic coloration in the plants they studied. While these few early mentions of visual aposematism in plants referred to poisonous ones, in papers published since 2001 more attention was given to visual aposematism of thorny, spiny and prickly ones (Lev-Yadun 2001, 2003a, b, 2006a, 2009a, b, c, 2011, 2014a; Midgley et al. 2001; Midgley 2004; Lev-Yadun and Ne'eman 2004, 2006; Rubino and McCarthy 2004; Ruxton et al. 2004; Speed and Ruxton 2005; Halpern et al. 2007a, b, 2011; Lev-Yadun and Halpern 2008; Fadzly et al. 2009; Lee and Gould 2009; Lev-Yadun and Gould 2009; Lev-Yadun et al. 2009a; Nassar and Lev-Yadun 2009; Ronel et al. 2009, 2010; Schaefer and Ruxton 2011; Ronel and Lev-Yadun 2012; Kavanagh et al. 2016; Midgley et al. 2016) and less attention to poisonous ones (Gould 2004; Lev-Yadun and Ne'eman 2004; Hill 2006; Lev-Yadun 2006a, 2009a, 2014a, d; Lev-Yadun and Gould 2007, 2009; Archetti 2009b; Archetti et al. 2009a; Lev-Yadun and Holopainen 2009; Cooney et al. 2012; Gerchman et al. 2012; Maskato et al. 2014; Hughes and Lev-Yadun 2015).

S. Lev-Yadun, *Defensive (anti-herbivory) Coloration in Land Plants*,
DOI 10.1007/978-3-319-42096-7_25

Cook et al. (1971), the earliest reference I managed to find, briefly proposed that aposematic coloration occurs in poisonous seeds of the plant *Eremocarpus setigerus* (Euphorbiaceae). Seeds of this plant are either camouflaged by mottling when less poisonous, or they are plain gray when they are much more poisonous, a color that does not camouflage them much on the soil surface. The mountain dove (*Zenaidura macroura*) rejected the gray seeds and eagerly ate the mottled ones, leading Cook et al. (1971) to propose that the gray seeds are aposematic. Rothschild (1972) proposed that some plants may develop a generalized warning system analogous to the aposematic coloration of distasteful animals to deter vertebrate herbivores but attract certain herbivorous insects that can sequester their toxic compounds. Hinton (1973), who was a zoologist, gave the first detailed hypothesis for a possible defense from herbivory of yellow, red and other types of vivid flower coloration. Hinton proposed that colorful poisonous flowers should be considered aposematic and that they probably have mimics. His review about deception in nature was published in a book about illusion which was not a biology book, but rather dealt with art. I had a copy of the book because of my interests as a photographer. This hypothesis, which was briefly referred to by the very influential Miriam Rothschild (1980) when she discussed the various roles of carotenoids, did not cause botanists or zoology-oriented ecologists to pursue this issue. It seems that the relevant community was willing to consider plant coloration mostly for physiological issues, or when plant-animal relations were considered, only for the attraction of pollinators and seed dispersers. Harper, who at about the same time (1977) wrote his comment about botanists that were reluctant to accept things that are commonplace for zoologists, did not discuss or explain why zoologists who dealt with animal aposematism and were also involved in research on plant-animal interactions had not recognized how common these phenomena are in many plant taxa and habitats. Wiens (1978) in his review of mimicry in plants, mentioned briefly that many examples of striking, contrasting, and often variegated or mottled patterns of coloration characterize plants, particularly leaves, and wondered if they serve aposematic functions. He proposed that if herbivores orient their feeding selection primarily visually, these patterns should function for aposematism. Wiens (1978) gave several examples of poisonous plant taxa with patterned leaf coloration, such as *Caladium* and *Dieffenbachia*, and colorful poisonous seeds of various plants including those of *Ricinus*, as good candidates for visual aposematism (Fig. 25.1). Wiens (1978) also mentioned personal communication with C. Dodson that suggested that red young leaves in tropical plants may also be aposematic. Eisner (1981), without using the term aposematic, actually described visual aposematism in the thorny plant *Schrankia microphylla*, which when touched folds its leaves, further exposing its thorns. Harborne (1982), in his book on chemical ecology, proposed that the brightly colored, purple-black berries of the deadly *Atropa belladonna* warn grazing mammals of the danger in consuming them. Williamson (1982) also briefly proposed that brightly colored (red or red and black) seeds lacking an arillate or fleshy reward (e.g., *Erythrina*, *Ormosia*, and *Abrus*) might be aposematically colored to warn seed eaters of their toxicity. Knight and Siegfried (1983) raised the question of whether green fruits signal unpalatability, and concluded that in the forest canopy, green does not provide

Fig. 25.1 Toxic variegated seeds of *Ricinus communis*. Depending on soil color, plant debris, illumination and distance these seeds may be cryptic, but at a closer range they are very conspicuous and shiny

enough contrast to be aposematic. Two decades later, Lev-Yadun and Ne'eman (2004) proposed that green can serve as an aposematic color by toxic or thorny plants in yellow-gray desert landscapes. Smith (1986) hypothesized that leaf variegation may theoretically function in aposematism, but concluded that for the vine species (*Byttneria aculeata*) he studied, the variegation was indeed related to visual defense from herbivory by mimicking leaf mining damage, but that it was not aposematic. Although Smith rejected the operation of aposematism in the plant species he studied, he gave a clear and detailed formulation of the aposematic hypothesis for poisonous plants: "The benefits to the plant of chemical defense against herbivores would be greater if herbivores avoided such plants altogether, rather than testing leaves for palatability, and so causing some damage. A distinct leaf color pattern linked with chemical defense might function in this way. Polymorphism for leaf color should then coincide with polymorphisms for chemical defense. Müllerian and Batesian mimicry could result in evolution of similar patterns of variegation, with or without associated toxicity, among other species which have herbivore species in common with the model species" (Smith 1986). Rothschild (1986) proposed that red may serve as an aposematic color in poisonous plants, without giving examples. Lee et al. (1987) concluded that anthocyanins in developing leaves of mango and cacao are not aposematic but rather have physiological functions. Givnish (1990) noted that Smith's (1986) rejected hypothesis regarding the aposematic potential of leaf variegation should be considered, but did not elaborate on this issue when he proposed that the understory herbs he studied in the forests of New England (USA) use leaf variegation for camouflage. Tuomi and Augner (1993) mentioned a possible association between bright colors in plants and toxicity. Augner (1994) modeled and discussed the conditions for the operation of aposematism in plants, focusing on chemical-based aposematism with no direct reference to a visual one, although it can be understood from the text that visual aposematism was not rejected. Augner and Bernays (1998) modeled the possibilities of the operation of plant defense signals and the existence of their mimics, and although they did not refer directly to visual aposematism, it is again clear from the text that they concluded that Batesian mimics of plant defense signals may be common (see proposed Müllerian and Batesian mimics in Lev-Yadun 2003a, b, 2006a, 2009a, b, c, 2014a, d; Lev-Yadun and Gould 2007, 2009). Archetti (2000), in his discussion on bright red and yellow autumn leaves that were proposed to signal aphids about the chemical defensive qualities of the trees, rejected the possibility that these leaves are aposematic. However, some years later, Archetti changed his mind and accepted aposematism as part of the explanation for autumn leaf coloration (Archetti et al. 2009a).

Another issue of importance concerning poison-related aposematism is the relativity of aposematism. Deciding that a certain branch, root, leaf, flower, fruit or seed is poisonous or unpalatable is a relative issue. Certain frugivores can consume fruits that are poisonous to other animals (Janzen 1979) and the same is true of any plant organ or tissue. Therefore, a chemically-based aposematic plant for certain animal taxa, may be edible, non-aposematic or even attractive for other taxa. Moreover, although visuality is important for decisions about plant palatability made by herbivores (Prokopy and Owens 1983; Reeves 2011), other senses are also involved (Provenza et al. 2000).

Chapter 26
Aposematic Coloration in Thorny, Spiny and Prickly Plants

The anecdotal status of studying and discussing plant aposematism that was revealed to me in the years 1996–2000 did not match what I found in the field. This was a typical scientific Déjà vu. It happened to me in my Ph.D. thesis about the regulation of ray formation in the secondary plant body (e.g., Lev-Yadun and Aloni 1995), it happened again when I struggled to convince that *Arabidopsis thaliana* can and should be an excellent model for wood and fiber formation (e.g., Lev-Yadun 1994, 1997), and when we argued for the origin of agriculture in the Near East about 10,600 years ago in a short, geographycally limited single concious event (e.g., Lev-Yadun et al. 2000). In all these cases the studied subjects were either considered by many as already resolved, considered to reflect a totally different solution, or in the case of *Arabidopsis* not to have the relevant tissues. I repeatedly identified an opportunity in such situations and following my previous experience, adopted pursuing plant aposematism as a personal crusade.

In English, there are three terms for sharp and pointed defensive plant appendages: thorns, when they are made of branches, spines, when they are made of leaves, and prickles, when they are made of cortical tissues (e.g., in roses). I will use the specific term whenever possible, and mention thorns or spines when generally discussing defense by sharp and pointed plant organs.

Thorns, spines and prickles provide mechanical protection against herbivory (Janzen and Martin 1982; Cooper and Owen-Smith 1986; Janzen 1986; Tomlinson 1990; Myers and Bazely 1991; Grubb 1992; Gowda 1996; Rebollo et al. 2002; Ronel and Lev-Yadun 2012) because they can wound mouths and digestive systems (Janzen and Martin 1982; Cooper and Owen-Smith 1986; Janzen 1986), and other body parts of herbivores, and even introduce dangerous microorganisms into the herbivore's tissues (Halpern et al. 2007a, b, 2011; Lev-Yadun and Halpern 2008). Spines and prickles have additional functions, such as reducing sun irradiation (Gibson and Nobel 1986; Mauseth 2006), taking part in seed dispersal by animals (Zohary 1962), vegetative dispersal by animals (Gibson and Nobel 1986; Allen et al. 1991; Bobich and Nobel 2001), and even camouflaging the plants (Benson 1982; Mauseth 2006).

S. Lev-Yadun, *Defensive (anti-herbivory) Coloration in Land Plants*,
DOI 10.1007/978-3-319-42096-7_26

From both evolutionary and ecological points of view, spininess and thorniness increase in various ecosystems following lengthy exposure to herbivory by large animals (Zohary 1983; Janzen 1986; Grubb 1992; Givnish et al. 1994; Nobel 1994; Ronel et al. 2007, 2010; Ronel and Lev-Yadun 2009, 2012). Even during the lifetime of an individual plant, thorniness and spininess may increase following animals' browsing as a way of induced defense (Milewski et al. 1991; Perevolotsky and Haimov 1991; Young et al. 2003).

Theoretically, once herbivores learn to identify thorns, spines and prickles (and their bright or contrasting colors, or their associated conspicuous markings should help in their recognition), they can avoid the harmful plants advertising them. The fact that thousands of thorny, spiny and prickly species have colorful sharp defensive structures or that they are otherwise conspicuous by association with white or colorful markings, somehow escaped the notice of botanists and zoologists, although cacti and other spiny taxa are found not only in the remote wilderness among scorpions and rattlesnakes in arid habitats of North America, but in the majority of botanical gardens. Interestingly, Gibson and Nobel (1986) in their book on cacti, wrote "When spines are young, they may range in color from white or nearly translucent to yellow, golden, brown, pink, orange, red, gray, or nearly black. As spines age they characteristically fade or weather to duller shades. Individual spines also can be multicolored. Unfortunately, no one has investigated the factors determining color. The red colors are likely to be produced by pigments, such as betalains; the red pigment in young spines is water soluble. Neither has anyone determined a function of spine coloration." The role for conspicuousness of thorns, spines and prickles in plants as a type of visual aposematism, and its probable ecological significance, was first proposed by Lev-Yadun (2001) in a study that included about 1000 cacti species and about 300 other species.

Since what is toxic to one animal might be harmless to another (Laycock 1978; Janzen 1979; Gleadow and Woodrow 2002), chemically based aposematism may not operate for all herbivores. Concerning sharp and pointed defensive organs, the situation is usually different. Naturally, there are differences in the sensitivity of herbivores to wounding by sharp objects, but even specialized mammalian herbivores like woodrats and collared peccaries, which are well adapted to deal with and exploit very spiny *Opuntia* plants, tend to choose the less spiny ones (Brown et al. 1972; Theimer and Bateman 1992). The need to touch and ingest sharp objects makes all large vertebrate herbivores sensitive to such plants to a certain extent. Visual aposematism towards vertebrate herbivores with conspicuous thorns, spines and prickles may therefore characterize a larger proportion of the thorny, spiny and prickly plants than the proportion of visual aposematism in poisonous ones.

Concerning defensive plant coloration I made my significant and target-oriented move only in the early winter of 1994/1995. At that time I decided that I had to understand why some local wild plant species look like green zebras. The specific trigger was the very spiny and common annual winter rosette plant *Silybum marianum* (Mary's thistle, milk thistle, Asteraceae). The leaves of this plant (and two other related local spiny species, *Notobasis syriaca* and *Scolymus maculatus*) are covered with a dense network of white stripes of various widths that make them

Fig. 26.1 *Silybum marianum*, the variegated wild-type growing mixed with *green*, non-variegated mutants. According to a legend, it became variegated when Mary nursed baby Jesus under a still non-variegated plant on her way from Bethlehem to Egypt, and a drop of milk spilled on the leaf and transformed it

very conspicuous from dozens of meters away (Figs. 1, 19.1). Thus, this coloration certainly cannot serve as camouflage. According to the legend, *S. marianum*, a ruderal plant common along unpaved roads, became variegated when Mary nursed baby Jesus under a still non-variegated plant on her way from Bethlehem to Egypt, and a drop of milk spilled on the leaf and transformed it (Fig. 26.1). The attempt to understand the functions of variegation in *S. marianum* followed previous developmental and genetic studies with other variegated species and genotypes that I had been conducting since the year 1987, when I found a variegated *Cupressus sempervirens* mutant tree (See Lev-Yadun et al. 2004b) while screening cypress populations all over Israel for another type of developmental mutant. It resulted in my permanent attention to various types of non-reproductive and non-physiological coloration in plants.

In the beginning, I thought that the white variegation in *S. marianum* leaves might mimic the damage of tunneling insects (Fig. 26.2), and by this be viewed by herbivores as if it were already infested, thus tricking insects into refraining from attacking them, or mammals from grazing on such infested plants [for the theoretical discussion of the hypothesis that leaves variegated in white appear to be infested see Smith (1986), Campitelli et al. (2008) for field data, and Soltau et al. (2009) for the most important experimental field study of this hypothesis so far]. As it turned out, my major hypothesis (but certainly not my only one), concerning the defensive function of the zebra-like leaf variegation was different.

As in many other cases in my scientific career, my understanding of the function of zebra-like leaf variegation arrived as a surprise, in an unexpected moment, after more than a year of documentation in the field and thinking. In February 1996 I was

Fig. 26.2 A leaf of *Sonchus oleraceus* infested by tunneling insects. My first hypothesis concerning the variegation of *Silybum marianum* was that it mimics such tunneling. I still think so, but I think that aposematic signaling of this very spiny plant is more important for defense than the tunneling mimicry

studying plant recovery from fire and grazing on a mountain ridge east of the Druze village Hurfeish in northern Israel (33⁰1′N; 35⁰22′E). When I had finished my measurements for that day and started to pay attention to plant coloration in that habitat, after only several minutes of slow walking, I came across two small, spiny, annual rosette plants growing only several centimeters apart. One was *Carduus argentatus* (Asteraceae) (Fig. 26.3) and the other was *Nonea obtusifolia* (Boraginaceae) (Fig. 26.4). The fact that these two spiny species that belong to two different plant families, separated from their last shared ancestor by several dozens of millions of years, had a similar type of coloration, i.e., white spots at the base of many of their small spines, struck me on the spot. I immediately understood that these plants do not hide their vicious nature from herbivores, but rather, like many dangerous animals, advertise their defensive qualities. I had a strong feeling that the enigma of the zebra-like appearance of *S. marianum* and the several other local wild spiny and variegated species was probably solved. Several days later, in order to see whether spine conspicuousness is a rule and not just an interesting but marginal adaptation of a small number of species, I examined thorn, spine and prickle conspicuousness in wild plants growing in a field located 150 km to the south, in the central coastal plain of Israel, about 15 km south of Tel Aviv. There, in a different ecology and flora, the phenomenon of conspicuousness of the defensive nature of spiny, thorny

Fig. 26.3 A spiny rosette of *Carduus argentatus*. Many of the spines have a *white marking* that makes them conspicuous. Seing a small plant of that species growing next to a spiny plant of *Nonea obtusifolia* with a similar *white marking* associated with its spines indicated to me that spiny plants may advertise their defense, i.e., are aposematic

Fig. 26.4 Leaves of *Nonea obtusifolia* with *white markings* associated with its small spines

and prickly plants was even more obvious, not only by white mottling, but also with colorful thorns, spines and prickles (Fig. 26.5), and therefore, when I understood that it is a general phenomenon, I began a broad and targeted study of the issue.

After realizing that the thorns, spines and prickles of many wild plants and some ornamentals growing in Israel are usually colorful or associated with conspicuous white or colorful markings, I decided to test if this principle is true in four very spiny taxa (cacti, and the genera *Agave*, *Aloe* and *Euphorbia*) that evolved

Fig. 26.5 A colorful, conspicuous very spiny *Eryngium creticum* plant from Israel. Several other spiny members of the genus are also colorful and conspicuous

independently in two different continents (cacti and *Agave* in America and *Euphorbia* and *Aloe* in Africa). Unlike cacti, that except for toxic spineless and camouflaged ones (Gibson and Nobel 1986) are mostly non-toxic, the genera *Agave*, *Aloe* and *Euphorbia* are characterized by toxic plants. First, I examined and documented the type, color and markings associated with thorns, spines and prickles in several cacti and other succulent taxa collections in Israel. Already during this survey, I realized that several mimicry types occur, an issue I studied in more detail several years later. In the second stage of study, I compiled detailed data by analyzing the pictures and text of published monographs on these four taxa (Reynolds 1969; Benson 1982; Gentry 1982; Preston-Mafham and Preston-Mafham 1994; Sajeva and Costanzo 1994). Since cacti have two classes of spines (regular spines and glochids), and their spines have several colors, the color of their spines and glochids was assigned separately. Within months I examined about 1300 spiny species, and it became clear that most spiny plants in these four taxa have a colorful spine system, commonly also marked by white mottling, and that being conspicuous is the dominant visual character in spiny, thorny and prickly plants everywhere (Lev-Yadun 2001).

Colorful spine systems, many of which are multi-colored, are found in many species of the Cactaceae (Figs. 26.6, 26.7–26.10). The spines are usually brown, yellow, red, white, gray, pink, black, or tan (Benson 1982). Benson (1982) listed 174 cacti species, 140 (80.5 %) of which have colored spines and 59 (33.9 %) have colored glochids (Figs. 26.11–26.12). Eight species (4.6 %) have one color in the spines, 39 (22.4 %) have two colors, 48 (27.6 %) three colors, 24 (13.8) four colors, 14 (8 %) five colors, five (2.9 %) six colors and two (1.1 %) seven colors. In 13 (7.5 %) species the glochids were brown, seven (4 %) species red, 28 (16.1 %)

Fig. 26.6 The multicolored spine system of a cactus

yellow, ten (5.7 %) tan, and one (0.6 %) gray (Benson 1982). Preston-Mafham and Preston-Mafham (1994) had color pictures of 973 cacti species, 862 (88.6 %) of which had white markings associated with the spines (71–72) and six (0.6 %) species had brown/black markings. The common white markings associated with spines of the Cactaceae are comprised of white wool or felt, and may deteriorate with time. They usually appear in the upper parts of the plant and are restricted to areoles, where they form well-arranged white spots or short stripes (Lev-Yadun 2001) (Figs. 26.13 and 26.14). These white markings are so conspicuous and typical that marcipan-made cacti in a marcipan museum in Budapest (Hungary) also have these markings (Fig. 26.15). The less colorful markings were not mentioned, but examination of cacti collections indicated that they are usually grayish but never green.

Agave species can have two types of spines in their leaves: spines at the distal end, or spines (teeth) along the margins. In addition to teeth along the margins, many *Agave* species also have colorful stripes along the leaf margins that enhance spine and teeth visibility (Gentry 1982). The spines and the teeth along the margins of the leaves are brown, reddish, gray, black, white, or yellow (Gentry, 1982). Gentry (1982) listed 194 *Agave* species, 112 (57.7 %) of which have apical spines,

Fig. 26.7–26.10 Typical conspicuous and colorful spines of various cacti taxa

86 (44.3 %) have teeth along the margins and 47 (24.2 %) species have colorful stripes along the margins (Figs. 26.16–26.17, 26.18 and 26.19).

In the genus *Aloe*, the colorful spines are white, red, black, or yellow. Many *Aloe* species have white markings on their succulent leaves and many species have both colorful spines and white marking (Reynolds 1969). Reynolds (1969) listed 137 *Aloe* species, 133 (97.1 %) of which have spines along the margins, 94 (68.6 %) species have colored spines, and 37 (27 %) species have white spines along the margins. Fifty species (36.5 %) have both white spots on the leaf surface and colored spines along the margins, 42 species (30.7 %) have colored thorns along the margins but no white spots, 13 (9.5 %) species have white spines along the margins but no white spots on the leaf surface, three species (2.2 %) have white spots on leaf surface but no spines, and two species (1.5 %) have both colored spines along the margins and colored spots (Reynolds 1969).

Fig. 26.11–26.12 Conspicuous glochides in cacti. Such glochides are formed in groups of hundreds (each colorful spot) of barbed detachable spines. From my experience since the 1960s in picking and eating *Opuntia ficus-indica* fruits I highly appreciate the irritating potential of such glochides

In the genus *Euphorbia*, colorful thorns and white or whitish variegation or white markings associated with thorns (Sajeva and Costanzo 1994) are also common. Sajeva and Costanzo (1994) listed 80 *Euphorbia* species, 48 (60 %) of which have colored thorns. Thirteen (16.3 %) species have white marking associated with thorns along their ribs and nine (11.3 %) other species have marking of other colors associated with thorns along the ribs (Sajeva and Costanzo 1994) (Figs. 26.20–26.21 and 26.22).

Lev-Yadun (2001) showed that two types of conspicuousness of thorns and spines are typical of many plant species: (1) colorful thorns and spines (commonly multicolored), and (2) white and colorful spots and stripes associated with thorns and spines in leaves, stems, and fruits. Both types of aposematic coloration dominate the thorn or spine system of taxa rich in thorny or spiny species: cacti, and members of the genera *Agave*, *Aloe*, and *Euphorbia*. Lev-Yadun (2001) proposed that since the coloration and markings of thorns, spines and prickles in plants are so common, it is probably not a neutral or a random phenomenon. It has been thus proposed that conspicuous thorns, spines and prickles are beneficial to plants since herbivorous vertebrates remember the signal and tend to avoid subsequent tasting of such conspicuous thorny or spiny plants. Furthermore, herbivores might pass over the aposematic individuals and eat their non-aposematic neighbors, thus reducing the competition between the aposematic and neighboring plants (Lev-Yadun 2001). Since aposematic signals associated with thorns, spines and prickles are found in plants of diverse geographical and taxonomic origins, Lev-Yadun (2001) proposed that this is an ancient signal that has been selected for many times. Rubino and McCarthy (2004) tested Lev-Yadun's (2001) aposematic hypothesis by examining the presence of putative aposematic coloration in thorny, spiny, and prickly vascular plants of southeastern Ohio, and because of their similar field results, reached the same conclusions.

Fig. 26.13 Typical *white markings* at the base of the spines of a cactus. Such markings can be seen even under moonlight

This phenomenon of visual aposematism in thorny, spiny and prickly plants, which seems to be very common, has been described and discussed at three levels: (1) the floristic approach, studying it across large taxa (Lev-Yadun 2001; Ronel et al. 2009), floras or ecologies (Lev-Yadun and Ne'eman 2004; Rubino and McCarthy 2004; Ronel and Lev-Yadun 2012); (2) the individual species level (Lev-Yadun 2003a; Lev-Yadun and Ne'eman 2006; Halpern et al. 2007a, b, 2011; Fadzly et al. 2009; Kavanagh et al. 2016; Midgley et al. 2016); and (3) mimicry of the phenomenon (Lev-Yadun 2003a, b, 2006a, 2009a, b, c, 2014a; Lev-Yadun and Gould 2009). Although Midgley et al. (2001) and Midgley (2004) did not use the word aposematic, they described the typical conspicuous white thorns of many African *Acacia* (= *Vachellia*) trees as visually deterring large herbivores, supporting the aposematic hypothesis. Ruxton et al. (2004) and Speed and Ruxton (2005), elaborated on the principle that unlike poisonous organisms, aposematic thorns and spines advertise their own dangerous quality (self-advertisement).

Lev-Yadun (2003a) showed that the rosette and cauline leaves of the very spiny winter annual plant species of the Asteraceae in Israel (*Silybum marianum*) resemble green zebras. The widths of typical variegation bands were measured and found to be highly correlated with leaf length, length of the longest spines at leaf margins and the total number of spines along leaf circumference. Thus, there was

Fig. 26.14 Typical *white markings* at the base of the spines of a cactus that form conspicuous white panels

a significant correlation between the spininess and strength of variegation (advertisement). Lev-Yadun (2003a) therefore proposed that this white variegation is a special case of aposematic (warning) coloration. The aposematic hypothesis concerning *S. marianum* gains further support by the fact that a variable proportion of its populations may accumulate a considerable amount of nitrate, which was found to be deadly to calves grazing on this plant (Kendrick et al. 1955). However, additional defensive roles of the variegation, such as mimicry of the tunnels of flies belonging to the Agromyzidae, reduced insect landing on the leaves in general (Lev-Yadun 2003a), just as the zebra stripes defend against tsetse flies (Waage 1981; Brady and Shereni 1988; Doku and Brady 1989; Gibson 1992; Egri et al. 2012), dazzle coloration (Lev-Yadun 2014c), and masquerade (Lev-Yadun 2014b) were also proposed.

Lev-Yadun (2001) proposed that if the cost of producing and maintaining the signal is lower than the cost of the damage it prevents, the gain will be large enough to favor selection for such mutations (see Johnstone 1995). Even annuals, but especially perennial plants, may survive damage caused by vertebrate herbivores (Crawley 1983; Karban 2015). Therefore, any herbivore reaction to the aposematic coloration and to the associated defenses (when they exist) by avoidance is of direct benefit to the individual plant that will suffer fewer repeated attacks. Hence, as with

Fig. 26.15 The *white markings* at the base of cacti spines are so typical that they were mimicked in a cactus made of Marcipane in a Marcipane museum in Budapest, Hungary

Fig. 26.16–26.17 Colorful *Agave* spines along their leaf margins

Fig. 26.18 A *dark* conspicuous *Agave* spine at the tip of its leaf. We used to trim them in our garden to avoid their risk, especially to the eyes

animals (Sillén-Tullberg and Bryant 1983), there was no need to propose kin/group selection, or altruism, as the evolutionary drive for the spread of this character.

Production of colorful thorns, spines and prickles, and white or colorful marking does require resources. However, when a conspicuous aposematic tissue serves more than one purpose, the relative cost of advertisement is reduced. For instance, in certain cacti, the white wool or felt (which mark spine groups and makes them more conspicuous) also reduces diurnal temperature extremes, thus protecting the shoot apex (Nobel 1978). Whether the aposematic signaling is honest is important for evaluating the evolutionary consequences of conspicuous thorns, spines and prickles. Evaluating honesty in biological signaling is complicated, especially in plants, since some other colorful or white plant parts are also used for non-aposematic signaling, e.g., signaling to pollinators and frugivores, or for physiological functions such as reduction of plant temperature. Hence, plant colors can signify not only that a plant is thorny or spiny, but rather signal "pay attention". The honesty of the aposematic signal is maintained by herbivores that taste the plants. If mimics are overly abundant, the signal may lose its deterrent quality because herbivores learn that it is unreliable (e.g., Lev-Yadun 2001, 2003a, b, 2009a). As I will show and discuss later, mimics of thorns, spines and prickles do exist.

Fig. 26.19 A panel that connects *Agave horrida* spines enhancing their visibility

Fig. 26.20–26.21 Conspicuous *Euphorbia* sp. thorns. They loose their color after a while

Fig. 26.22 A *white* panel connects *Euphorbia* thorns enhancing their visibility, a phenomenon found also in cacti and *Agave*

Table 26.1 The number of species and frequency of common colors of spins, thorns and prickles in 167 plant species in the flora of Israel[a]

Color	Yellow	Red	Orange	White	Purple	Black
Annuals	19	13	6	9	1	0
Perennial herbs	30	6	13	10	4	1
Dwarf shrubs	8	7	9	7	1	2
Shrubs	7	12	11	8	0	3
Trees	2	5	4	4	0	1
Total	66 (39.5 %)	43 (25.7 %)	43 (25.9 %)	38 (22.8 %)	6 (3.6 %)	7 (4.2 %)

After Ronel and Lev-Yadun (2012)

[a]Numbers may add up to more than the number of species as many plant species have more than one color in their thorns, spines and prickles

Ronel and Lev-Yadun (2012) examined the aposematic thorn, spine and prickle color hypothesis in the whole wild flora of Israel. Out of the 294 thorny, spiny and prickly species in the whole flora (about 11 % of all vascular species), they could determine thorn, spine and prickle colors in 167 species (56.8 %). The common colors were yellow (66 species), red or reddish (43 species), orange (43 species), white (38 species), black (7 species) and purple (6 species) (Table 26.1), supporting the hypotheses about spines being visually aposematic. Interestingly, in the flora of Israel, annuals do not have black spines and trees do not have purple ones.

I conclude this part of the discussion by positing that the recent proposals that thorny, spiny and prickly plants may be visually aposematic (e.g., Lev-Yadun 2001, 2009a; Rubino and McCarthy 2004; Ruxton et al. 2004; Speed and Ruxton 2005; Halpern et al. 2007a; Fadzly et al. 2009; Lev-Yadun et al. 2009a; Schaefer and Ruxton 2011; Ronel and Lev-Yadun 2012; Farmer 2014; Kavanagh et al. 2016; Midgley et al. 2016), were based on the fact that thorns, spines and prickles are usually colorful or are also conspicuous because they are marked by various types of associated coloration in the tissues that form them, or positioned next to them, including white markings. Similarly, it has also recently been proposed that many spiny animals have colorful spines and that they are thus aposematic (Ruxton et al. 2004; Inbar and Lev-Yadun 2005; Speed and Ruxton 2005), a fact that was discussed only briefly in the classic monograph by Cott (1940).

Chapter 27
Fearful Symmetry in Aposematic Spiny Plants

Symmetry has been proposed to increase the efficiency of visual aposematic displays in animals (Forsman and Merilaita 1999; Forsman and Herrström 2004), although the role of symmetry in aposematism is not global (e.g., Langridge 2006; Stevens et al. 2008a). This hypothesis was contested with the hypothesis that asymmetry impairs the efficiency of visual aposematism while increasing the defensive value of cryptic color patterns (e.g., Forsman and Merilaita 2003). Forsman and Merilaita (2003) compared intra-individual asymmetry in color patterns of cryptic patterns on forewings and aposematic patterns on hindwings of three moth species. They found somewhat larger asymmetry in cryptic patterns but it was not significant, and therefore they concluded that there were probably developmental or genetic constraints on the evolution of pronounced asymmetry in the system they studied. Langridge (2006) found in the cuttlefish *Sepia officinalis* that color changes that enhanced crypsis were symmetrical while those that enhanced anti-predator signaling were asymmetrical, contrary to the theoretical predictions. These results indicate that understanding the role of symmetry in both crypsis and aposematism is not straightforward and may also be context related. Later, Stevens et al. (2009) showed that asymmetry in warning signals may not be costly to prey compared with other features of the aposematic signal, such as color and overall size. However, there is an innate preference for symmetry in the visual system of animals, probably because of the need to recognize objects (Enquist and Arak 1994), a well known fact from pollination biology (Neal et al. 1998).

Lev-Yadun (2011) reviewed various cases of proposed/putative visual aposematism in plants and showed that symmetry characterizes many of them. A careful examination of the major plant taxa and plant organs proposed to be visually aposematic shows a good agreement with the zoological hypothesis for symmetry increasing the effect of aposematism in animals. However, this was by no means proof that symmetry in these plants evolved because of aposematic signaling, but rather that such signaling has a potential to be stronger when associated with symmetry that exists for various developmental or physiological reasons. These cases were: (I) Toxic colorful flowers (Fig. 27.1), which were one of the first plant parts proposed

S. Lev-Yadun, *Defensive (anti-herbivory) Coloration in Land Plants*,
DOI 10.1007/978-3-319-42096-7_27

Fig. 27.1 The toxic colorful and symmetric flower of the Mediterranean species *Anemone coronaria*. These toxic plants are avoided by most grazing mammals and therefore dominate heavily grazed areas

to be aposematic (Hinton 1973; Lev-Yadun 2009a), are typically symmetric (Stebbins 1974; Neal et al. 1998). (II) Similarly, putative aposematic toxic fruits (Hill 2006; Lev-Yadun et al. 2009a) and spiny unripe red fruits that were also proposed to be aposematic (Lev-Yadun et al. 2009a) are symmetric. (III) Many putatively aposematic spiny plants are wholly symmetric, or their aposematic organs are symmetric. The spiniest plant taxa (cacti, *Aloe* spp., *Agave* spp., *Euphorbia* spp.), which have been proposed to be aposematic because of their colorful spine system (Lev-Yadun 2001), their shoots, and in cacti also the very spiny, colorful and conspicuous fruits, are commonly radially symmetric. Moreover, in the genera *Agave* and *Aloe*, the individual spiny leaves are commonly bilaterally symmetric. In many cacti species, the spines are arranged along the symmetrically arranged stem ribs or formed in groups evenly distributed over flat cladode areas, and in *Euphorbia* spp. they are arranged in columns along the stem ribs in symmetric pairs. The same is true for the spiny leaf rosettes and flowering spiny heads of the many east Mediterranean species of the Asteraceae (Ronel et al. 2009) that are symmetric (Fig. 27.2).

Thus, Lev-Yadun (2011) proposed that like in certain animals (Forsman and Merilaita 1999; Forsman and Herrström 2004), visual aposematism in spiny and poisonous plants also seems to be commonly associated with symmetry. This association may be the outcome of developmental constraints, or in flowers and fruit, selected to be symmetrical along with other signaling systems aimed for gene dispersal by animals, as well as with perceptual exploitation of animal sensing *sensu* Schaefer and Ruxton (2009), because animals probably pay more attention to symmetric shapes (e.g., Enquist and Arak 1994; Neal et al. 1998).

Fig. 27.2 The very spiny colorful and symmetric inflorescence of *Silybum marianum*

There is an obvious need for experimental testing of the role of symmetry and asymmetry in visual plant aposematism and other types of defensive plant coloration. This need is part of the general great need to test by experiments the various current hypotheses of plant aposematism (see Ruxton 2014). However, it is probable that since aposematism in general and symmetry-related in particular, is only one of several defensive components of plants, it will be very difficult to single out the role symmetry plays in direct defense from herbivores.

Chapter 28
Color Changes in Old Aposematic Thorns, Spines, and Prickles

Of the various colorful plant/animal communication systems, adaptive color changes are known to take part in the two extensively studied gene dispersal systems: pollination by both invertebrates and vertebrates and seed dispersal *via* frugivory. Young and still unrewarding animal-pollinated flowers and young and unripe fleshy fruits are usually green and cryptic. Many flowers become colorful and visually conspicuous only towards anthesis, when they open and offer nectar and/or pollen, or other rewards to pollinators, because their still immature colorful petals are commonly covered by bracts or by green sepals. Many flowers retain their conspicuous advertising colors till they wilt. However, post-pollination floral color change is a well-documented phenomenon in various taxa and life forms on all continents except for Antarctica (Weiss 1991, 1995; Weiss and Lamont 1997; Bradshaw and Schemske 2003). Color change of individual flowers in an inflorescence may reduce their advertising intensity, and thus their detectability by pollinators. Retaining the attractive coloration after pollination, or after they turn unreceptive, may reduce pollinator visits to un-pollinated flowers because the pollinators learn that there is no reward and fly to other plants, thus diminishing the plant's reproductive success. By the simultaneous reduction of both the reward after pollination and of attractiveness (advertisement) by changing their color, plants direct their pollinators to un-pollinated flowers within the same inflorescence or plant.

Fleshy fruit aimed at attracting and rewarding seed dispersers usually become colorful (yellow, pink, white, orange, red, brown, blue, purple or black) only toward ripening, when they also become edible following the lowering of the content of protective, poisonous, and otherwise harmful or repelling secondary metabolites, and by increasing their sugar, protein and fat content as well as flavor and softness (Ridley 1930; van der Pijl 1982; Snow and Snow 1988; Willson and Whelan 1990; Schaefer and Schaefer 2007), a phenomenon also considered to be at least partly adaptive (e.g., Willson and Whelan 1990; Schaefer and Ruxton 2011).

While the adaptive significance and the broad occurrence of color change in flowers (Weiss 1991, 1995), fruits (van der Pijl 1982; Willson and Whelan 1990)

S. Lev-Yadun, *Defensive (anti-herbivory) Coloration in Land Plants*,
DOI 10.1007/978-3-319-42096-7_28

and leaves (Archetti 2000; Hamilton and Brown 2001; Hoch et al. 2001; Lev-Yadun and Gould 2007; Archetti et al. 2009a) has been widely discussed, the phenomenon of color change in thorns, spines and prickles has been described as a widespread signaling phenomenon and discussed as such only by Lev-Yadun and Ne'eman (2006).

Color changes of aposematic thorns, spines and prickles make them less conspicuous and they lose most or even all their visual aposematic character (Lev-Yadun and Ne'eman 2006). For instance, in *Rosa* species, when the branches are young and green the growing prickles are also green (Fig. 28.1), the yellow, orange, red, brown, or black prickles are very conspicuous (Fig. 28.2). Later, when the branch changes its color from green to brown or gray, the prickles that have lost their original coloration and have changed their color to white (Fig. 28.3) are much less conspicuous and thus less aposematic. In other species (e.g., *Araucaria araucana, Washingtonia filifera, Carissa grandiflora*) (Figs. 28.4, 28.5, 28.6, 28.7 and 28.8), the color changes make them much less conspicuous and the spines and thorns lose all or most of their aposematic function (e.g., the spines along petiole margins of the large leaves of the palm *Washingtonia filifera*, which have the same brown or gray color as do the dry and dead petioles). *Chorisia speciosa*, an American tree grown in Israel as an ornamental, has large, red prickles, several cm. long, on its young green stems and branches. These original prickles change their color to whitish after several years. As these fast-growing *Chorisia* trees expand in girth, new red, conspicuous prickles are formed between the groups of old whitish prickles that have lost their color (Fig. 28.9). The tree trunk, however, with its new groups of red prickles, remains conspicuous and aposematic. The scale of this phenomenon of thorn, spine and prickle color change on a taxon, flora, continent or global scale is still unknown.

Lev-Yadun and Ne'eman (2006) emphasized that color change in thorns, spines and prickles is not mandatory. The color changes and the aposematic character loss typically occur when the defended organs become less edible to large herbivores because of their increased size, mechanical rigidity or chemical defense, or when there is no need for defense. Reducing the cost of defense seems to be the reason for the ephemeral nature of conspicuousness of many plant thorns, spines and prickles (Lev-Yadun and Ne'eman 2006). The adaptive value may lie in reducing the investment in coloration, since a thin ephemeral coloration layer demands fewer resources than a thick permanent one. Keeping thorns, spines, or prickles colorful for a long time is more costly and the benefit of being aposematic is smaller in older, larger, or otherwise better protected and less nutritious organs. The tendency of plants to lower the cost of defense by thorns, spines and prickles is a well-known phenomenon. For instance, African acacias and other woody plants have longer thorns on the lower branches than on the higher ones (Cooper and Owen-Smith 1986; White 1988; Milewski et al. 1991; Brooks and Owen-Smith 1994; Young and Okello 1998; Gowda and Palo 2003). Certain trees (e.g., various citruses and palms) have large thorns or spines only when young and fewer or none when mature (e.g., Kozlowski 1971; Cooper and Owen-Smith 1986; Cornett 1986; Clement and Manshardt 2000). Moreover, like several other types of induced defenses, thorns and spines are known

Fig. 28.1 Green, immature prickles in the wild rose *Rosa canina* growing in Mount Hermon, Israel

to increase in size and number following herbivory (e.g., Milewski et al. 1991; Perevolotsky and Haimov 1991; Young et al. 2003). There is therefore no theoretical difficulty in proposing that color changes in thorns, spines and prickles also reflect conservation of resources (Lev-Yadun and Ne'eman 2006). However, a simple alternative theoretical explanation exists: the thorns, spines, and prickles are colorful simply because the hard polymers composing them are colorful by nature. Lev-Yadun and Ne'eman (2006) dismissed this possibility because the thorns, spines and prickles that lose or change color remain hard and mechanically functional. The layer of coloration seems not to have a significant, or even any role, in producing their sharpness or strength. The broad taxonomic distribution of color changes in thorns, spines and prickles (Table 28.1) indicates that this character has evolved repeatedly and independently (convergent character) in both Gymnosperms and Angiosperms, monocotyledons and dicotyledons, probably in response to selection by visually oriented herbivores combined with conservation of resources.

The cost to the plant of producing colored organs has three aspects. First, it requires allocation of resources for the synthesis of the pigments and a mechanism to regulate the color pattern formation. Second, any color of a plant's photosynthetic organ, other than green, is linked in many cases (but not all!) to lower photosynthesis and productivity, and many young and growing thorns, spines and prickles are

Fig. 28.2 Colorful conspicuous rose prikles

green and photosynthetic (Fig. 28.1). Third, conspicuousness may attract herbivores and this risk has to be compensated for by better defense. Lev-Yadun and Ne'eman (2006) proposed that the discussed color changes of thorns, spines and prickles are adaptive, providing some advantage to the plant, and that the expression of color loss depends on the reduced vulnerability of the defended organ to large herbivores correlated with its growth or maturity.

Fig. 28.3 White, old prickles in the wild rose *Rosa canina* growing in Mount Hermon, Israel

Fig. 28.4 Thorns of *Carissa grandiflora* with typical *red* marking of their sharp ends

Fig. 28.5 Old thorns of *Carissa grandiflora* that have lost their typical *red* marking of their sharp ends

Fig. 28.6 Colorful conspicuous spines along the petioles of the green leaves of the fan palm *Washingtonia filiferra*

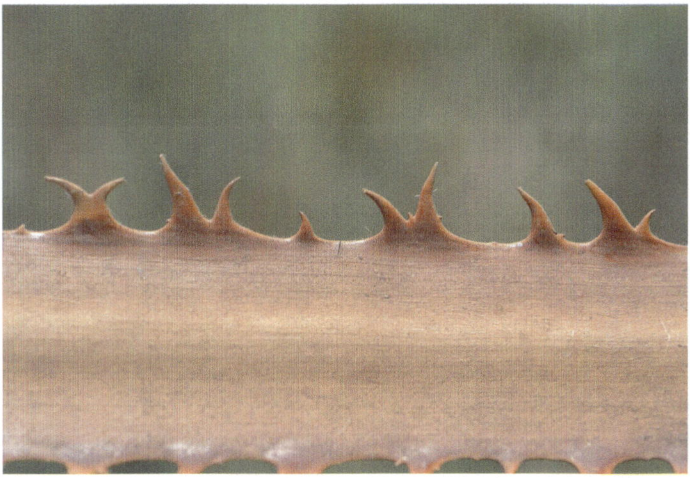

Fig. 28.7 Brown spines along the petioles of a dry and dead leaf of the fan palm *Washingtonia filiferra*

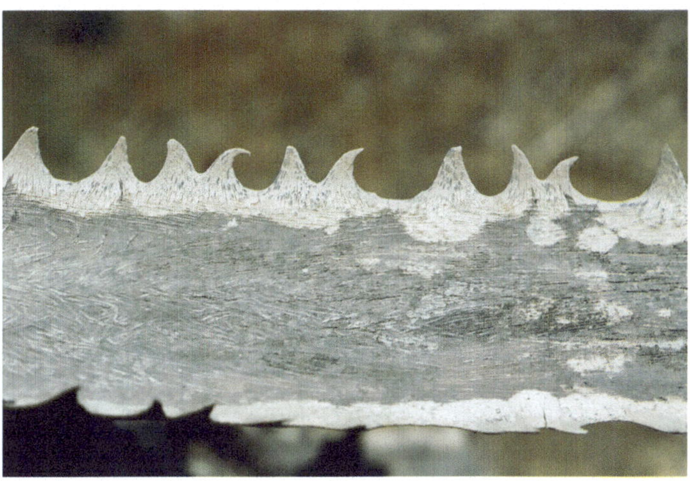

Fig. 28.8 White spines along the petioles of a 30–40-year-old dead leaf of the fan palm *Washingtonia filiferra*. The spines are still functioning at that stage as I know from repeated personal experience

Fig. 28.9 New *reddish-brown* prickles emerging among older whitish prickles on the trunk of *Chorisia speciosa* tree grown as an ornamental in Israel

Table 28.1 Differential discoloration in young vs. old thorns, spines, and prickles

Taxon	Color of mature thorns, spines, and prickles and background tissue	Color of old thorns, spines, and prickles and background tissue	Comments
Gymnosperms			
Pinus taeda	Brown thorns on green	Gray thorns on brown	In seed cones (T) [A]
Pinus contorta	Brown thorns on green	Gray thorns on brown or gray	In seed cones (T) [A]
Araucaria araucana	Brown thorns on green	Brown thorns on brown	In leaves (T) [A]
Angiosperms (monocotyledons)			
Phoenix dactylifera	Yellow and black on green	Brown on brown	Leaflets (T) [AF, AS]
Washingtonia filifera	Orange or brown on green	Whitish spines on gray	Along petioles (T) [A]
Yucca aloifolia	Dark brown on green	Gray on gray	End of leaves (T) [A]
Angiosperms (dicotyledons)			
Calicotome villosa	Black on green	Gray on gray	Branch tips (S) [AS]
Capparis spinosa	Yellow or orange on green or red	Gray on gray	Along branches (S) [AS, E]
Carissa grandiflora	Orange or red on green	Gray on gray	Along branches (S) [AF]
Cereus peruvianus	Orange and brown on green	Gray on gray	Along stems (T) [A]
Chorisia speciosa	Red on green	Whitish or gray on green or gray + growth of new red prickles among the old ones	Along trunk and branches (T) [A]

(continued)

Table 28.1 (continued)

Taxon	Color of mature thorns, spines, and prickles and background tissue	Color of old thorns, spines, and prickles and background tissue	Comments
Crataegus aronia	Red	Gray	Thorns (T) [AS]
Erythrina corallodendrum	Black on green or gray	Whitish on gray	Along branches and trunk (T) [A]
Euphorbia milii	Reddish brown on gray	Gray on gray	Along trunk and branches (S) [AF]
Rosa species	Yellow, orange, red brown or black on green	Whitish on brown	Along stems (S) [A, AS, E]
Rubus sanguineus	Yellow, orange or red on green or red	Gray on gray	Along stems (C) [AS]
Ziziphus spina-christi	Orange-brown on gray	Gray on gray	Along branches (T) [AF. AS]
Ziziphus lotus	Pink, red, brown on gray	Gray on gray	Along branches (S) [AF. AS]

After Lev-Yadun and Ne'eman 2006
(*C*) climber, (*S*) shrub, (*T*) tree, [*A*] America, [*AF*] Africa, [*AS*] Asia, [*E*] Europe

Chapter 29
Pathogenic Bacteria and Fungi on Thorns, Spines and Prickles

As a child that used to walk barefoot in the summer and step on various sharp objects including thorns, spines and prickles, and because of the wounds inflicted on me, several times I got emergency anti-tettanus injections. After I understood that thorns, spines and prickles are commonly visually aposematic, I wondered how common are bacterial pathogens found on these sharp pointed structures and with our enthusiastic departmental microbiologist Malka Halpern I began to examine this. Four publications (Halpern et al. 2007a, b, 2011; Lev-Yadun and Halpern 2008) showed that thorns, spines and prickles regularly harbor an array of pathogenic bacteria and fungi and discussed its potential significance in the evolution and functionality of plant aposematism. Spines from date palm (*Phoenix dactylifera*) trees, thorns from common hawthorn (*Crataegus aronia*) trees and two thorny shrub species, thorny burnet (*Sarcopoterium spinosum*) and manna tree (*Alhagi graecorum*), were sampled for aerobic and anaerobic bacteria in Israel. Every typical mature individual of these trees and shrubs carries hundreds or even thousands of conspicuous and therefore potentially aposematic spines or thorns. The common knowledge about the severity and frequency of aggressive bacterial infections among date-palm orchard workers in Israel following spine wounding, has necessitated the costly practice of removal of all the millions of spines from many of the orchards by mechanical saws. This was a very good indication that an important aspect of defensive ecology was hidden there.

Indeed, even the small number of spines and thorns studied (dozens) resulted in the cloning and identification of several very pathogenic aerobic and anaerobic bacteria species including *Clostridium perfringens*, *Bacillus anthracis* and *Pantoea agglomerans* (Halpern et al. 2007a, b, 2011). *Clostridium perfringens* is known to be a flesh-eater in that it can produce a necrotizing infection of the skeletal muscle called gas gangrene (Shimizu et al. 2002). *Bacillus anthracis* is the etiological agent of anthrax, a notoriously acute fatal disease in both domesticated and wild animals, particularly herbivorous ones, and humans (Jensen et al. 2003). The cutaneous form of the disease is usually acquired through injured skin or mucous membranes, a typical thorn, spine or prickle injury. *Clostridium tetani* is the etiological agent of

S. Lev-Yadun, *Defensive (anti-herbivory) Coloration in Land Plants*,
DOI 10.1007/978-3-319-42096-7_29

tetanus, a serious disease in humans and animals can be fatal when left untreated. Thorn, spine and prickle injuries have been known to cause tetanus in the USA, Ethiopia, and Turkey (Hodes and Teferedegne 1990; Ergonul et al. 2003; Pascual et al. 2003). In their review of the medical literature, Halpern et al. (2007b) found that septic inflammation caused by plant thorn, spine or prickle injury can result not only from bacteria, but also from pathogenic fungi. Dermatophytes that cause subcutaneous mycoses are unable to penetrate the skin and must be introduced into the subcutaneous tissue by a puncture wound (Willey et al. 2008). Thus, the physical defense provided by thorns, spines and prickles against herbivores might be only the tip of the iceberg in a much more complicated story. These sharp plant structures may inject bacteria into herbivores by wounding, enabling the microorganisms to pass the animal's first line of defense (the skin) and in so doing may cause severe infections that are much more dangerous and painful than the mechanical wounding itself (Halpern et al. 2007a, b, 2011; Lev-Yadun and Halpern 2008).

An obvious question concerning the potential defensive role of pathogenic microorganisms on plant surfaces is about those found on surfaces other than thorns, spines or prickles. The positive answer in many cases is theoretically simple. Thousands of plant species have sharp microscopic alternatives to thorns, spines and prickles that can insert the bacterial and fungal pathogens into the tissues of the herbivores that attack them. Given that microorganisms are generally ubiquitous, there is no reason to assume that only specific plants or specific plant organs will be rich in microorganisms. Despite this ubiquitous occurrence, however, certain plants or plant organs may have specific chemical components or structures on their surfaces that either reduce or increase the possibility of survival for microorganism taxa. Microorganisms can grow on plant surfaces in biofilms, which are assemblages of bacterial cells attached to a surface and enclosed in adhesive polysaccharides excreted by them. Within the biofilm matrix, several different microenvironments can exist, including anoxic conditions that facilitate the existence of anaerobic bacteria. Considering the findings of Halpern et al. (2007a, b, 2011) in regard to spines and thorns, it is clear that not only aerobic, but also pathogenic anaerobic bacteria can survive on these defensive structures. Although it is assumed that an array of biofilm types is formed on plant surfaces, this issue should be systematically studied in relation to defense from herbivory in order to gain a better understanding of the anti-herbivory role of microorganisms. Although limited in its scope, the study by Halpern et al. (2011) gave good indications that in the palm *Washingtonia filifera* there are significant differences in the composition of the bacterial flora of the colorful spines, *versus* green, flat and photosynthetic tissues of the same leaf.

Another theoretical aspect is the delay between the thorn's contact and wounding and the microorganism's action. While the pain from contacting thorns, spines or prickles is immediate, the microorganism's action is delayed. However, the same is true for the delayed action of poisons in aposematic poisonous organisms, and yet there is general agreement that mant poisonous and colorful organisms are aposematic (e.g., Cott 1940; Edmunds 1974; Gittleman and Harvey 1980; Harvey and Paxton 1981; Ruxton et al. 2004). Therefore, there is no reason to view a microorganism's contamination and its delayed action any differently.

None of the published medical data concerning infections caused following thorn, spine, or prickle injuries discussed ecological or evolutionary issues or aposematism, but rather was published only in the interests of medical practice. However, the data presented in Halpern et al. (2007a, b, 2011) showed that plant thorns, spines and prickles may regularly harbor various toxic or pathogenic bacteria. Halpern et al. (2007a, b) and Lev-Yadun and Halpern (2008) proposed that thorn-, spine- and prickle-injected microorganisms have a considerable potential role in anti-herbivory, actually serving as a biological warfare agent, and that they may have uniquely contributed to the common evolution of aposematism (warning coloration) in thorny, spiny and prickly plants.

The question of whether simultaneous olfactory aposematism is involved in bacteria-infected plants has never been addressed. Moreover, the possibility that the pathogenic bacterial biofilms omit attractive molecules that cause herbivores to approach the plants and be wounded and infected to benefit the bacteria that by this will contaminate new victims and in the long run also benefit the plants because of reduction of herbivore populations has also never been addressed.

Chapter 30
Aposematism in Plants with Silica Needles and Raphids Made of Calcium Oxalate

Lev-Yadun and Halpern (2008) proposed that many plant species that do not have thorns, spines, or prickles have an alternative of one out of two types of usually internal, but sometimes external, sharp, microscopic defensive structures: silica needles and raphids (which are needles made of calcium oxalate). Plant silica bodies (phytoliths) are formed by an ordered biological deposition of silicon that enters the plant *via* the roots (Richmond and Sussman 2003). Silica bodies have several known functions including structural, optical, serving as co-factors in the detoxification of heavy metals, and defense from herbivory (e.g., Richmond and Sussman 2003; Wang et al. 2004; Katz 2015; Meharg and Meharg 2015). Lev-Yadun and Halpern (2008) discussed their specific potential biological warfare defensive function: enabling the penetration of microorganisms into the bodies of herbivores.

Thousands of plant species belonging to many families produce raphids (Franceschi and Horner 1980). Usually, raphids are formed in specific parenchymatic cells that differ from their neighboring cells and are called idioblasts (Fahn 1990). The raphids are formed in large numbers and are packed compactly, aligned parallel to each other, and spread when the tissue is wounded. Raphids are always elongated, needle-shaped, and usually have two sharp pointed ends. This, however, is not their whole structural story. Studies conducted with a scanning and transmission electron microscope have revealed that in many cases, the raphids may be barbed or may have deep grooves along them. The grooves serve as channels through which plant toxins are introduced into the tissues of the herbivores (Sakai et al. 1972; Franceschi and Horner 1980). Like silica bodies in plants, calcium oxalate bodies have several functions, including tissue calcium regulation, transmission of light into deeper tissues, defense from herbivory, metal detoxification, and structural functions (Franceschi and Horner 1980; Ruiz et al. 2002; Nakata 2003; Franceschi and Nakata 2005; Gal et al. 2012).

In addition to the ability of both types of internal microscopic spines (raphids and silica needles) to introduce plant toxins into the wounded tissues of the herbivore by causing a mechanical irritation, Lev-Yadun and Halpern (2008) proposed that they are also able to introduce pathogenic microorganisms into the herbivore's

S. Lev-Yadun, *Defensive (anti-herbivory) Coloration in Land Plants*,
DOI 10.1007/978-3-319-42096-7_30

Fig. 30.1 *Dieffenbachia* sp. leaves with their typical white variegation. Since they are toxic, it is probable that the *white marking* serves for visual aposematism

tissues. Because of their small size, raphids and silica needles can internally wound the mouth and digestive system not only of large vertebrates, but also of insects and other small herbivores such as snails that usually manage to avoid thorns, spines and prickles by passing between them. Like thorns, spines and prickles, the raphids and silica needles actually inject the pathogenic microorganisms found on the plant surfaces themselves, as well as those found in the mouths and digestive tracts of the herbivores and cause infection.

Certain well known, well-defended plants with a combination of raphids and toxins seem to be visually aposematic. The poisonous American genus *Dieffenbachia* is a very good example, demonstrating the probably broad, but overlooked phenomenon of association and mutual defensive function of raphids and white (as well as non-white) leaf variegation. Leaves of wild members of the genus *Dieffenbachia* growing in the forests of Central and South America are typically variegated in white (Lee 2007) (Fig. 30.1). The plants are well defended from herbivores by the large amounts of sharp grooved raphids that, in addition to wounding, administer various plant toxins into the wounded tissues of the herbivores (or humans) that eat them (Arditti and Rodriguez 1982; Evans 1987; Gardner 1994; Bradbury and Nixon 1998). The combination of visual conspicuousness with unpalatability in the genus *Dieffenbachia* points to the probable aposematic effect of their leaf variegation.

Chapter 31
Müllerian and Batesian Mimicry Rings of Aposematic Thorny, Spiny and Toxic Plants

An obvious question concerning visual aposematism is that of the existence of Müllerian and Batesian mimics. When in a given habitat the proportion of aposematic thorny and spiny plants increases for a period long enough for an evolutionary change, Müllerian mimicry may lead to the establishment of defense guilds (see Waldbauer 1988). This question was examined at both a specific flora (of Israel) (Lev-Yadun 2003a, 2009b; Ronel et al. 2009; Ronel and Lev-Yadun 2012) and with a global perspective (Lev-Yadun 2009c, 2014d) and indeed, Müllerian mimicry seems to be common within the group of thorny, spiny and prickly plants.

Lev-Yadun (2009b) showed that twenty-one wild thorny, spiny or prickly plant species growing in Israel that are conspicuous because of white stripes and spots found on their leaves have all that is required to be considered a Müllerian mimicry ring or rings. Eight Israeli spiny species of the Asteraceae were found to have white variegation. The first three: *Silybum marianum*, *Notobasis syriaca* and *Scolymus maculatus*, are the spiniest of the eight (spines ranging from several mm. to several cm. in length). All three species typically have annual rosettes that have a white network of stripes (several cm. to dozens of cm. long, and several mm. wide) on the upper surface of their spiny leaves, and in *S. maculatus*, also in the spiny wings formed along its stems and branches, which gives them a zebra-like appearance (Lev-Yadun 2003a) (Fig. 1, 19.1, 19.4). White spots (ca. 3–20 mm. size) are found in the rosette, cauline leaves, and in the spiny wings formed along the stems and branches of the spiny annual *Carduus argentatus*, which has moderately long (2–10 mm.) spines. *Launaea spinosa*, a small thorny desert shrub, has leaves with white-tipped lobes. Two perennial herbs of the same genus, *L. nudicaulis*, and *L. capitata* (= *L. glomerata*), also have leaves with small spines along their white spotted margins (Feinbrun-Dothan 1978). *Helminthotheca echioides*, a spiny annual, has short spines (2–5 mm. long) and white spots (ca. 1–4 mm. size) on the rosette and cauline leaves. Eight annual species of the Boraginaceae are characterized by white spots (1–3 mm. size) found around the base of the short and somewhat soft spines (ca. 1–4 mm long) that characterize leaves of *Nonea obtusifolia*, *N. ventricosa*, *Echium judaeum E. rauwolfii*, *Anchusa aegyptiaca*, *A. strigosa*, *A. undulata*

S. Lev-Yadun, *Defensive (anti-herbivory) Coloration in Land Plants*,
DOI 10.1007/978-3-319-42096-7_31

and *Cerinthe palaestina*. Two annuals of the Brassicaceae also have white spots on their bristle-covered leaves. *Raphanus aucheri* has conspicuous white spots on its spiny rosette leaves. Some of the plants of *Brassica tournefortii* also have white spots on their somewhat spiny leaves, but they are much less conspicuous than those of *R. aucheri*. Two tree species of the Cupressaceae (*Juniperus drupacea* and *J. oxycedrus*) have white stripes along their needle-like sharp adult leaves 1.2–2.0 cm. long. These white longitudinal stripes are formed along the adaxial side of the spiny leaves. In the Liliaceae, many individuals of *Smilax aspera*, a poisonous climber, have white variegated leaves (variegation several mm. to 40 mm. in size), and very sharp prickles (several mm. to 1 cm. long) on their branches and spines on some of their leaves. These prickles and spines are usually colorful (yellow, brown, red, black and their combinations).

In addition to the 21 thorny, spiny or prickly species with white variegation or mottling, four non-spiny *Launaea* species have white variegation on their leaves. *Launaea resedifolia*, a perennial herb, has dentate leaves that are white calloused at the tip. *Launaea tenuiloba*, a perennial herb, has dentate leaves with white-tipped lobes. *Launaea mucronata*, a perennial herb, has dentate leaves with white callous-tipped lobes. *Launaea angustifolia* is an annual that has incised-dentate leaves ending in a hard white mucro (Feinbrun-Dothan 1978).

Twenty of these spiny and white variegated species occupy open habitats, and only the climber *Smilax aspera* is found in both shady and open habitats. In order to form mimicry rings, the relevant species must have an overlapping distribution. To examine the possibility that the 21 variegated plant species form mimicry rings, their overlapping distribution was tabulated (Table 31.1) from the data in the distribution maps for each species given by Danin (2004). The compiled data of the geographical distribution of the 21 variegated and spiny or prickly plant species clearly showed such overlap. Moreover, the same is true for the overlapping geographical distribution of the non-spiny white variegated *Launaea* species with the variegated spiny species. Therefore, there is no distribution-based limitation on the possibility that these two plant species groups form Müllerian and Batesian mimicry rings respectively. The geographical distribution of the 21 variegated and spiny plant species (Table 31.1) showed that in most regions of Israel, the Palestinian Authority and Jordan, at least 10 such species are found together or in close proximity. The smallest number of such species growing in a single zone (the Southern Negev) was six, and the highest number was 15 (in Samaria) (Table 31.1).

An important issue is the possibility of testing the hypothesis. Intuitively it seems simple, but in reality it is very complicated and there are contradicting theoretical considerations. (1) If white variegation has an aposematic role, then it should theoretically deter herbivores with previous foraging experience, but not naïve ones. The problem is that since aposematism is a very old and common phenomenon, there are no genetically naïve animals. Moreover, at least in *Silybum marianum*, which was proposed to use visual aposematic markings by white stripes (Lev-Yadun 2003a), Rothschild and Moore (1987) proposed that it uses olfactory aposematism *via* pyrazine. It is likely that both types of aposematism operate simultaneously not only in the *Silybum* case, but also in some other spiny and white variegated species. The

Table 31.1 Overlapping and non-overlapping distribution of white variegated spiny plant species in Israel according to the 30 territories of Israel and Jordan (compiled from Danin 2004).

	Sm	Ns	Scm	Ca	Ls	Ln	Lc	He	No	Nv	Ej	Er	Aa	As	Au	Cp	Ra	Bt	Jd	Jo	Sa	Lr	Lt	Lm	La
Coastal Galilee	+	+	+	+	−	−	−	+	−	−	−	−	−	+	+	−	−	+	−	−	+	−	−	−	-
Acco Plain	+	+	+	+	−	−	−	+	−	−	−	−	−	+	+	−	−	+	−	−	+	+	+	−	−
Carmel Coast	+	+	+	+	−	−	−	+	−	−	+	−	+	+	+	−	−	+	−	−	+	−	−	−	−
Sharon Plain	+	+	+	+	−	−	−	+	−	+	+	−	+	+	+	+	−	+	−	−	+	+	+	−	−
Philistean Plain	+	+	+	+	−	+	−	+	−	+	+	−	+	+	+	−	−	+	−	−	+	+	+	−	−
Upper Galilee	+	+	+	+	−	−	−	+	+	−	+	−	+	+	+	+	+	−	−	+	+	−	−	−	−
Lower Galilee	+	+	+	+	−	−	−	+	+	−	+	−	+	+	+	+	+	−	−	−	+	−	−	−	−
Mt. Carmel	+	+	+	+	−	−	−	+	+	−	+	−	+	+	+	+	−	+	−	−	+	−	−	−	−
Esdraelon Plain	+	+	+	+	−	−	−	+	+	−	+	−	+	+	−	−	+	−	−	−	−	−	−	−	−
Samaria	+	+	+	+	−	−	−	+	+	+	+	−	+	+	+	+	+	+	−	−	+	−	−	−	−
Shefela	+	+	+	+	−	−	−	+	+	+	+	−	+	+	+	−	−	−	−	−	+	−	−	−	−
Judean Mountain	+	+	+	+	−	−	−	+	+	+	+	−	+	+	+	−	−	+	−	−	+	−	−	−	−
Northern Negev	+	+	+	+	−	+	−	−	+	−	+	−	+	+	−	−	−	+	−	−	−	+	+	+	−
Western Negev	+	+	+	+	−	+	+	−	−	+	+	−	+	−	−	−	−	+	−	−	−	+	+	+	−
Negev Highlands	−	+	−	+	−	+	+	−	+	−	+	+	+	+	−	−	−	+	−	−	−	+	+	+	−
Southern Negev	−	−	−	−	+	+	−	−	−	−	+	+	+	−	−	−	−	+	−	−	−	+	+	+	+
Hula Plain	+	+	+	+	−	−	−	+	+	+	+	−	−	+	+	−	+	−	−	−	+	−	−	−	−

(continued)

Table 31.1 (continued)

Kinrot Valley	+	+	+	+	−	−	−	+	−	−	+	−	+	+	+	+	+	+	−	−	−	−	−	−	−
Beit Shean Valley	+	+	+	+	−	−	−	−	−	−	+	−	+	+	−	−	+	−	−	−	−	−	−	−	−
Mt. Gilboa	+	+	+	+	−	−	−	+	+	−	+	−	+	+	+	−	+	−	−	−	−	−	−	−	−
Samarian Desert	+	+	+	+	−	−	−	−	−	−	+	−	+	+	+	+	−	−	−	−	−	−	−	+	−
Judean Desert	+	+	+	+	−	+	+	−	−	−	+	+	+	+	+	−	−	−	−	−	−	−	−	+	+
Lower Jordan Valley	+	+	+	+	−	+	−	−	−	+	+	−	+	+	−	+	+	−	−	−	−	−	−	+	−
Dead Sea Valley	+	+	+	+	+	+	−	−	−	−	+	+	+	+	−	−	−	−	−	−	−	−	−	+	+
Arava Valley	−	−	−	−	+	+	+	−	+	−	+	+	+	−	−	−	−	+	−	−	−	+	+	+	+
Golan & Mt. Hermon	+	+	+	+	−	−	−	+	+	−	+	−	+	+	+	+	+	−	+	−	+	−	−	−	−
Gilead	+	+	+	+	−	+	−	−	+	−	+	+	+	+	+	+	+	−	−	−	+	−	−	−	−
Ammon	+	+	+	+	−	+	−	−	+	−	+	−	+	+	−	+	−	−	−	−	+	−	−	−	−
Moav	+	+	+	−	−	+	+	−	−	+	+	−	+	+	−	+	−	−	−	−	−	−	−	+	+
Edom	−	+	−	+	+	+	+	−	−	−	+	+	+	+	−	+	−	+	−	−	+	−	−	+	+

The 21 spiny variegated species: (Sm) *Silybum marianum*, (Ns) *Notobasis syriaca*, (Scm) *Scolymus maculatus*, (Ca) *Carduus argentatus*, (Ls) *Launaea spinosa*, (Ln) *L. nudicaulis*, (Lc) *L. capitata*, (He) *Helminthotheca echioides*, (No) *Nonea obtusifolia*, (Nv) *N. ventricosa*, (Ej) *Echium judaeum*, (Er) *E. rauwolfii*, (Aa) *Anchusa aegyptiaca*, (As) *A. strigosa*, (Au) *A. undulata*, (Cp) *Cerinthe palaestina*, (Ra) *Raphanus aucheri*, (Bt) *Brassica tournefortii*, (Jd) *Juniperus drupacea*, (Jo) *J. oxycedrus*, (Sa) *Smilax aspera*. The four non-spiny *Launaea* species that have white variegation on their leaves: (Lr) *Launaea resedifolia*, (Lt) *L. tenuiloba*, (Lm) *L. mucronata*, (La) *L. angustifolia*

possibility that thorny, spiny and prickly plants simultaneously use visual and olfactory aposematism was never studied systematically (see Lev-Yadun 2009a, 2014a) and it complicates the analysis of whatever results one gets. Moreover, as already pointed out (Lev-Yadun 2003a, 2006a, 2009a, 2013a, 2014a, b, c, d), white variegation may have several simultaneous defensive and physiological functions that may further complicate the analysis of experimental results. (2) If white variegation has an aposematic role, then it should be more widespread in ecosystems with heavy grazing pressure than in areas with less grazing. This hypothesis will certainly be difficult to study in the Near East or in any other part of the Mediterranean Basin because heavy grazing has been practiced everywhere there for millennia. However, field work I conducted in other regions with much less grazing (Canada, Estonia, Finland, Russia), but with different plant and animal taxa, clearly indicated a dramatically lower level of spininess in the ecosystem, along with much lower levels of white leaf variegation. (3) If non-spiny plants with white leaf variegation act as Batesian mimics of protected plants, then their abundance should be low compared with their models, because Batesian mimicry is expected to be effective only if mimics are much rarer than the models. This hypothesis is generally supported, since spiny plants with white variegation are much more common and conspicuous in the Near East, e.g., the very common *Silybum marianum*, *Notobasis syriaca* and *Scolymus maculatus*, than non-spiny white-variegated plants.

In conclusion, the hypothesis presented in Lev-Yadun (2003a, 2009b, 2014a) that white leaf variegation may be aposematic (along with other defensive functions) and that Batesian mimics for this character also exist, seems to be supported by the data presented there. It seems that both Müllerian mimicry defense guilds and Batesian mimicry defense guilds exist in white variegated wild plants in Israel and other Mediterranean countries. The long history of the considerable impact of grazing in this arid region (Zohary 1962, 1983; Thirgood 1981; Perevolotsky and Seligman 1998) seems to have selected for these characters. It has been shown that increasing the number of Müllerian mimics helps to reduce the cost of learning by herbivores even if the various species differ in their strength of defense (Rowland et al. 2007; Sherratt 2007), and the common existence of well-defended white variegated plants in the Near East seems to support their conclusions. The possibility that a parallel phenomenon of chemically defended plants with white variegated leaves or fruits exists in the Near-Eastern and in other floras has not escaped my attention, and this subject was under investigation (e.g., Lev-Yadun 2013a, 2014a, d), but is also still only partly described and understood.

After discussing Müllerian mimicry rings from a Mediterranean point of view, I wish to discuss it from a global perspective. To consider the existence of Müllerian mimicry rings in aposematic organisms, two factors are needed: (1) a similar signal, and (2) an overlapping distribution of the mimics in respect to the territory of predators in animals, or herbivores in plants. Lev-Yadun (2009b, c) showed that for the plant taxa proposed there to form Müllerian mimicry rings, both criteria are fulfilled.

The accumulating data about the common association of plant defenses by spines with visual conspicuousness, along with the fact that many such species overlap in their habitat, raises the possibility of the global broad phenomenon of the existence of Müllerian defensive mimicry rings in plants. Even from the limited number of publications proposing visual aposematism in spiny plants, the operation of vegetal Müllerian mimicry rings seems to be obvious. The phenomenon can be traced to both the Old World (Asia, Africa, and Europe) and the New World (South- and North America). Lev-Yadun (2009c) proposed that many species of visually aposematic thorny, spiny and prickly plants of the following taxa: (1) Cactaceae (Lev-Yadun 2001), (2) the genus *Agave* (Lev-Yadun 2001) (both are American taxa), (3) the African genus *Aloe* (Lev-Yadun 2001), (4) African thorny members of the genus *Euphorbia* (Lev-Yadun 2001), (5) African acacias with white thorns (Midgley et al. 2001, 2016; Midgley 2004), (6) spiny vascular plants of southeastern Ohio (Rubino and McCarthy 2004), (7) spiny Near Eastern plants with white variegated leaves (Lev-Yadun 2003a, 2009b), (8) Near Eastern members of the Asteraceae with yellow spines (Ronel et al. 2009; Ronel and Lev-Yadun 2012), all form Müllerian mimicry rings of spiny plants.

In the four very spiny taxa (Cactaceae and the genera *Agave*, *Aloe* and *Euphorbia*) that were the first to be proposed as visually aposematic (Lev-Yadun 2001) there is a very strong morphological similarity. In cacti, there are two types of conspicuousness of spines that are typical of many species: (1) colorful spines, and (2) white spots, or white or colorful stripes, associated with spines on the stems. These two types of aposematic coloration also dominate the spine system of the genera *Agave*, *Aloe* and *Euphorbia*. The fact that many species of three of these four spiny taxa (*Agave*, *Aloe* and *Euphorbia*) are also poisonous (Reynolds 1969; Gentry 1982; Nobel 1994) further indicates their potential to form defensive Müllerian mimicry rings.

Lev-Yadun (2009c) proposed that each of these groups for itself and some of these groups (e.g., Cactaceae and the genus *Agave* in North America; *Aloe*, *Euphorbia* and acacias in eastern and southern Africa) among themselves, when they have overlapping distribution and share at least some of the vertebrate herbivores, form a upper tier of Müllerian mimicry rings. Cacti and *Agave* that have an overlapping distribution over large areas in North America (Gentry 1982; Benson 1982) form the first Müllerian mimicry ring. Most of the large herbivores in North America disappeared not so long ago in evolutionary time scales and seem to have shaped the spiny defense of these plant taxa (Janzen 1986). The second Müllerian mimicry ring proposed by Lev-Yadun (2009c) is of the spiny and thorny members of the African genera *Aloe*, *Euphorbia* and certain acacias with very conspicuous white thorns, which partly overlap in distribution and share various large mammalian herbivores (Midgley et al. 2001, 2016; Midgley 2004; Reynolds 1969; Sajeva and Costanzo 1994). The third Müllerian mimicry ring proposed by Lev-Yadun (2009c) is the outcome of the common presence of aposematic coloration in spiny vascular plants of southeastern Ohio (Rubino and McCarthy 2004), with color patterns in thorns, spines and prickles similar to those of Cactaceae and the genera *Agave*, *Aloe* and *Euphorbia* described in Lev-Yadun (2001). The next case of potential operation of Müllerian mimicry ring of spiny plants with overlapping territories that suffer from the same large herbivores, but on a much smaller geographical

Table 31.2 Species with yellow thorns, spines and prickles in the flora of Israel.

Family and species (annuals)	Habitat	Chorotype	Spiny organ/Comments and spine color
Amaranthaceae			
Bassia muricata (L.) Asch.	D	IT/SA	R- spiny fruiting perianth (yellow)
Zygophyllaceae			
Tribulus terrestris L.	M/D	M/IT/ES	R- mericarps with prickles (yellowish/reddish/brownish)
Asteraceae			
Xanthium spinosum L.	M/D	T	S, R- prickly fruit (yellow)
Carlina racemosa L.	M	M	S, R- spiny bracts (yellow-orange)
Atractylis prolifera Boiss.	D	SA	S, R- spiny involucral bracts (yellow-orange)
Carduus argentatus L.	M/D	M	S, W, R- spiny involucral bracts (yellow)
Notobasis syriaca (L.) Cass.	M/D	M	S, R- spiny inflorescence (yellow/red)
Picnomon acarna (L.) Cass.	M/D	M/IT	S, W, R- spiny involucral bracts (yellow)
Silybum marianum (L.) Gaertn.	M/D	M/IT	S, R- spiny involucral bracts (yellow/orange)
Centaurea solstitialis L.	M	M/IT	R- spiny involucral bracts (yellowish)
Centaurea verutum L.	M/D	M	R- spiny involucral bracts (yellow)
Centaurea sinaica DC.	D	SA	R- spiny involucral bracts (yellow)
Centaurea hyalolepis Boiss.	M/D	M/IT	R- spiny involucral bracts (yellow)
Centaurea procurrens Spreng.	M/D	M	R- spiny involucral bracts (yellow)
Centaurea pallescens Delile	D	SA	R- spiny involucral bracts (yellow)
Carthamus persicus Willd.	D	IT	S, R- spiny involucral bracts (yellow)
Carthamus nitidus Boiss.	D/M	SA	S, R- spiny involucral bracts (yellow)
Carthamus glaucus MB.	M/D	M	S, R- spiny involucral bracts (yellowish)
Carthamus tenuis (Boiss. et Blanche) Bornm.	M/D	M	S, R- spiny involucral bracts (whitish/yellowish)

(continued)

Table 31.2 (continued)

Family and species Species (perennial herbs)	Habitat	Chorotype	Spiny organ/Comments
Acanthaceae			
Blepharis ciliaris (L.) B.L. Burtt	D	SA/S	S, R- spiny flower bracts (yellow/orange)
Blepharis attenuata Napper	D/M	IT/SA	S, R- spiny flower bracts (yellow/orange)
Asteraceae			
Gundelia tournefortii L.	M/D	IT	S, R- spiny bracts of head (in leaves yellow/orange; in flowers yellow/orange/purple
Echinops viscosus DC. subsp. *macrolepis* (Boiss.) Feinbrun	M	M	S, R- spiny head (yellowish)
Echinops gaillardotti Boiss.	M/D	M	S, R- spiny head (yellowish/reddish)
Echinops adenocaulos Boiss.	M/D	M	S, R- spiny head (yellow/orange)
Echinops glaberrimus DC.	D	SA	S, R- spiny head (yellow)
Echinops polyceras Boiss.	D/M	IT	S, R- spiny head (yellow)
Cousinia moabitica Bornm. et Nábělek	D	IT	S, R- spiny bracts (yellow)
Cousinia postiana Winkl.	D	IT	S, R- spiny involucral bracts (yellow)
Cirsium phyllocephalum Boiss. et Blanche	M	M	S, R- spiny involucral bracts (yellow/orange/brown)
Cirsium alatum (S.G.Gmel.) Bobrov	M	IT	S, W, R- spiny involucral bracts (yellow)
Ptilostemon diacantha (Labill.) Greuter	M	M	S, R- spiny involucral bracts (yellow)
Cynara syriaca Boiss.	M	M/IT	S, R- spiny involucral bracts (yellow with purple base)
Onopordum cynarocephalum Boiss. et Blanche	M/D	M	S, W, R- spiny involucral bracts (yellow)
Onopordum blancheanum (Eig) Danin	M	M	S, W, R- spiny involucral bracts (yellow)
Onopordum carduiforme Boiss.	M/D	M	S, W. R- spiny involucral bracts (yellow)
Onopordum ambiguum Fresen.	D	IT/SA	S, W, R- spiny involucral bracts (yellow)
Onopordum anisacanthum Boiss.	D	IT	S, W, R- spiny involucral bracts (golden yellow)

(continued)

Table 31.2 (continued)

Family and species Species (perennial herbs)	Habitat	Chorotype	Spiny organ/Comments
Onopordum alexandrinum Boiss.	D	IT/SA	S, W, R- spiny involucral bracts (yellow/orange)
Onopordum transjordanicum Eig	D	IT	S, W, R- spiny involucral bracts (yellow)
Onopordum jordanicolum Eig	D/M	SA	S, W, R- spiny involucral bracts (yellowish)
Onopordum palaestinum Eig	D/M	M/IT	S, W, R- spiny involucral bracts (yellow)
Onopordum floccosum Boiss.	M	M	S, W, R- spiny involucral bracts (yellow)
Centaurea rigida Banks et Sol.	M/D	IT	R- spiny involucral bracts (yellow)
Centaurea lanulata Eig	D	SA	R- spiny involucral bracts (yellow)
Centaurea aegyptiaca L.	D	SA	R- spiny involucral bracts (purple/yellow)
Scolymus hispanicus L.	M/D	M	S, R- spiny involucral bracts (yellow/brown)
Family and Species (dwarf -shrubs)	Habitat	Chorotype	Spiny organ/Comments
Fabaceae			
Alhagi graecorum Boiss.	M/D	M/IT	T (yellow/black)
Ononis spinosa L.	M/D	M/IT	T (whitish/yellowish with reddish tip)
Zygophyllaceae			
Fagonia bruguierei DC.	D	SA	S- stipular spines (yellow/orange)
Fagonia arabica L.	D	SA	S- stipular spines (yellow/orange/brown)
Fagonia mollis Delile	D	SA	S- stipular and leaf spines (yellow)
Asteraceae			
Atractylis serratuloides Cass.	D/M	SA	S, R- spiny involucral bracts (yellow/orange)
Launaea spinosa (Forssk.) O. Kuntze	D	SA	T (yellow/brown)
Asparagaceae			
Ruscus aculeatus L.	M	M	S (yellow/brown/black)
Family and Species (shrubs)	Habitat	Chorotype	Spiny organ/Comments
Brassicaceae			
Capparis sicula Veill.	M/D	M/IT/SA	S- stipular spines (yellow/orange)
Capparis spinosa L.	M/D	M	S- stipular spines (yellow/orange)
Rosaceae			
Rubus canescens DC.	M/D	M	R, P- prickly branches and pedicle (yellow/orange/red)
Rubus sanguineus Friv.	M/D	M/IT	S, R, P- prickly branch, petiole and pedicle (yellow/orange/red)

(continued)

Table 31.2 (continued)

Family and Species (shrubs)	Habitat	Chorotype	Spiny organ/Comments
Fabaceae			
Calicotome villosa (Poiret) Link	M/D	M	T (yellow/black)
Asteraceae			
Iphiona mucronata (Forssk.) Asch. et Schweinf.	D	SA	S (yellow/orange)
Smilacaceae			
Smilax aspera L.	M	M	S, P- prickly stems (yellow/red/black)
Family and Species (trees)	Habitat	Chorotype	Spiny organ/Comments
Zygophyllaceae			
Balanites aegyptiaca (L.) Delile	D	S	T (yellow/orange)
Rhamnaceae			
Ziziphus spina-christi (L.) Desf.	M/D	S	S- stipular spines (yellow/orange/red when young; whitish when old)
Arecaceae			
Phoenix dactylifera L.	D	SA	S (yellow/red/black when young)

After Ronel and Lev-Yadun (2012)
Habitat: *D* desert, *M* Mediterranean. Chorotype: *A* African, *C* Cosmopolitan, *ES* Euro-Sibirian, *M* Mediterranean, *IT* Irano-Turanian, *S* Sudanese, *SA* Saharo-Arabian, *T* Tropical. Spiny organ: *P* prickles, *R* reproductive structures, *S* spiny leaves, *T* thorny branches, *W* spiny wings on stem (Spine, thorn and prickle color)

scale, has recently been proposed for several Mediterranean spiny species with white-variegated leaves (Lev-Yadun 2003a), and later for more than 20 spiny species in the flora of Israel that have white markings associated with their spines (Lev-Yadun 2009b). Another case of a probable Müllerian mimicry ring of spiny plants was described by Ronel et al. (2009) and Ronel and Lev-Yadun (2012), who while studying the spine system of Near Eastern spiny members of the Asteraceae, found 46 spiny species with yellow spines, and 19 additional species with yellow spines that belong to other taxa (Ronel and Lev-Yadun 2012) (Table 31.2). Since some of these species and others with yellow spines also grow in southern Europe, and north Africa, it is clear that the same phenomenon is also common there. The last proposed case of Müllerian mimicry rings is that of the white latex, a toxic and sticky defense substance produced by about 20,000 plant species and secreted following wounding. Lev-Yadun (2014d) proposed that since white and toxic latex evolved independently many times in various plant taxa in several continents, it forms many defensive Müllerian mimicry rings.

Lev-Yadun (2009c) concluded that defensive Müllerian mimicry rings seem to be very common in plants, and that it is probable that many other spiny or toxic plants that form Müllerian mimicry rings are waiting to be studied.

Chapter 32
Batesian Mimicry and Automimicry of Aposematic Thorns, Spines and Prickles

Mimicry of defended and aposematic animals by both defended and non-defended animals is very common (Bates 1862; Müller 1879; Wallace 1889; Poulton 1890; Cott 1940; Wickler 1968; Edmunds 1974; Ruxton et al. 2004), and several classic authors have already proposed that defensive visual mimicry also operates in plants as an anti-herbivore mechanism. Wiens (1978) estimated that about 5% of land plants (some 20,000 species) are mimetic (an underestimation I think), listing several types of defensive plant mimicry. Since there are so many colorful or otherwise visually marked (aposematic) thorns, spines and prickles, their Batesian mimics are expected.

Cases of automimicry, i.e., mimicry of some defensive parts by other parts of the same individual, have rarely been described. Most described cases of automimicry in animals have a chemical (toxic) basis (Guilford 1994). Visual weapon automimicry of physical defenses such as horns or canines has been shown in several mammalian species (Guthrie and Petocz 1970) and has been proposed to be of intraspecific threatening value along with the defense from predators. However, this intriguing idea has received very limited attention in plants, e.g., Lev-Yadun (2003b).

Spine Automimicry

During my field work with conspicuous spines of several *Agave* species cultivated in Israel, I came across clear cases of visual spine automimicry, and later searched for such mimics in various taxa and characterized them. After I identified the phenomenon of spine automimicry during fieldwork, *Agave* handbooks (Gentry 1982; Sajeva and Costanzo 1994) were consulted in order to examine how common this character of potential visual defensive automimicry is. Lev-Yadun (2003b) described two types of Batesian thorn, spine or prickle mimicry: (1) A unique type of weapon (spine) automimicry (within the same spiny or prickly individual), a phenomenon previously known only in animals (e.g., Guthrie and Petocz 1970), and (2) classic

S. Lev-Yadun, *Defensive (anti-herbivory) Coloration in Land Plants*,
DOI 10.1007/978-3-319-42096-7_32

Batesian mimicry of aposematic colorful thorns, spines and prickles, by colorful elongated and pointed plant organs (buds, leaves and fruits), which despite their appearance, are not sharp. Interestingly, some insects, such as the tree-hopper *Umbonia* sp., mimic colorful aposematic plant thorns to escape predation (Purser 2003).

Agave species can have two types of spines in their leaves: (1) spines at the distal end, or (2) teeth along the margins (Gentry 1982), and many of these are colored and were proposed to be visually aposematic (Lev-Yadun 2001). *Aloe* species typically have spiny teeth along their leaf margins. Colorful spines along leaf petiole margins are also common in various palm species. My search revealed several dozens of *Agave* species that show spine automimicry (Figs. 26.16, 32.1 and 32.2). The mechanism that causes it is very simple: developing *Agave* leaves are rolled one around the other and around themselves and are strongly pressed against one another. The teeth that develop along the margins press against the surface of the same leaf or of another adjacent leaf, and the pattern of the teeth is copied and retained along the non-spiny parts of the leaves when the leaves unfold. For instance, in *Agave americana*, a common ornamental in Israel, the teeth copies are seen in many leaves. The *Agave* species showing the most remarkable teeth mimicry is *A. impressa*, in which the teeth mimicry print is visually enhanced by white material

Fig. 32.1 Leaves of *Agave americana* showing in their middle many prints (automimicry) of the real spines found at their margins

Fig. 32.2 Leaves of *Agave* sp. showing many prints (automimicry) of their real spines

and is therefore very conspicuous. However, the specimens of *A. impressa* I saw in Israel were still young and thus smaller than the ones described in Gentry (1982). The pictures of several dozens of other *Agave* species that were published in Gentry (1982) also show teeth mimicry. The other species are *A. abrupta*, *A. applanata*, *A. avellanidens*, *A. bovicornuta*, *A. cerulata*, *A. colorata*, *A. cupreata*, *A. decipiens*, *A. deserti*, *A. flexispina*, *A. fortiflora*, *A. franzosini*, *A. gigantensis*, *A. gracilipes*, *A. guadalajarana*, *A. horrida*, *A. kerchovei*, *A. macroculmis*, *A. maximiliana*, *A. mckelveyana*, *A. obscura*, *A. pachycentra*, *A. palmeri*, *A. parrasana*, *A. polyacantha*, *A. potatorum*, *A. pumila*, *A. salmiana*, *A. scabra*, *A. seemanniana*, *A. shawii shawii*, *A. sobria*, *A. victoriae-reginae*, *A. wercklei*, *A. wocomahi* and *A. zebra*. Additional *Agave* species with such teeth mimicry are *A. ferox*, *A. lophantha*, *A. macroacantha*, *A. marmorata*, *A. parryi*, *A. triangularis*, *A. utahensis* and *A. xylonacantha* (see figures in Sajeva and Costanzo 1994). The same type of colorful teeth along the margins and their mimicry by impression is obvious in the American palm *Washingtonia filifera* (Palmaceae), a common ornamental and a feral tree in Israel, and in *Aloe* sp. (Liliaceae) (Lev-Yadun 2003b).

Regular Spine and Thorn Mimicry

Colorful thorn-like, spine-like, or prickle-like structures were found in several wild and ornamental species growing in Israel. For instance, in several *Erodium* spp. (Geraniaceae), the elongated unripe fruits, which are several cm long, beak-like, pointed, and self-dispersing (by drilling into the soil, see Abraham and Elbaum 2013), are red or have other conspicuous colors, but this red fruit coloration is not aimed at attracting seed dispersers because they are self-dispersers (Fig. 32.3). In the annual *Erodium laciniatum* subsp. *laciniatum*, the elongated, several-cm.-long beak-like, pointed fruits are red. In the small perennial desert herb *E. crassifolium* (= *E. hirtum*), the elongated, several-cm.-long beak-like pointed fruits are partly or completely red. Another small perennial desert herb of this genus, *E. arborescens* also has red, elongated, several-cm.-long beak-like, pointed fruits that look like thorns although they are soft. In *Sinapis alba*, an annual of the Brassicaceae, the distal part of the elongated, pointed fruit, when fully developed but not yet ripe, is colorful (yellow, red, purple or various combinations of these) (Fig. 32.4). The pointed but soft distal ends of the new leaves of the perennial geophytes *Urginea maritima* and *Asphodelus ramosus* of the Liliaceae are red or orange and look like

Fig. 32.3 Unripe but fully grown fruits of *Erodium lacinlatum* from the coastal plain of Israel. They look like *red* (aposematic) thorns but they are soft and not sharp. Several days later they will dry up, disperse themselves and drill themselves into the soil. The *red* coloration is therefore not aimed to attract dispersing animals

Fig. 32.4 The fruits of *Sinapis alba*, an annual of the Brassicaceae, with their typical colorful elongated pointed but soft distal fruit part

spines. In *Limonium angustifolium* (= *Limonium* meyeri), a wild and domesticated perennial of the Plumbaginaceae, the distal part of its large leaves is red and looks like a spine although it is soft. The pointed but soft distal ends of leaves of the wild and domesticated shrub *Myrtus communis* of the Myrtaceae are red. In *Aegilops geniculata* (= *A. ovata*), an annual of the Poaceae, the elongated, and several-cm.-long, thorn-like but soft awns are red-purple (Lev-Yadun 2003b). The pointed distal part of the fleshy leaves of *Carpobrotus acinaciformis* (= *Mesembryanthemum acinaciforme*), a clonal perennial of the Aizoaceae, is red in many cases. Similarly, the pointed distal part of the fleshy leaves of *Aeonium decrum* (= *Sempervivum decorum*), a small shrub of the Crassulaceae, is red in many cases, as it is in several other unidentified species belonging to the genus *Aeonium*. Roses (*Rosa* sp.) have colorful (whitish, yellow, orange, red, brown, black) prickles along their stems. When buds develop along stems following their release from suppression by the apical bud, while they are still small and soft, they temporarily mimic the prickles in size, spacing, and color (Fig. 32.5). It is very easy to mistake these soft buds for prickles (Lev-Yadun 2003b).

Lev-Yadun (2006a) and Lev-Yadun and Gould (2009) proposed that there are two possible evolutionary routes toward mimicry of colorful thorns, spines, or prickles. In the first, an aposematic spiny, thorny or prickly plant may have lost its

Fig. 32.5 A rose branch with both prickles and soft young buds. It is difficult to distinguish visually the buds from the prickles

costly defensive character but retained the shape and aposematic signal becoming a secondary Batesian mimic. In the second, a non-aposematic and non-spiny, thorny or prickly plant can acquire the signal, becoming a primary Batesian mimic. Theoretically, such Batesian mimics may later in their evolution develop thorns, spines and prickles from the non-sharp but colorful tissues and become visually aposematic. Alternatively, the structure and coloration may have a different, unknown function. Williamson (1982) proposed that because plants are sessile, plant mimics are less likely to be mistaken for their models than animal mimics, which are mobile, a view which is only partly correct as plants, especially their leaves and branches, move in the wind, as do light spots within the canopy or beneath it. Williamson (1982) also proposed that the autonomy of plant appendages implies that warning mimicry provides less advantage to plants than to animals, because plants, which can usually regenerate, suffer less than animals from sampling by naive herbivores. On the contrary, Augner and Bernays (1998) studied the theoretical conditions for an evolutionary stable equilibrium of defended signaling plants and plants mimicking these signals. The modeling showed that mimicry of plant defense signals may be common, and even imperfect mimics could invade a population of defended signaling plants. Theoretically, selection allows the success of even poor and imperfect mimics (e.g., Edmunds 2000; Penney et al. 2012). In any case, Williamson (1982) seems to have considerably underestimated the role of mimicry in plant defense from herbivory, and the botanical reality supports the hypothesis presented by Augner and Bernays (1998).

Conspicuous thorns, spines and prickles were proposed to be beneficial for plants as herbivorous animals would remember the visual (aposematic) signal and tend to avoid subsequent tasting of such marked plants (e.g., Lev-Yadun 2001, 2009a). Lev-Yadun (2003b) proposed that the plants that visually mimic colorful thorns, spines and prickles may benefit from this mimicry even when this mimicry is imperfect. These plants are typical Batesian mimics in which non-protected organisms mimic dangerous ones. Lev-Yadun (2003b) also proposed that the plants that automimic spines and have fewer real spines than the amount they appear to show, may benefit from spine automimicry even if the mimicry of their own spines is imperfect. This is a special case of Batesian mimicry in which non-protected tissues mimic protected ones within the same organ.

The spine automimicry shown in Lev-Yadun (2003b) is a vegetal parallel for "weapon automimicry" of horns or canines known in several mammalian species (Guthrie and Petocz 1970). In both automimetic plants and animals, visually oriented animals were the selective agent that selected for such mutants, because animals see the visual patterns, change their behavior and thus by attacking the non-mimetic genotypes more, select for the mimics.

Chapter 33
Additional Cases of Defensive Visual Batesian Mimicry Among Plants

Leaf shape is known to partly explain herbivorous insect richness on various plant taxa (Jones and Lawton 1991), and this is probably related to more than one mechanism. The little known, and even less tested visual Batesian leaf mimicry will be briefly discussed in this short chapter. Shimshi (1979/1980) proposed that the hill ecotype of *Iris atrofusca* from the heavily grazed desert/steppe transition zone in Israel with its erect (ensiform) leaves that differ from the regular curved (falcate) leaves of the common type in other ecosystems is a mimic of the poisonous common geophyte *Asphodelus ramosus* (= *A. microcarpus*) of the Liliaceae that grows in the same habitat, which is not grazed in the winter and spring because of toxicity (Seligman et al. 1959). Brown and Lawton (1991) postulated that the two non-spiny species *Celmisia lyalli* and *C. petriei* (Asteraceae) growing in New Zealand look rather like spiny members of the genus *Aciphylla* (Apiaceae). Brown and Lawton (1991) also described the close resemblance of general plant morphology and leaf shape of the European white dead-nettle (*Lamium album*) to the well-defended stinging nettle (*Urtica dioica*) as a case of Batesian mimicry. This nettle mimic and other species that visually mimic nettle (*Lamium purpureum*, *Lamiastrum galeobdolon*, *Ballota nigra*, *Galeopsis tetrahit* and other *Galeopsis* species) are named dead nettles because they lack the stings of stinging nettle (Wheeler 2004). Interestingly, in November 2015, a group of my first-year biology students wrongly identified a *Ballota saxatilis* plant growing on Mount Carmel, Israel as nettle during a field class when they had to collect and identify several plant species. Recently, morphological Batesian mimicry was proposed to exist between the chemically defended leaves of the model *Pseudowintera colorata* by the leaves of the non-defended *Alseuosmia pusilla*, two species from New Zealand (Yager et al. 2016). Yager et al. (2016) showed that these two species had leaves that were morphologically distinct from all neighboring species, and that *A. pusilla* individuals were more similar to neighboring than to distant *P. colorata* plants. Moreover, 90 % of the leaf shape variation in these two species varied similarly across an altitudinal gradient. These authors used a quantitative geometric morphometric approach that allows for comparisons based on actual values and not only on general visual similarity. This

S. Lev-Yadun, *Defensive (anti-herbivory) Coloration in Land Plants*,
DOI 10.1007/978-3-319-42096-7_33

is expected to allow better quantitative critera for mimicry and for comparative studies on the strength of evolution in cases of visual mimicry. These few examples probably reflect a much broader phenomenon, partly discussed above, for instance when mimicry of host leaf shape by mistletoes was described. These few examples probably reflect a much broader phenomenon, partly discussed above, for instance when mimicry of host leaf shape by mistletoes was described.

Chapter 34
When May Green Plants Be Aposematic?

Green plants dominate most productive terrestrial ecosystems (Hairston et al. 1960). Therefore, green can neither be conspicuous nor aposematic under productive conditions. Knight and Siegfried (1983) raised the question of whether unripe green fruits signal unpalatability, and concluded that in the forest, green does not provide enough contrast to be aposematic. Lev-Yadun and Ne'eman (2004) posited, however, that in deserts there are conditions when green can be visually aposematic.

Water availability is the main limiting factor for plant growth in hot deserts. In the hot deserts of the Middle East, the rainy season is restricted to several months in winter and spring and the typical rains are scattered and usually provide only a few mm of rainfall per rain event (Lev-Yadun et al. 2009c). Therefore, desert plants have developed various adaptive strategies to survive during the long, hot and dry summer (Zohary 1962; Evenari et al. 1982; Archibold 1995). Annual desert plants survive the dry season as dormant seeds in the soil, and geophytes and hemicryptophytes pass the summer as dormant underground corms, bulbs or roots. Shrubs and dwarf-shrubs drastically reduce their leaf area and canopy size in summer by shedding leaves and branches, or by replacing large winter leaves with smaller summer ones to lower their water loss. Their summer leaves are covered in many cases by dense white trichomes that reflect sun radiation, thus reducing heating and transpiration (Orshan 1963, 1986; Evenari et al. 1982). Other plants have succulent leaves or stems with a very thick waxy cuticle layer that considerably reduces transpiration during the daytime, when their stomata are closed. Many desert succulents have adopted the CAM photosynthetic pathway, opening their stomata at night rather than during the day (Winter and Smith 1996). Winter and spring are the main growing seasons for the vast majority of the annual and perennial Middle Eastern desert plants. Therefore, on the landscape scale, but only in rainy years, the dominant landscape color during winter and spring is green. In spring or at the beginning of the summer, annuals may complete their lifecycle and in any case they dry up, as do the aboveground parts of geophytes and hemicryptophytes (Zohary 1962). Shrubs and dwarf-shrubs, which retain live canopies during the summer, lose most or all their green parts. As a result, the landscape turns yellow or gray. Surprisingly, there

S. Lev-Yadun, *Defensive (anti-herbivory) Coloration in Land Plants*,
DOI 10.1007/978-3-319-42096-7_34

are some conspicuously green desert plant species in the summer that contradict the general phenological and visual pattern.

Heavily grazed areas in the west, north, east and central parts of the Negev Desert, Israel were inspected repeatedly, in all seasons for many years. Special attention was given to regions densely populated by Bedouins, who own large flocks of sheep and goats (cattle, donkeys and camels are much less abundant). Following this vast field experience, Lev-Yadun and Ne'eman (2004) presented several cases of common poisonous or thorny plants that keep their vital green color during the summer in the heavily grazed, nearly plant-less, summer-yellow or gray landscape in the Negev Desert in Israel. They proposed that the conspicuously green color of these well-defended plant species is probably aposematic based on both visual and olfactory aposematism.

Grazing by flocks of animals owned by Bedouin almost completely harvests the low plant biomass of the desert. In the summer, in heavily grazed areas, the remaining plants become extremely sparse. When viewed from a distance of a dozen meters, it looks as if the grazing herds are eating stones. However, even under such drastic grazing pressure, there are several common plant species that are conspicuously green but are not eaten by the herds. It is clear that the flocks see the green plants, since they eat all other dry but palatable plants around them, but leave these green plants untouched. All these plants are poisonous or thorny.

The poisonous group includes the green desert Gymnosperm shrub *Ephedra aphylla* (= *E. alte*) (Ephedraceae), which contains ephedrin or related defensive molecules (Caveney et al. 2001). The shrub *Pergularia tomentosa* [= *Daemia tomentosa*] (Asclepiaceae) and the small tree *Calotropis procera* (Apple of Sodom), growing in hot but spring-irrigated or otherwise locally humid habitats, are poisonous, like most members of the family, which are rich with saponins and alkaloids (Waisel et al. 1977). The aposematic grasshopper *Poekilocerus bufonius* even sequesters the poisons of *Calotropis procera* for self-protection (Euw et al. 1967). Three very common species, *Anabasis setifera*, *A. syriaca* and *A. articulata* (Amarantaceae), are also alkaloid-rich summer-green shrubs that are not eaten by the flocks (Zohary 1980; Feinbrun-Dothan and Danin 1991). The common geophyte *Asphodelus ramosus* (= *A. aestivus*=*A. microcarpus*) of the Liliaceae has poisonous green leaves that are not grazed in the winter or in spring, but are eaten when dry and brown in the summer (Seligman et al. 1959; Shmida and Darom 1986). The perennial herb *Peganum harmala* of the Zygophyllaceae is very poisonous and not eaten by the flocks. Therefore, it is very common near Bedouin camps where the flocks eat its competitors (Danin 1977). Like *P. harmala*, the shrub *Thymelaea hirsuta* of the Thymelaeaceae, which is also poisonous, is not eaten by the flocks, and is abundant near Bedouin camps (Shmida and Darom 1986). The summer annuals *Chrozophora tinctoria* and *C. obliqua* (Euphorbiaceae) are also avoided by the flocks. Another member of the Euphorbiaceae, the tall shrub/small tree *Ricinus communis* (Castor oil plant), also remains green but is not grazed. Of the Solanaceae, the introduced shrub *Nicotiana glauca*, the native shrub *Withania somnifera*, and the introduced summer annuals *Datura innoxia*, *D. stramonium* and *Solanum nigrum* are all alkaloid-rich poisonous plants (Waisel et al. 1977) with a very strong deterrent odor, especially when wounded, and they are not eaten by the flocks.

The thorny summer-green plants group includes the shrubs *Alhagi graecorum* (= *A. maurorum*) (Papilionaceae) and *Zilla spinosa* (Brassicaceae), which have green stems and branches even in the summer, after their summer leaf shed. Both species are extremely thorny. The shrubs *Capparis aegyptia* and *C. sinaica* (= *C. cartilaginea*) (Capparaceae), found in rocky habitats in the desert, are both very spiny, as is the shrub *Asparagus stipularis* (Liliaceae), which retains green stems during the summer.

Thus, the small number of common plants that are conspicuously green in the dry summer in the Negev Desert (Israel), when most other plants are either totally dry (annuals, geophytes and hemicryptophytes) or are summer deciduous, have good protection against herbivory either by being poisonous or thorny. Sheep and goats that consume all other plants around them do not graze on these plants. The selective impact of millennia of grazing by the flocks of nomad and semi-nomad people resulted in the dominance of such well defended plants in the summer vegetation in that region. It is tempting to suggest that these plants have shifted their season of activity along with the development of their protective characters. Thus, their being green in summer is an analog for an evolutionary color change in conventional aposematic organisms that are red, yellow, etc. Interestingly, and in support of this hypothesis, plants that are evergreen in predominantly deciduous forests, plants leafing or flowering especially early in spring in the temperate region, or spring ephemerals in Great Britain, i.e., plants that are green when their neighbors are not, have a high incidence of toxicity (Grubb 1992).

Lev-Yadun and Ne'eman (2004) proposed that in various arid ecosystems or other types of ecosystems, when most plants are dry or leafless, green color is conspicuous and contrasts with the yellow-gray desert or leafless background and can act as an aposematic color. However, it is clear that green cannot be aposematic in environments where it does not contrast with the background. Thus, green fruits posted against a green canopy cannot be aposematic, as explained by Knight and Siegfried (1983). Moreover, most of the proposed aposematic summer-green plants are green also during winter, when their color cannot be aposematic, because then they are surrounded by many other green plants that are winter annuals or seasonal perennials such as hemicryptophytes. Secondary compounds are usually found at lower levels in deciduous leaves with a short life span and in ephemeral plants than in evergreen leaves of perennial plants with a long life span (Price 1997). As a result, during winter, the summer-green aposematic plants of the desert are protected from herbivory by their many more palatable neighbors, which will be probably eaten first (see Milchunas and Noy-Meir 2002). The levels of secondary metabolites in these plants are subject to seasonal changes, lower during winter and higher in summer (Price 1997). Consequently, Lev-Yadun and Ne'eman (2004) hypothesized that also in desert plants the concentration of the secondary metabolites increases at the beginning of the summer when the background turns from green to yellow/gray and the plants' green color becomes aposematic.

Some of the poisonous summer-green plant taxa (i.e., *Calotropis*, *Datura*, *Nicotiana*, *Ricinus*, *Thymelaea*) also have specific strong deterrent odors, especially when damaged. This fact supports the hypothesis of olfactory aposematism in poisonous plants (i.e., Eisner and Grant 1981; Launchbaugh and Provenza 1993; Provenza et al. 2000) and also supports the proposal that multimodal signals are more effective than simple ones (Lindström et al. 2001). It's possible that olfactory aposematism in poisonous plants is more common than previously realized. In an almost plant-less arid environment, a strong deterrent odor combined with a conspicuous color may be an easy-to-remember aposematic signaling combination. The higher concentration of volatile secondary metabolites during summer increases both the repellent odor and the level of poisonous substances (Price 1997). Therefore, during the summer, there is an increase in the repellant odor of the plants, an increase in the concentration of poisonous substances (Price 1997), and an increase in the visual contrast provided by the green color (Lev-Yadun and Ne'eman 2004).

Lev-Yadun and Ne'eman (2004) concluded that green plants may be aposematic during the almost plant-less summer in hot deserts and that some of these poisonous plants also use olfactory aposematism to communicate with herbivores for their mutual protection. Their ability to avoid even causal tasting by grazers is critical for their survival in the highly grazed Negev Desert or in similar ecosystems in the Near East and elsewhere. The green aposematic plant hypothesis of plants growing in barren habitats is supported by the fact that plants that are more apparent (do not match local soil or rock color) are more defended than cryptic ones growing in the same habitat (Strauss et al. 2015).

Chapter 35
Spiny East Mediterranean Plant Species Flower Later and in a Drier Season Than Non-spiny Species

From a global point of view, thorny, spiny and prickly plants that are protected by such physical defenses from mammalian or avian herbivores (Cooper and Owen-Smith 1986; Janzen 1986; Grubb 1992) are more common in arid regions of the world than in humid ones (Carlquist 1974; Parsons and Moldenke 1975; Shmida 1981; Milton 1991; Grubb 1992). These trends are in accordance with the general pattern of stronger anti-herbivory defense found in slow-growing plants in habitats with limited resources (McKey et al. 1978; Coley et al. 1985; Endara and Coley 2011). Ronel et al. (2010) tested if this global perspective is mirrored concerning the flowering phenology of spiny *versus* non-spiny plants in the flora of Israel.

The Mediterranean climate is characterized by a short, mild and wet winter contrasting with a long, dry and hot summer (Zohary 1973). Lush green herbaceous plants dominate the Mediterranean landscape during late winter and spring. Hot and dry spells of 35–45 °C characterize the late spring in Israel, and under these extreme weather conditions the green landscape changes its color to light green and then to yellow within a couple of days and the annual vegetation enters its dry summer state. The typical hot and dry Mediterranean summer presents significant physiological challenges to plants. As a result, early spring is the peak flowering season in the Mediterranean basin (Petanidou et al. 1995; Bosch et al. 1997). Zohary (1962) compiled the flowering time of the whole flora of Israel and found that the peak of flowering occurs in April, when about 1600 species (out of 2700) flower simultaneously. This makes spring a wonderful season for nature lovers, but a harsh one for insect-pollinated flowers that must compete strongly for their pollinators. Competition for pollinators has been well documented and reviewed (Mitchell et al. 2009), and several hypotheses and models have been proposed about the outcome of competition for pollinators. It may lead to the rapid elimination of a minority species (Levin and Anderson 1970), to the evolution of autogamy (Levin 1972), to specialization to exploit different pollinating agents (Grant and Grant 1965), or it may cause a shift in the flowering time of some species (Mosquin 1971). Seasonal segregation in blooming time of related plant species, or plants that belong to the same pollination guild, is well documented in many biomes: Arctic (e.g., Hocking

S. Lev-Yadun, *Defensive (anti-herbivory) Coloration in Land Plants*,
DOI 10.1007/978-3-319-42096-7_35

1968), temperate (e.g., Mosquin 1971; Heinrich 1975; Lack 1976) and neotropical (e.g., Heithaus 1974; Frankie et al. 1974; Feinsinger 1978). This segregation has been interpreted as an evolutionary result of competition for pollination, and in some cases competition for pollinators was also experimentally proven to be the evolutionary driver (Waser 1978; Pleasants 1980).

Ronel et al. (2010) tested the hypothesis that thorny and spiny plants of the Israeli flora may also represent such a case. However, in addition to the risk of drought, these species are also exposed to increased risk from large herbivores (e.g., Seligman et al. 1959; Zohary 1962, 1983; Gutman and Seligman 1979). Under heavy grazing, a considerable part of the grass and many other types of non-woody vegetation disappears shortly after the end of the rainy season (Seligman et al. 1959; Gutman and Seligman 1979; Henkin et al. 1998). Thus, delaying growth and flowering phenology from April to May or June, exposes the plants to a combination of severe drought and increased grazing risk because the group defense by the lush green spring vegetation (the vegetal parallel of schooling in animals) has disappeared. Ronel et al. (2010) thus examined the hypothesis that in the semi-arid flora of Israel non-spiny plants usually flower in spring, while spiny plants that are better defended from mammalian herbivores flower later, at the beginning of summer. This was done by studying the flowering phenology of spiny and non-spiny species of the Asteraceae, Fabaceae, and Lamiaceae families, which include the largest number of spiny species expressing visual aposematism in the flora of Israel (see Ronel and Lev-Yadun 2012).

Data about the flowering season (given in the flora in months) was compiled from the analytical flora (Feinbrun-Dothan and Danin 1991) of all wild native species of the three families that include the largest number of spiny species: Asteraceae with 92 spiny out of 287 species, 78 of which have colorful spines (e.g., Ronel et al. 2009; Ronel and Lev-Yadun 2012), Fabaceae with 28 spiny out of 298 species, 20 of which have colorful thorns or spines, and Lamiaceae with 13 spiny out of 227 species, 6 of which have colorful thorns or spines (Ronel and Lev-Yadun 2012). The months in which flowering was reported was tabulated, and the number of non-spiny and spiny species that flowered in each month was counted. The median of the distribution of the monthly number of flowering species was used as an indicator for the peak flowering season for each group. The distribution of the number of non-spiny and spiny species that flowered in the various months differed significantly for the Asteraceae, Fabaceae and for all families together, but not for the Lamiaceae (Ronel et al. 2010). The 195 non-spiny species of the Asteraceae have their flowering peak at the beginning of April, while the 92 spiny species have it in mid-June. The 270 non-spiny species of the Fabaceae have their flowering peak in mid-April, while the 28 spiny species have it in mid-May. The 114 non-spiny species of the Lamiaceae have their peak of flowering at the beginning of May, while the 13 spiny species have it in mid-May. A comparison of all 133 spiny species of these families with all 589 non-spiny species of these families showed that the peak in flowering of the spiny species is at the beginning of May, while that of the non-spiny species is in late March; the peak of flowering (when the highest number of species flowered) was 4–8 weeks later for spiny species than for non-spiny species (Table 35.1).

Table 35.1 The number of spiny and non-spiny species of the Asteraceae, Fabaceae and Lamiaceae in the flora of Israel that flower in each month

	Jan	Feb	Mar	Apr	May	Jun	Jul	Aug	Sep	Oct	Nov	Dec
Asteraceae												
Spiny	1	3	19	44	51	52	52	35	19	7	1	0
Non-spiny	14	29	112	156	101	49	39	37	31	21	11	12
Fabaceae												
Spiny	2	8	11	16	17	17	10	8	4	1	0	0
Non-spiny	7	70	212	250	152	53	20	16	10	7	2	3
Lamiaceae												
Spiny	0	0	2	6	8	8	6	7	5	3	0	0
Non-spiny	9	20	42	68	79	68	50	35	24	13	3	3
All families												
Spiny	3	11	32	66	76	77	68	50	28	11	1	0
Non-spiny	30	119	366	474	332	170	109	88	65	41	15	18

After Ronel et al. (2010)

Ronel et al. (2010) thus found a clear phenological difference in flowering time between spiny and non-spiny plant species, in which the functional group of spiny species (their majority is visually aposematic) grows and flowers later in the season, when the weather is warm and dry, while the non-spiny species grow and flower earlier in the spring under more favorable conditions, which might be a consequence of a long and repetitive evolutionary process. Their data clearly indicated that the physical defense of the plants by spines is associated with a flowering season postponed to the dry summer, a period with a much greater impact of mammalian grazers and drought stress. As the summer progresses, in many areas of the Near East the dry non-spiny and non-poisonous plants gradually disappear as the result of intensive grazing activity (Seligman et al. 1959; Gutman et al. 1990; Noy-Meir 1990). Even in the green spring (March-April) and conspicuously throughout the summer, spiny thistles such as *Echinops viscosus*, *Notobasis syriaca*, *Silybum marianum*, *Onopordum* sp., *Centaurea* sp., *Carthamus* sp. and *Scolymus maculatus* dominate many areas that are under intensive cattle and sheep grazing, when the favored edible plants can hardly be seen. I stress that the published stocking rates of sheep, goats and cattle in studied rangelands are not very high because studies of modern range management include supplemental feeding and high economic returns. However, traditional grazing does not include supplemental feeding or only to a small extent and its stocking rate is much higher, with the consequent higher historical, millennia-long impact on the vegetation. I saw these higher levels of grazing impact many times in areas under traditional grazing regime in the Palestinian Authority, southern Lebanon, the northern Sinai (Egypt) and in the Bedouin rangelands in the Arad region in the Negev desert (Israel). Therefore, the published values of the remaining plant biomass and taxa in controlled experiments (e.g., Seligman et al. 1959; Gutman and Seligman 1979; Gutman et al. 1990, 1999; Noy-Meir 1990) represent only the moderate levels of grazing typical to modern herding, which is not as severe as the traditional one.

The seasonal shift in flowering time from the main season, when most Mediterranean plants bloom, to the end of the flowering season, when fewer species bloom, might be the evolutionary result of a change in phenology reducing the competition for pollinators. The results presented by Ronel et al. (2010) clearly indicated that spiny plants in the semi-arid east Mediterranean region express a delayed flowering season, shifted to the beginning of the dry summer, when most of the herbaceous vegetation is already dry. During this season, mammalian grazers consume any edible herbaceous vegetation, selecting for late flowering species that allocate more resources for anti-herbivore defenses than early flowering species. There is a well-known global geographical trend where the occurrence of spiny plants is higher in arid regions than in humid ones (Grubb 1992). In parallel to the global trend, Ronel et al. (2010) showed a seasonal one, that non-spiny plants grow and flower in the spring, which is the main flowering season in the Mediterranean basin, while spiny plants commonly flower later, in the hot and dry summer. Since many of these thorny and spiny plants are visually aposematic because of their colorful thorns, spines and other markings (Lev-Yadun 2009b; Ronel et al. 2009; Ronel and Lev-Yadun 2012), I posit that in addition to the needed physiological adaptations, these plants are well defended by thorns and spines that express aposematic coloration.

There are two alternative scenarios that could have led to the association of defense by spines and summer flowering: (1) the later flowering exposed the plants to an increased risk of grazing, and this pressure selected for better defense, or (2) plants that were better defended could evolve towards later flowering to improve pollination because by being well defended they were damaged less by the herbivores. Without a thorough comparative study of the genetic basis for early *versus* late flowering, and defense by spines *versus* the lack of such a defense, in several pairs of species per discussed family (Asteraceae, Fabaceae and Lamiaceae), it is impossible to determine the evolutionary route of the seasonal shift.

Chapter 36
Biochemical Evidence of Convergent Evolution of Aposematic Coloration in Thorny, Spiny and Prickly Plants

A very strong, indirect evidence for the operation of aposematic coloration in thorny, spiny and prickly plants and its convergent evolution is the fact that conspicuous thorn, spine and prickle coloration is found in Angiosperm taxa that have mutually exclusive biochemical pathways of coloration. For instance, taxa belonging to the Caryophyllales (e.g., Cactaceae, Caryophyllaceae, Amaranthaceae) produce yellow, orange, red, purple and black pigments *via* the betalain pathway (Stafford 1994). Most other Angiosperm families use anthocyanins or carotenoids for similar patterns of coloration. The fact that spines of cacti are usually conspicuous because of their coloration (Lev-Yadun 2001), commonly including yellow, orange, red and black coloration resulting from betalain derivatives, indicates that this group of pigments is involved, among their various important physiological functions (e.g., Jain and Gould 2015), in aposematic coloration. By contrast, in Rosaceae, Asteraceae and Fabaceae as well as in many other Angiosperm families that use anthocyanins for orange, pink, red, blue, purple and black coloration of thorns, spines and prickles, the chemical origin of the aposematic coloration (Lev-Yadun 2001, 2006a; Lev-Yadun and Gould 2009) is different. It seems therefore that the aposematic coloration of thorny, spiny and prickly plants is a good case of convergent evolution.

S. Lev-Yadun, *Defensive (anti-herbivory) Coloration in Land Plants*,
DOI 10.1007/978-3-319-42096-7_36

Chapter 37
Aposematic Coloration in Poisonous Flowers, Fruits and Seeds

Flower and fruit colors and their chemical and physical defenses were commonly discussed as mechanisms for filtering non-legitimate or less functional pollinators and seed dispersers rather than concerning aposematism (e.g., Ridley 1930; Faegri and van der Pijl 1979; Herrera 1982; Willson and Whelan 1990; Weiss 1995; Clegg and Durbin 2003; Irwin et al. 2004; Schaefer et al. 2004, 2007; Schaefer and Ruxton 2011). Yet, already in the year 1973, Hinton proposed that the bright colors of poisonous flowers are not only attractive, but also aposematic. The duality of plant reproductive signaling systems, serving in attracting certain animals and repelling others at the same time did not get much research attention.

Herrera et al. (2002) proposed that plants that possess a particular combination of traits that simultaneously enhance pollination and defend from herbivores enjoy a disproportionatly higher fitness advantage over plants possessing individual traits of such combinations. There is, however, a large body of evidence (that will be reviewed here only partly and briefly) for the simultaneous operation of olfactory and visual aposematism in both flowers and fruits, although the authors of these studies referred to filtering of pollinating and of dispersing animals rather than to aposematism. For instance, Pellmyr and Thien (1986), in a broad theoretical study on the origin of angiosperms, proposed that floral fragrances originated from chemicals serving as deterrents against herbivore feeding. In a much more focused study of flower defense in the genus *Dalechampia*, Armbruster (1997) and Armbruster et al. (1997) proposed that defensive resins have evolved into a pollinator-reward system, and that several defense systems have evolved from such advertisement systems. Pollen odors in certain wind-pollinated plants that certainly are not aimed of attracting pollinators are rich in defensive molecules, such as α-methyl alcohols and ketones (Dobson and Bergström 2000). The de-aromatized isoprenylated phloroglucinols may visually attract pollinators of *Hypericum calycinum* by their UV pigmentation properties, but at the same time the plant may use this pigmentation as a toxic substance against caterpillars, defending the flowers from herbivory (Gronquist et al. 2001). The dual action of attracting pollinators while deterring other animals was also found in other taxa, e.g., *Catalpa speciosa* and *Aloe*

S. Lev-Yadun, *Defensive (anti-herbivory) Coloration in Land Plants*,
DOI 10.1007/978-3-319-42096-7_37

vryheidensis (Stephenson 1981; Johnson et al. 2006; Hansen et al. 2007). Thus, certain floral scents may have a defensive role (Knudsen et al. 2006; Junker et al. 2007) in addition to their well-known attracting function. A similar double strategy of using signals to attract certain animals and repel others occurs in fruits (Cipollini and Levey 1997; Tewksbury and Nabhan 2001; Izhaki 2002). Thus, in many cases, the combination of visual signaling and chemical defense and unpalatability of flowers and fruits should have led to the view that they are aposematic (e.g., Hinton 1973). Interestingly, Schaefer and Ruxton (2011) in their very detailed review of plant-animal communication, posited that aposematic fruit signals are not likely to be common and did not mention flower aposematism. I will describe and discuss the meager direct information concerning aposematic reproductive structures in plants and I will do it while considering the wealth of supporting information that was not considered as such by the authors probably because of the general ignorance of plant aposematism by many.

As described earlier in this book, poisonous seeds were probably the first plant parts that were proposed to be visually aposematic because they are both poisonous and visually conspicuous or colorful (Cook et al. 1971; Wiens 1978; Williamson 1982). The experimental data presented in Cook et al. (1971) supported the aposematic hypothesis. However, aposematism was mentioned only briefly in each of these three papers and to the best of my knowledge there was no further or later research done on aposematic seeds in the wild.

Historically, the second reproductive plant organ proposed to be poisonous and colorful and therefore aposematic was the flower (Hinton 1973; Rothschild 1980), but again, this hypothesis was not pursued much further. Lamont (1993) proposed that red-flowered cyanogenic *Grevillea* species may be a cue for higher animals of their poisonous nature, without referring to aposematism and of course, a cue is not a signal. Irwin et al. (2003) showed that red flower morphs of the wild radish, which were also richer with secondary defensive metabolites than white morphs, were better defended from various herbivores. While Irwin et al. (2003) did not use the term aposematic, they described it operating. Hanley et al. (2009) proposed that red cyanogenic flowers of several Australian *Hakea* species deter florivores, but again without using the term aposematic. The metabolic association of conspicuous color with chemical defense was discussed by Fineblum and Rausher (1997). Later, Schaefer and Rolshausen (2006a) proposed a new hypothesis of defensive plant coloration, the "defense indication hypothesis", based on the common biochemical pathways for certain plant colors and defensive molecules, focusing on anthocyanins. This hypothesis predicted that fewer herbivorous insects [and Lev-Yadun and Gould (2007) proposed any sensitive herbivore] would feed on plants that have strong anthocyanin coloration because it correlates with the strength of a chemical defense. The biochemical basis for this correlation is that anthocyanins and a number of defense chemicals such as tannins stem from the same biosynthetic pathways. Lev-Yadun and Gould (2007, 2009) and Lev-Yadun (2009a) proposed that the association of conspicuous coloration with defensive chemicals should be considered aposematic. Gerchman et al. (2012) tested the question of whether visual signaling for pollinators can simultaneously serve for aposematism. They used as their experimental system the conspicuous purple tufts of leaves ("flags"), which often terminate vertical inflorescences in the Mediterranean annual *Salvia viridis*. These flags were shown to attract insect pollinators to the flowering patch. They have determined the aposematic potential of *S. viridis* flags on three levels: (1) concentrations of anthocyanins, suggested to function as aposematic visual signals, in flags *versus* leaves, (2) spectrometry to estimate whether the color-vision system of two common but very different Mediterranean generalist herbivores (locusts and goats) can discriminate colorful flags from green leaves, and (3) choice experiments to determine food preferences of the same herbivores. Anthocyanin concentrations in flags were found to be more than 10-times higher than in leaves. Flags exhibited peak reflectance at 450 and 700 nm wavelengths, while leaves reflected maximally at 550 nm. Goats preferred feeding on clipped inflorescences over intact control inflorescences. Locusts preferred leaves over colorful flags. To test whether this was due to deterrence from the flags' coloration, they also offered them a choice between leaves and a rare, white morph of the flags. The locusts chose both equally immediately after presentation, but the leaves attracted more individuals after 5 min of feeding. The locusts also preferred green cabbage over anthocyanin-rich red cabbage. Their results supported the possibility of a secondary function of colorful extra-floral displays as warning signals.

A special case of proposed flower aposematism is that of bee or wasp mimicry by orchid flowers. For many decades, bee or wasp mimicry by orchid flowers was considered to serve pollination. Lev-Yadun and Ne'eman (2012) discussed this issue and proposed, following a century-old hypothesis that predated the understanding about orchid pollination by bee or wasp mimicry, that since attracting solitary male bees or wasps, which are their species-specific pollinators, is mostly (not only!) based on olfactory signaling, that in addition to pollination, such plants exhibit another, rarely mentioned and practically forgotten, non-exclusive function of defensive, anti-herbivory bee- or wasp mimicry (Batesian mimicry). This mimicry may deter large mammalian herbivores, and possibly also insects, from the plants and especially from their flowers by a type of visual and possibly also olfactory deceptive aposematism. This issue will be discussed in more detail in a later chapter.

Another aspect of probable flower aposematism involves nectar. Colored nectar (yellow, red, brown, black, green, blue) is known in 68 taxa belonging to 20 genera in 15 families (Hansen et al. 2007). These authors showed data indicating that the colors and their associated nectar chemistry serve as honest signal and as floral filters against nectar thieves and inefficient pollinators. However, aposematism was not discussed although signaling by plants to repel animals should be considered as aposematism.

Concerning fruits, Harborne (1982) proposed that the brightly colored, purple-black berries of the deadly *Atropa belladonna* (deadly nightshade) warn grazing mammals of the danger in consuming them. Aposematism in fruits mimicking thorns (Lev-Yadun 2003b) or aposematic caterpillars (Lev-Yadun and Inbar 2002; Aviezer and Lev-Yadun 2015; Lev-Yadun 2015a) are discussed in depth in other chapters. Schaefer and Schmidt (2004), without using the term aposematic, actually described visual aposematism in chemically defended fruits. Hill (2006) was the only one that directly reffered to fruit aposematism and experimentally examined the aposematic function of poisonous and colorful fruits and gave good indications for the warning function of the coloration. Lev-Yadun et al. (2009a) proposed that certain red unripe toxic or prickly/spiny fruits may be aposematic.

Unripe Red Fruits May in Many Cases Be Visually Aposematic

Multicolored fruit displays, where fruits first change their color from green to a conspicuous color when they have reached full size, but are still unripe and later change their color to a second conspicuous one upon ripening, have been studied in plants of several ecosystems (Stiles 1982; Willson and Whelan 1990; Janson 1987; Morden-Moore and Willson 1982; Willson and Thompson 1982; Willson and Melampy 1983; Wheelwright and Janson 1985; Fuentes 1995; Traveset and Willson 1998; Cramer et al. 2003; Schaefer and Ruxton 2011). Some studies indicated that bicolored fruit displays enhance seed dispersal (Stiles 1982; Janson 1987; Morden-Moore and Willson 1982; Willson and Thompson 1982; Willson and Melampy 1983; Wheelwright and Janson 1985; Fuentes 1995; Cramer et al. 2003; Greig-Smith 1986; Izhaki 1998) but other studies did not (Janson 1987; Morden-Moore and Willson 1982; Willson and Thompson 1982; Fuentes 1995; Izhaki 1998). Thus, it seems that promotion of dispersal can only partly explain multicolored fruit displays.

The similarity between the aposematic colors in poisonous insects and colors of certain ripe fruits has already been demonstrated (Herrera 1985a; Gamberale-Stille and Tullberg 2001), but an aposematic role for colorful fruits was not considered by Herrera (1985a) and was not found to operate in artificial fruit-like objects introduced to domestic chicks (Gamberale-Stille and Tullberg 2001). Several types of defensive coloration or other visual aspects that possibly or probably repel large herbivores have been proposed to operate *via* aposematism or otherwise in fruits: (1) brightly colored poisonous fruits (Wiens 1978; Schaefer and Schmidt 2004; Hill 2006; Schaefer and Schaefer 2007), (2) pods of several wild annual legumes (*Lathyrus ochrus*, *Pisum humile*, *P. elatius* and *Vicia peregrina*) that have conspicuous reddish spots arranged along them that appear to mimic aposematic lepidopteran caterpillars and may repel various herbivores (Lev-Yadun and Inbar 2002; Aviezer and Lev-Yadun 2015; Lev-Yadun 2015a), (3) colorful (yellow, red, purple or various combinations of these) aposematic thorn-like unripe soft fruits in several wild *Erodium* species and in *Sinapis alba* growing in Israel (Lev-Yadun 2003b) (Figs. 32.3 and 32.4), and (4) white dots on unripe and non edible or even toxic or unpalatable fleshy fruits belonging to many species originating from several continents and various taxa (Lev-Yadun 2013a) (Figs. 13.7, 13.8 and 13.9). Lev-Yadun et al. (2009a) examined several species of the Israeli flora (listed below) that have conspicuous unripe red fruit that are poisonous, unpalatable or defended by prickles, and proposed that in such cases fruit color may serve as an aposematic signal that deters herbivores and frugivores from the dangerous or unpalatable unripe fruit. Thus, colorful aposematic unripe fruits may increase plant fitness by deterring herbivores and frugivores from consuming such fruits and by this consume or just ruin their immature seeds.

Nerium oleander is a tall evergreen multi-stemmed shrub or small tree (Apocynaceae), common along seasonal streams, rivers and in other wet habitats. Its fruit (typically 5–15 cm long and 6–10 mm in diameter) consists of two follicles. The young unripe small fruit is green but it commonly turns red when fully grown but still unripe (Fig. 37.1). Upon maturation, the fruit dries, turns brown, and its follicles twist to liberate the many small, plumed, wind-dispersed seeds (Herrera 1991) (Fig. 37.2). The whole plant, including its fruit, is extremely toxic, containing about 2 % cardiac glycosides (Langford and Boor 1996). More than ten glycosides with known chemical structure were reported in *N. oleander*. The cardiac effects of the glycosides are due to direct cardiotoxicity and an indirect effect *via* the vagal nerve. The direct effect is due to the inhibition of the Na-K ATP-ase pump enzyme system. This specific action increases intracellular sodium ions and serum potassium concentration (Osterloh et al. 1982). The lethal dose of leaves reported for

Fig. 37.1 Red and very toxic fruit of *Nerium oleander*. Since its seeds are dispersed by the wind (Fig. 37.2), the red coloration is not aimed to attract dispersing animals

Fig. 37.2 *Nerium oleander* seeds dispersed by the wind

several animal species is about 0.5 mg leaf tissue per 1 kg body mass (Pearn 1987). Since *Nerium oleander* disperses its seeds by wind, the red advertisement of the unripe and very poisonous fruits is not aimed at attracting frugivores in order to promote seed dispersal. The combination of red fruits with their very poisonous latex (e.g., Herrera 1991; Langford and Boor 1996) is sufficient to characterize their color as aposematic.

Anabasis articulata is a poisonous shrub (Chenopodiaceae = Amaranthaceae) common in the Negev desert. Its unripe winged fruits are purple, red or pink. After ripening, they are wind dispersed. The plant is rich in alkaloids and is not eaten by grazers (Zohary 1980; Feinbrun-Dothan and Danin 1991; Lev-Yadun and Ne'eman 2004). The wind-dispersed poisonous purple, red or pink fruits of *Anabasis articulata* need no advertisement to promote seed dispersal and thus may also be characterized as aposematic.

Emex spinosa is an annual plant (Polygonaceae), common in semi-arid open areas and grasslands. Its fully grown unripe fruits are red and spiny. When they dry, their spines turn light brown and stick to the fur or hooves of passing animals, thereby dispersing the seeds over distances. *Hedysarum spinosissimum* (= *H. pallens*) is an annual plant (Fabaceae), typical of sandy soils of the steppe. Very conspicuous red prickles cover its unripe pods (Fig. 37.3), which turn light brown when

Fig. 37.3 The mature very prickly red fruits of *Hedysarum spinosissimum*. When they are ripe they turn *brown* and they are dispersed by adhering to the skin or fur of passing animals

ripe and stick to the fur of passing animals, dispersing the ripe seeds. The spiny unripe red fruits of *Emex spinosa* and the prickly ones of *Hedysarum spinosissimum* are unpalatable for large mammalian herbivores and need no advertisement to attract frugivores because they are epizoochorous. Therefore, the red color of their unripe fruits can also be regarded as aposematic.

Rhamnus alaternus (Rhamnaceae) is a tall evergreen shrub or small tree, common in the Mediterranean maquis. Its young small fruits are green and turn red when fully grown but still unripe (Fig. 37.4). The red fruits contain 7.8 ± 2.8 ppm of the anthraquinone emodin and are poisonous to some extent. This concentration decreases to 2.5 ± 0.9 ppm in ripe black fruits (Tsahar et al. 2002). A similar fruit color change also occurs in *R. lycioides*, in *R. punctata*, and probably also in other *Rhamnus* species. The fleshy-fruited *Rhamnus* species disperses its seeds *via* frugivores only after the fruits change color from red to black. It has been demonstrated for the Old World's *Rhamnus alaternus* and *R. lycioides* (= *R. palaestinus*) and for the New World's *R. cathartica*, that bird species do not consume unripe red fruits (Sherburne 1972; Maw 1981; Tsahar et al. 2002). Emodin is the predominant secondary metabolite in *Rhamnus* fruits and recently it has been demonstrated that birds and small mammals are unable to detoxify high concentrations of emodin efficiently (Izhaki 2002). There are good indications for considerable predispersal seed predation in *R. alaternus* (Bas et al. 2005), which may trigger the evolution of defense.

Rubus sanguineus (=*R. sánctus*) (blackberry) (Rosaceae) is a prickly climber, often with a shrubby form, common along streams and in other wet habitats. Blackberries are characterized by compound drupes carried in groups. The color of young small fruits is green (Fig. 37.5). After an initial growth period, they reach their final size and turn bright red (Fig. 37.6). However, at this stage, the fruit is still

Fig. 37.4 Unripe *red* and ripe *black* fleshy fruits of the small Mediterranean tree *Rhamnus alaternus* common in Israel

Fig. 37.5 Young, not fully grown *green* fruits of *Rubus sanguineus* (blackberry)

Fig. 37.6 Fully grown but still not ripe and very sour non-edible *red* fruits of *Rubus sanguineus*

Fig. 37.7 A mixture of non ripe *red* and ripe and edible *black* fruits of *Rubus sanguineus*

unripe and very sour. Many times, the conspicuous groups of unripe red compound drupes dominate the plants. Only later in season, when the fruits turn black and sweet (Fig. 37.7), frugivorous birds consume them and disperse the seeds (Jordano 1982, 1984). We measured the sugar content of fruits at the three color stages. The green fruits contained 7.5 %, the red 9 % and the black 20 % sugars. In parallel, the acidity decreased from pH 2.6–2.9 in green and red unripe fruit to pH 3.6 in black ripe fruits.

Species of the woody genus *Pistacia* (e.g., the trees *P. atlantica* and *P. palaestina*, and the shrub *P. lentiscus*), grow in various habitats in the Mediterranean, steppe or in the case of *P. atlantica* even in desert districts of Israel and Syria (Zohary 1973). In *Pistacia* spp. the young green fruits turn red (or yellow in some *P. lentiscus* individuals) when fully grown but still unripe with immature seeds, and are unpalatable (Izhaki 1986). Later, upon fruit ripening and seed maturation, the fruits turn dark blue or black. Seed ripening as well as structural and chemical changes in the fruit pulp accompany this color change. The fruit softens, the concentrations of its secondary repellent compounds decrease while that of lipids and proteins increase, and it turns palatable to frugivores (Herrera 1984; Izhaki 1986; Jordano 1989). The red immature fruits are less rewarding and less palatable than the mature dark ones; consequently, frugivorous birds prefer to consume dark fruits and not red ones (Izhaki 1998). *Pistacia* also has typical red seedless fruits that are not aborted but retained on the tree. These empty fruits contribute to the visual attraction of fruit-bearing trees and shrubs, but are consumed by frugivores only when the dark seed containing fruits are not available (Izhaki 1998; Jordano 1989; Traveset 1994).

In the bird-dispersed *Rhamnus alaternus*, *Rubus sanguineus* and *Pistacia* sp., the fruits express three color stages. The first green stage is common to all unripe small fruits. At this stage the fruit is cryptic and photosynthetically active, contributing to its own production costs and defended by both crypsis and unpalatability. Producing and bearing non-green, fully grown or almost fully grown but yet unripe fruits for substantial periods, adds to the costs of fruit production. Such a trait is supposed to be compensated for by another contribution to the fitness of the plant. Increasing the long-range advertisement of a fruit-bearing plant and its consequent attractiveness to frugivores, which also increases the probability of ripe fruit removal, can provide such compensation. Indeed, it has been demonstrated that red and unripe or empty fruits of *Pistacia* contribute to the removal rate of the black ripe fruits by increasing the attraction of frugivores (Jordano 1989).

Lev-Yadun et al.'s (2009a) proposal that certain unripe red fruits are aposematic contributed an additional aspect to the understanding of the ecology of fruit color changes and did not contradict any other defensive or physiological explanation. Signaling *via* color change in fruits can be regarded as analogous to signaling *via* color change of flowers (e.g., Weiss 1995; Weiss and Lamont 1997) and to color changes in aposematic thorns, spines and prickles (Lev-Yadun and Ne'eman 2006). Pre- and post-color change organs (fruits, flowers, leaves, spines) contribute to the overall attraction of pollinators and frugivores, or to the repulsion of herbivores. However, when approaching the plant, pollinators usually visit the pre-color change rewarding flowers, while frugivores commonly turn to the post-color-change palatable, rewarding, non-poisonous or more edible fruits. This strengthens the concept that color itself has only a limited direct meaning unless coupled with reward or punishment. Thus, the red color can be attractive when related to rewarding fruits or aposematic when related to poisonous, sour, bitter or spiny fruits. Similarly, floral color and odor function as pollinator attractors when coupled with a reward (Faegri and van der Pijl 1979), but there is also good evidence for their direct defensive function (Irwin et al. 2003; Strauss and Irwin 2004; Lev-Yadun et al. 2009b; Schaefer and Ruxton 2011; Lev-Yadun and Ne'eman 2012).

The possible benefits for the signaler plant are: (1) increase in the advertisement of the fruit-bearing plant and attraction of more frugivores per fruit unit from longer distances (Stiles 1982; Willson and Melampy 1983; Wheelwright and Janson 1985; Cramer et al. 2003; Greig-Smith 1986), (2) reduction in the consumption of immature seeds, (3) reduction in the abortion of damaged fruits, before seed maturation, because wounding of fruits can stimulate ethylene production and consequent abscission (Abeles et al. 1992), and (4) reduction in the possible damage to fruits, which might later decrease seed dispersal because frugivores avoid eating damaged fruits (especially in large fruits) (Janzen 1977; Herrera 1982; Manzur and Courtney 1984; Borowicz 1988; Buchholz and Levey 1990; Cipollini and Stiles 1993; Marchand and McNeil 2004). The frugivores benefit from the signal because: (1) they avoid poisonous fruits that contain harmful substances or spiny ones, and (2) they refrain from damaging future fruit resources in their home range.

Lev-Yadun et al. (2009a) concluded that there are certain red-colored unripe fruits that are poisonous or spiny, which are probably aposematic, as demonstrated from the flora of Israel, and they also proposed that this is probably a common but largely overlooked worldwide syndrome. As with other color-dependent defensive strategies, they expected that aposematic unripe red fruits were subjected to mimicry by young unripe red fruits that contain no poison or no strong physical defenses.

Altogether, in spite of the huge body of research conducted to characterize visual and chemical signaling by plants to animals in flower and fruit biology, the aposematic hypothesis for these very important plant organs, which are commonly visually and chemically conspicuous, has received much too little attention. The data reviewed above demonstrates that aposematism by defended plant reproductive organs is probably quite common but mostly overlooked.

Chapter 38
Aposematic Trichomes: Probably an Overlooked Common Phenomenon

Trichomes, the unicellular and multicellular appendages of the epidermis (Fahn 1990) are well known for their multiple functions. Trichomes may serve in protecting plants from excess sun irradiation of various wavelengths, including UV (Fahn and Cutler 1992; Manetas 2003), secrete toxic ions, especially in saline habitats (Fahn 1988), function in water absorption (Fahn and Cutler 1992), reduce transpiration (Fahn and Cutler 1992; Werker 2000), be involved in temperature regulation (Wagner et al. 2004), defend from insect or other herbivorous invertebrates by reducing accessibility or by the actual trapping of their legs or by various chemical means (Levin 1973; Fahn 1979, 1988; Werker 2000) and defend from large herbivores when they sting for instance in *Urtica* (Thurston and Lersten 1969; Levin 1973; Fahn 1990; Wheeler 2004; Fu et al. 2006). In addition, in certain carnivorous plants such as *Drosera* and *Dionea*, they may take part in the attraction, capture and digestion of insects (Juniper et al. 1989; Fahn 1990).

Many plant trichomes are colorful (yellow, orange, red, brown, blue, purple, white) and very conspicuous, especially on the green background of photosynthetic tissues and by shining when back illuminated. In certain cases, such as in cotton plants, colorful trichomes produce toxins that defend from caterpillars (Agrawal and Karban 2000). In addition, in various plants the trichomes have conspicuous red markings at their base, e.g., *Echium angustifolium* (Boraginaceae) and *Echinops adenocaulos* (Asteraceae) that enhance their visibility at close range. Thorns, spines and prickles are large and usually spaced and their ability to defend from insects is limited although they may sometime entrap and kill caterpillars (e.g., Potter and Kimmerer 1988; Yamazaki 2014), whereas trichomes, because of their size, density, structure, and chemical composition may commonly defend plants from insects including by entrapping them (e.g., Levin 1973; Fahn 1979, 1988; Werker 2000; Krimmel and Pearse 2013).

Lev-Yadun (2009a) proposed that many colorful or otherwise conspicuous trichomes are aposematic. Since many insects see UV (Briscoe and Chittka 2001) the possibility that trichomes may signal and therefore deter insects in the UV channel should be considered and studied. The possibility that trichomes produce olfactory

S. Lev-Yadun, *Defensive (anti-herbivory) Coloration in Land Plants*,
DOI 10.1007/978-3-319-42096-7_38

aposematic signals in addition to visual ones should also be considered, in light of the secretive nature of many trichomes (e.g., Fahn 1979, 1988). Lev-Yadun (2014e) proposed that plants with colorful sticky or hooked trichomes may deter insects and serve as aposematic coloration not only directly, but also by using the dead or entrapped insects. LoPresti (2014) studied the defensive role of salt bladders in several salt-secreting members of the Chenopodiaceae (four *Atriplex* species, *Blitum californicum*, ten *Chenopodium* species, and *Oxybasis glauca*) against a variety of insects and found in choice, no choice and field tests that generalist insect herbivores preferred to feed on leaves without salt bladders or consumed less from intact leaves than from leaves in which the salt bladders were removed. A different mode of trichome aposematism was proposed by Yamazaki and Lev-Yadun (2015). They proposed that dense, thread-like white trichomes of some plant species from Estonia, Greece, Israel and Japan visually mimic spider webs, lepidopteran and spider-mite web nests and plant-pathogenic fungi, and that it may result in reduced herbivory, since various herbivores avoid spider- or other arthropod webs to circumvent predation or toxic attacks, or refrain from colonizing plants that have already been occupied by other herbivores and pathogens. Yamazaki and Lev-Yadun (2015) did not dismiss the possibility that these web-like structures may also have other defensive or physiological functions or that they are not classic mimics but rather exploit the herbivore's perceptual state concerning the avoidance of potentially risky objects.

Altogether, there is sufficient evidence to consider plants with conspicuous trichomes to be visually aposematic on both a chemical and physical basis, but this issue was mostly overlooked.

Chapter 39
Why is Latex Usually White and Only Sometimes Yellow, Orange or Red? Simultaneous Visual and Chemical Plant Defense

Latex is a common chemical and physical plant defense that exudates following wounding from canals called laticifers. It has evolved repeatedly, and is known from over 20,000 species belonging to more than 40 families (Agrawal and Konno 2009). Latex protects plants from herbivores, especially insects, but also from fungi, bacteria and vertebrate herbivores, as well as physically sealing wounds (Fahn 1979; Konno 2011). Latex contains bioactive compounds including alkaloids, cardiac glycosides, terpenes, digestive proteins, sugar-mimicking substances and many other defensive components (Agrawal and Konno 2009; Konno 2011; Mithöfer and Boland 2012), and the full repertoire of the compounds that repel or damage herbivores and the identity of these herbivores are enormous. While these toxic substances repel both invertebrates and vertebrates, certain specialized herbivores (especially various arthropods), may not be deterred, but rather attracted to them (e.g., Zalucki et al. 1990; Agrawal et al. 2012), and some even sequester them (Nishida 2002; Opitz and Müller 2009) for their own defense. Moreover, latex commonly sticks insects to the plant and by this causes their death, or glues their mouth parts shut, preventing them from feeding (Dussourd 1995; Konno 2011). The fact that various insect types regularly sever the leaf's veins and drain the latex before they consume plant tissues (e.g., Dussourd and Eisner 1987) is an excellent demonstration of its defensive role (Mithöfer and Boland 2012). Latex exudetes from plants only after wounding, a situation well known to involve the simultaneous emitting of defensive volatiles, including ones that attract predators and parasitoids (e.g., De Moraes et al. 1998; Kessler and Baldwin 2001) and volatiles that repel various herbivores (Birkett et al. 2000; De Moraes et al. 2001). Altogether, the combination of toxicity and mechanical defense causes many plants that produce latex to be considered toxic or otherwise non-palatable or even deadly to many insects, and when highly toxic (e.g., *Nerium oleander*) even to various large vertebrate herbivores (Knight and Walter 2001).

A striking fact about latex is that the majority of plants produce it white, e.g., *Papaver* spp., *Ficus elastica*, *Euphorbia* spp., *Calotropis procera* (Konno 2011; Mithöfer and Boland 2012) (Figs. 39.1–39.2), although yellow and orange, e.g.,

S. Lev-Yadun, *Defensive (anti-herbivory) Coloration in Land Plants*,
DOI 10.1007/978-3-319-42096-7_39

Figs. 39.1–39.2 White latex that dripped from a wounded branch on a leaf and on another branch of *Ficus elastica*

Nerium indicum, and red latex, e.g., *Cannabis* spp., *Croton lechleri*, *Daemonorops* spp., *Dracaena* spp., *Pterocarpus* spp., also exist (Langenheim 2003; Gupta et al. 2008; Agrawal and Konno 2009; Konno 2011; Jura-Morawiec and Tulik 2015). The white color is the outcome of the rubber particles dispersed in the latex (Agrawal and Konno 2009), rubber being an important component of latex stickiness. There are three theoretical options concerning the function of white latex color: (1) that there is no function or importance to its color, (2) that there are chemical constraints for the production of latex in other colors, and (3) that the white color has a visual advantage, a specific function. The fact that latex can also be yellow, orange or red indicates that there is no inherent reason to produce it almost only in white. Since white is its dominant color and since white latex has evolved independently many times in various plant lineages (e.g., Hagel et al. 2008), there should have been a strong selection for white latex. This possibility has only recently been addressed by Lev-Yadun (2014d). Because latex (or resin) color was till recently never considered to have a significant, if any function, there is no data on the relative frequency of various latex colors or about the ecological affinities of plants that produce latex in these colors.

The High Visibility of White Markings

As was discussed earlier in this book, a white signal, when contrasted with a darker tissue, may have a visual advantage over a colorful one, because even color-blind animals can see it (Givnish 1990). In addition, it is still visible even under low light levels and under various spectra regimes, e.g., close to sunrise, near sunset (e.g., Troscianko et al. 2009), at the bottom of a dense forest or within a dense canopy, and under very cloudy conditions (Givnish 1990; Midgley 2004; Lev-Yadun 2009a,

2013a, 2014a). It is therefore no surprise that many road markings are white (yellow road markings – a typical aposematic color – are also common), as the price for road mistakes by injury, jailing, financial losses and even death (for both drivers and pedestrians) can be very high.

Lev-Yadun (2014d) proposed that latex is predominantly white because white is the best solution for visual aposematism under the lighted conditions within and under plant canopies, and on the typical green color of leaves, young stems or unripe fruits, and on the typical gray-brown colors of mature bark, or towards color-blind animals. Thus the white color of plant latex allows it to serve as a visual aposematic signal about the defensive qualities of such plants. Lev-Yadun (2014d) also proposed that because latex signals both chemical and physical defenses, many Müllerian mimicry rings of plants with white latex probably exist, and that yellow, orange and red latex types should also be considered to be visually aposematic, especially in well-illuminated habitats. Latex in all colors also has a chemically-based potential for aposematism because of its olfactory component (see below). Unfortunately, with the current low level of understanding of plant aposematism compared to animal aposematism, it is impossible to relate specific visual aspect (specific colors that are not white), notwithstanding specific latex-associated chemical aposematic signaling *via* volatiles to specific environments, and examining such possible relationships is an open question.

Chemically Based Olfactory Aposematism in Plants

Olfactory aposematism, whereby poisonous plants deter mammalian or insect herbivores, received the early attention of students of plant aposematism (e.g., Eisner 1964; Eisner and Grant 1981; Rothschild 1986; Rothschild and Moore 1987; Moore et al. 1990; Launchbaugh and Provenza 1993; Provenza et al. 2000) and has continued since then, e.g. (Massei et al. 2007; Lev-Yadun 2009a; Lev-Yadun et al. 2009b). The olfactory component of latex-related aposematism is probably of significant importance especially against nocturnal, blind, or subterranean herbivores.

Since latex contains so many types of toxic secondary metabolites, including various volatiles (e.g., Oliveira et al. 2010) and because latex is exuded following wounding that by itself involves the emission of various additional defensive volatiles, there is a very strong direct and indirect chemical aspect to defense by latex. The potential for a chemically based olfactory aposematism role of such volatiles, especially towards nocturnal or blind herbivores or even towards subterranean ones, is an open area of research.

Sticky Latex as an Indirect Aposematic Plant Defense, a Case of an Extended Phenotype

Lev-Yadun (2014e) proposed that in addition to the well-known direct physical/chemical defenses by trichomes, resins and latex, insect carrion and trapped live insects that are attached to the surfaces of many sticky plant species *via* trichomes, latex or resin may defend plants by being indirectly aposematic. This type of indirect aposematism (an extended phenotype) is based on cueing visually to other herbivores that the plants are already occupied, and on cueing and signaling both visually and by rotting carrion or stress volatiles emitted from trapped insects, that such plants are dangerous or even deadly. This issue will be addressed in more detail later.

Do Members of the Genus *Euphorbia* Mimic Oozing Latex as Defense?

Mimicry of chemically defended butterflies by non-defended ones was the first type of proposed defensive mimicry (Bates 1862), now known as Batesian mimicry (Wickler 1968). Since mimicry of defended organisms or defensive mechanisms are so common in animals (Wickler 1968; Cott 1940; Ruxton et al. 2004) and is also found in plants (Lev-Yadun 2003a, b, 2009a, b, c), theoretically, latex mimicry is expected to occur. It seems that a number of succulent species belonging to the genus *Euphorbia* (e.g., *E. buruana*, *E. dauana*, *E. fluminis*, *E. horwoodii*, *E. knobelii*), a genus typically defended by white and very toxic latex, express conspicuous variegation *via* white coloration of parts of their green tissues (see figures in Sajeva and Costanzo 1994) that was proposed by Lev-Yadun (2014d) to look as if white latex was oozing there. This is again a case of automimicry (white variegation mimicking white latex), adding to the cases of spine automimicry (Lev-Yadun 2003b) described above.

Since there are so many common plants that produce white latex as defense, Lev-Yadun (2014d) proposed considering the ones that overlap in geographical distribution and are attacked by the same generalist herbivores as Müllerian mimicry rings. When there is a partial geographical overlap of plants from one white latex-based Müllerian mimicry ring with plants of other similar mimicry rings, a chain of Müllerian mimicry rings of such defended plants may exist. Such Müllerian mimicry rings concerning defended spiny plants marked with white or other types of defensive coloration were also proposed to exist (Lev-Yadun 2009b, c, 2014a). Because various plant taxa produce latex of different levels of toxicity and because of the variable sensitivity of various herbivores towards the latex, there are probably complex networks of regional mosaics of quasi-Müllerian or quasi-Batesian mimicry rings concerning the chemical (olfactory) and visual aposematic defense by latex.

Chapter 40
Visual Defenses Basically Operating by Various Mechanisms That Have an Aposematic Component

Several different visual defenses may serve as visual aposematism even though they mostly operate by other defensive, signaling, and physiological mechanisms. I give a short account of some of them here, since an in-depth treatment of each of them is given elsewhere in this book.

(1) Undermining insect camouflage: a case of habitat aposematism. It has been suggested that many patterns of plant coloration may undermine the camouflage of small invertebrate herbivores (Lev-Yadun et al. 2004a). Lev-Yadun (2009a) concluded that since insects, like many other animals, tend to avoid surfaces that don't match their coloration, morphology or pattern (e.g., Cott 1940; Kettlewell 1973; Endler 1984; Stamp and Wilkens 1993; Carrascal et al. 2001; Purser 2003; Ruxton et al. 2004), plant coloration that undermines their camouflage can be viewed as habitat aposematism.

(2) Delayed greening as unpalatability-based aposematism. The common phenomenon of delayed greening of young leaves in the tropics is associated with their low nutritive value, and it seems that this property defends them from herbivory (Kursar and Coley 1991, 1992, 2003; Coley and Barone 1996; Numata et al. 2004). Lev-Yadun (2009a) therefore proposed that the associations of being unpalatable with the conspicuous colors (e.g., whitish, pink, red) associated with delayed greening may signal herbivores about the lower nutritive value, a typical aposematism.

(3) Colorful autumn leaves. The liveliest discussion on defensive plant coloration so far in the 21st century has centered on the phenomenon of red and yellow autumn leaves, discussed in detail later in this book. Archetti (2000), in the first paper about the anti-herbivory function of yellow and red autumn coloration, specifically rejected the possibility that autumn leaf coloration is aposematic. Lev-Yadun (2006a) and Lev-Yadun and Gould (2007) however, proposed that the function of the bright autumn leaf coloration may in many cases represent aposematism or its Batesian mimicry. Lev-Yadun and Gould (2007, 2009) proposed that if the "Defense Indication Hypothesis" posited by Schaefer and Rolshausen (2006a) concerning red autumn leaves is accepted, it directly follows that plant parts rich in anthocyanins may serve in many cases as aposematic (warning) coloration because of

S. Lev-Yadun, *Defensive (anti-herbivory) Coloration in Land Plants*,
DOI 10.1007/978-3-319-42096-7_40

chemically-based unpalatability. Species with either red or yellow toxic autumn leaves should naturally be considered aposematic (Lev-Yadun 2006a; Lev-Yadun and Gould 2007). When red and yellow autumn leaves have a low nutritive value (a case of unpalatability) (Sinkkonen 2006a, b), they should also be considered aposematic for this reason alone (Lev-Yadun and Gould 2007). Later, in Archetti (2009b) and Archetti et al. (2009a), the overlap between the co-evolution signaling hypothesis was recognized as a case of aposematism, an issue further discussed in Lev-Yadun (2010).

(4) Animal mimicry and herbivore damage mimicry may serve as visual aposematism with or without coloration. It is probable that various types of defensive mimicry by plants may trick animals into behaving according to the plant's interests, just as they are tricked by bee mimicry of orchid flowers during pollination (e.g., Dafni 1984; Jersáková et al. 2006). Defensive visual animal mimicry by plants exists or at least proposed in several forms: (1) butterfly egg mimicry, (2) ant mimicry, (3) aphid mimicry, (4) caterpillar mimicry, (5) beetle mimicry, (6) spider web mimicry, (7) snake mimicry, (8) winged arthropod mimicry, (9) bee and wasp mimicry, (10) animal chewing or tunneling damage mimicry, (11) gall midge mimicry, (12) eye spot mimicry, (13) visual feces mimicry, and (14) feces and carrion odor mimicry. All these and the various modes by which they participate in plant defense will be discussed in detail in the relevant chapters. In any case, since some of the mimicry cases are of dangerous animals (ants, bees, wasps, toxic aposematic caterpillars, spider webs, snakes), and others cause plants to look as if they are already damaged or occupied, this can be considered a type of aposematism even if based on deception.

A related phenomenon, the use of aposematic insects that occupy various plants as defense of these plants from large herbivores, was proposed by Rothschild (1972, 1986). Various poisonous aposematic insects aggregate on poisonous plants, adding to the plant's aposematic odor and possibly also to its coloration. This type of plant-insect mutualism *via* aposematism following an extended phenotype deserves much more descriptive, theoretical and experimental studies.

Chapter 41
Plant Aposematism Involving Fungi

The understanding that plants have mutualistic relationships with various fungi including pathogenic ones is not new. Most suggestions for such relations were based on the chemical defenses provided by endophytic or parasitic fungi (Clay 1990; Bush et al. 1997; Omacinl et al. 2001; Clay and Schardl 2002), but these studies did not discuss aposematic issues. For instance, fungal endophytes in the genus *Neotyphodium* (Ascomycetes: Clavicipitaceae) form mutualistic associations with a variety of grasses (Clay 1990; Breen 1994). The fungal hyphae grow intercellularly in leaf and stem tissues, causing infections that are transmitted to the next plant generation exclusively through host plant seeds. The fungus benefits from access to plant nutrient and photosynthetic resources, while the plant benefits from enhanced resistance to insect herbivores or vertebrate grazers (Clay 1990; Breen 1994; Elmi and West 1995). A series of fungal endophyte-mediated-alkaloids commonly provides the basis for the acquired chemical defense against herbivory (Porter 1994; Justus et al. 1997). In addition, there were two suggestions, one that plants expressing mimicry of fungal coloration, and the second that plants with actual colorful fungi are aposematic, with or without known toxins: (1) when plants by coloration look as if they are infested (Lev-Yadun 2006b, 2009a, 2014a), and (2) when they are infested by toxic fungi (Lev-Yadun and Halpern 2007). Some of the above (when actual fungi are involved) are cases of extended phenotypes used by plants for defense, and at the same time, these fungi defend their own habitat – the infected plant. Ergots are classic examples of such relationships and I will elaborate on this below.

Ergot, a fungal disease of rye and other cereals, caused by the fungus *Claviceps purpurea*, is one of the most notorious toxic fungi in human history (Matossian 1989; Bennett and Bentley 1999; Alm 2003). There are more than 40 species in the genus *Claviceps* (Pažoutová and Parbery 1999) that attack grasses, rushes, and sedges. Most ergot species have colorful and conspicuous sclerotia that form on the ears (inflorescences) of their host (Tenberge 1999). Plant/*Claviceps* (ergot) interactions may be viewed simply as a disease, or else as a mutualistic relationship. Ergot fungi parasitize rye and other grasses, reducing their reproductive capacity because

S. Lev-Yadun, *Defensive (anti-herbivory) Coloration in Land Plants*,
DOI 10.1007/978-3-319-42096-7_41

their sclerotia replace some of the host's seeds. The fungal sclerotia are rich in a wide array of highly toxic alkaloids (Mantle 1969; Lorenz 1979; Matossian 1989; van Dongen and de Groot 1995). Ergotism, the toxic condition in humans and animals that eat *Claviceps*-infected grain, is often accompanied by psychotic delusions, nervous spasms, abortion, convulsions, gangrene, and even death (Mantle 1969; Lorenz 1979; Matossian 1989; van Dongen and de Groot 1995). The best-known ergot species, *C. purpurea*, forms conspicuous purple-black sclerotia up to several-cm-long on the ears of rye, where they are easily seen. When infected rye (a staple for humans in European countries with cold wet climates) is ground and used to produce bread, non-lethal levels of ergot poisoning may cause severe hallucinations or intense burning pain (St Anthony's Fire), and a stronger poisoning causes gangrene of feet, hands, and entire limbs, because of the vasoconstrictive action of the ergot alkaloids (De Costa 2002). Historically, ergotism has directly killed many thousands of people. For instance, over 40,000 deaths attributed to ergot poisoning were recorded in France in 943 (Prescott et al. 2005). In addition, many witch-trials, and subsequent executions over centuries in medieval Europe, and possibly some in North America involved women who did not behave according to acceptable moral norms, most likely as a result of intoxication by ergot alkaloids (Matossian 1989; Alm 2003). The pharmacological activities of the fungus are due to various components that include lysergic acid diethylamide (LSD) (Matossian 1989; Bennett and Bentley 1999; Alm 2003; Eadie 2004). People learned to refrain from eating infected rye, and it is likely that herbivores that consume grasses learn to avoid it too. The antiherbivory functions of ergot alkaloids, against both invertebrates and vertebrates, have led to it being regarded as mutualism: the plants are defended from herbivory by the poisonous substances of the fungi, while the fungi receive nutrition from the host (Clay 1988).

Lev-Yadun and Halpern (2007) proposed that the very poisonous purple-black sclerotia of the infamous fungus *Claviceps purpurea* (ergot) and the colorful sclerotia of many other *Claviceps* species are visually aposematic. The very toxic fungal sclerotia of the genus *Claviceps* typically have conspicuous colors (black, yellow, purple, reddish, brown, violet, white and combinations of these) and severely harm herbivores that consume the infected plants, thus meeting the basic criteria for chemically-based visual aposematism. Pažoutová (2006) provided detailed data on the colors of sclerotia for most *Claviceps* species. Eighteen species have black or blackish sclerotia, in eight the sclerotia are of various shades of brown, in seven they are yellow, four have red or purple sclerotia, one has green, and one is partly white, all are typical colors of poisonous aposematic organisms. Thus, the majority of the *Claviceps* species have conspicuous sclerotia that are poisonous. These fungi, which on the one hand only moderately reduce the reproductive capacity of their hosts, can protect the host plants from herbivory on the other, compensate for the reduced reproduction, and weaken the evolutionary tendency of their hosts to evolve better resistance towards their infection. Moreover, by doing so, ergots defend their own habitats, i.e., the host plants. Ergot infests cereals which grow in open habitats where grazers usually feed during the daytime, optimal conditions for visual aposematism.

Since *Claviceps* species need plant hosts, the more toxic the ergots are to herbivores, the better for both ergot and host. Chemicals in the sclerotia may also directly prevent their consumption by the herbivores or defend them from attacks by other fungi and microorganisms. Indeed, large herbivores usually learn to avoid toxic plants (e.g., Howe and Westley 1988; Lev-Yadun and Ne'eman 2004), resulting in a reduced tendency to consume infected hosts. For instance, opossums (*Didelphis virginiana*), are known to develop an aversion towards the very poisonous mushroom *Amanita muscaria*, in part due to the toxin muscimol (Camazine 1983). Alternatively, ergot fungi may induce abortions or even kill the animals that eat ergot-infested species, and by this decrease grazing pressure. Since low levels of ergot consumption are not lethal, but cause sickness (Clay 1988; Matossian 1989), such conditions are appropriate for the development of food aversion towards ergot-infected grasses. The large array of alkaloids in ergots makes it difficult for herbivores to evolve resistance to ergot toxicity. Lev-Yadun and Halpern (2007) proposed that the association between *Claviceps* and grasses is a fine-tuned ecological tactic that fits the definition of "dangerous liaisons" *sensu* van Baalen and Jansen (2001). Lev-Yadun and Halpern (2007) proposed that odor, in addition to color, might be involved in the aposematic signaling of ergots even in their open, well-illuminated habitats, as was found for various fungal taxa that grow in dark habitats, as described by Camazine (1983, 1985), Guevara and Dirzo (1999) and Sherratt et al. (2005). Field data from Finland for ergot (*Claviceps purpurea*) mutualism in *Festuca rubra* plants in grazed areas, showed that mammalian grazers indeed refrain eating infected plants (Wäli et al. 2013), thus supporting the hypothesis posited by Lev-Yadun and Halpern (2007) about probable ergot aposematism.

The hypothesis presented by Lev-Yadun and Halpern (2007) may theoretically either apply only to the dozens of *Claviceps* species that form colorful toxic ergots, or have a broader scope. There are indications for the latter since other toxic fungal taxa express potential aposematic coloration. For example, *Aspergillus* species produce aflatoxin, a fungal metabolite that is a very potent toxin if consumed by animals (Payne and Brown 1998) and has toxic and colored sclerotia (black in *A. flavus* and *A. tamarii*) (Raper and Fennell 1965; Goto et al. 1996; Chang et al. 2001). A similar picture emerges in the genus *Fusarium* (e.g., Toussoun and Nelson 1968; Bottalico 1998), as various animals are repelled by food contaminated with its toxins (Mirocha et al. 1976). Finally, the insect pathogenic fungus *Cordyceps* produces several types of pigments, some of which are bright red (Unagul et al. 2005) and seem to fit into the general picture of toxicity associated with bright or contrasting coloration.

A second type of fungal-related plant aposematism was proposed by Lev-Yadun (2006b). According to this, whitish-colored plants may appear to be infested by fungal disease. Because there are very good indications that infested plant parts, or ones that may just be infested by fungi are rejected by animals, i.e., frugivores avoid eating damaged fruits (Janzen 1977; Herrera 1982; Manzur and Courtney 1984; Borowicz 1988; Buchholz and Levey 1990; Ruxton et al. 2015), Lev-Yadun (2006b) proposed that white plant surfaces that mimic fungal-infested plants may reduce the tendency of herbivores to consume such plants. This is also a type of visual aposematism, practically a case of Batesian mimicry.

Chapter 42
Do Plants Use Visual and Olfactory Carrion-Based Aposematism to Deter Herbivores?

Sticky plant surfaces made of trichomes were shown experimentally not only to defend plants directly by killing, damaging, or slowing down insects (*sensu* Levin 1973) but also to enhance indirect defense by attracting predaceous arthropods to such carrion-carrying plants because it provides these predators with food (Krimmel and Pearse 2013; LoPresti et al. 2015). Concerning a possible role of carrion odor in plant defense from herbivory, Lev-Yadun et al. (2009b) proposed on theoretical grounds that plants' carrion or dung odors produced by flowers of certain species may simultaneously serve both the long-known attraction of pollinators and also act as an olfactory aposematism towards mammalian herbivores. The anti-herbivory (defensive) carrion odor hypothesis was supported by field data showing that cattle repeatedly refrained for many years from grazing in very productive paddocks used for the disposal of cattle carcasses (Lev-Yadun and Gutman 2013).

There are two types of sticky plant surfaces. The first and better studied sticky plant surface is the one covered by sticky or hooked trichomes (Levin 1973; Krimmel and Pearse 2013). Although trichomes are usually a constitutive component of the epidermis in the many taxa and genotypes that produce them during normal ontogeny, additional trichomes may be induced in many plant species following damage (Agrawal 1999; Karban 2015), further indicating their defensive role. The second type of sticky plant surface is temporary, made of fresh latex or resins that cover areas close to wounds as a type of defense expressed only after wounding (Fahn 1979). Insects are trapped in latex and resins oozing after wounding (Konno 2011) and if the trapped insects are not covered altogether by the latex or resin, they "decorate" the plant surface as do those entrapped by trichomes.

S. Lev-Yadun, *Defensive (anti-herbivory) Coloration in Land Plants*,
DOI 10.1007/978-3-319-42096-7_42

Predators Exploit Both Living and Dead Insects Stuck to Plant Surfaces

The opportunistic exploitation by predators of both live and dead insects stuck to or entrapped by plant trichomes has been known for a long time (Sugiura and Yamazaki 2006; Romero et al. 2008; and citations therein), but was considered only recently as regularly enhancing an indirect defense against invertebrate herbivores for plants (Krimmel and Pearse 2013; LoPresti et al. 2015). Because sticky plants as the outcome of trichomes (Duke 1994) and resin/latex secretion (Konno 2011) amount together to about a third of the land plant species, for small arthropods, both herbivorous and predaceous, and for small predatory birds, reptiles and mammals, the fact that many plants are sticky is potentially of considerable ecological and evolutionary importance and their avoidance or attraction may commonly be critical for the herbivorous insects, their potential host plants and for the predators.

Arthropods Trapped on Plant Surfaces May Serve for Visual and Olfactory Plant Aposematism

Lev-Yadun (2014e) proposed that the attached dead or trapped living insects may serve as billboards: (1) cueing visually to other herbivores that the plants are already occupied, and (2) cueing and signaling them both visually and by rotting carrion odors or stress volatiles emitted from trapped insects that such plants are dangerous and even deadly. This is a type of an extended phenotype (*via* insect bodies and volatiles), "second hand" plant aposematism, based on non-self, but still plant-associated signals. The suggestion that insects may visually and by odors serve plant aposematism, dates back to the early years of discussing plant aposematism. Accordingly, aposematic insects were proposed to defend plants from large herbivores when various poisonous aposematic insects aggregate on poisonous plants, adding to the plant's aposematic odor and coloration (Rothschild 1972, 1986).

In order to propose that live or dead arthropods stuck to plant surfaces have the potential to function as plant aposematism, several issues had to be addressed: (1) is this a common or anecdotal phenomenon, (2) do herbivorous insects indeed respond by avoidance to the presence of other insects or non-insect arthropods (herbivorous or predacious) on plants, (3) do herbivorous insects indeed respond by avoidance to the presence of the arthropod's death-related odors, (4) if avoidance behavior of herbivorous insects towards arthropod infestation exists, is it inherited or learned, (5) are the entrapped insects and their alarm and death odors just cues or are they also signals? and (6) if avoidance behavior of herbivorous insects exists, is it based on aposematism, on perceptual exploitation *sensu* Schaefer and Ruxton (2009), or on both?

From the studies cited above it was clear that the phenomenon of live or dead arthropods stuck to plant surfaces is common, and thus there is a strong likelihood that it allowed evolutionary responses to it, as did indeed occurr concerning the regular attraction of predaceous arthropods to such insect-carrion-carrying plants, because they provide predators with food (Krimmel and Pearse 2013).

The Supporting Evidence for the Hypothesis

There is considerable evidence that herbivorous insects indeed respond by avoidance to the presence of other insects or non-insect arthropods (herbivorous or predacious) on plants in the process of decision making during host selection (e.g., Williams and Gilbert 1981; Bernays and Chapman 1994; Sendoya et al. 2009). This avoidance seems to be inherited, but learning by long-lived herbivorous insects should also be considered.

Various arthropods are known to pay special attention to dead conspecifics identified by volatiles, and take special care to evacuate them from their habitats (Trumbo et al. 1997; Franks et al. 2005). Therefore, it is probable that similarly to the way that insect carrion attracts various predaceous arthropods (Krimmel and Pearse 2013), it may deter herbivorous ones.

The question of whether the known avoidance behavior of herbivorous insects towards arthropod infestation is inherited or learned is a very complicated issue. Even for the much better studied animal-animal aposematism, the relative role of inherited *versus* learned avoidance is not well known (Ruxton et al. 2004). When it comes to the dramatically less studied plant aposematism we are simply still completely in the dark.

The question of whether the trapped insects and their emitted odors function just as cues or also as extended phenotype signals was formulated by Lev-Yadun (2014e), but this issue has never been studied in depth. In this connotation, the question of whether avoidance behavior of herbivorous insects from plants "decorated" by dead or dying insects exists, and if it is based on aposematism, or on perceptual exploitation *sensu* Schaefer and Ruxton (2009), or on other principles is not less intriguing, but again it has never been studied.

Eisner et al. (1998) showed that glandular trichomes may sometimes trap not only herbivores, but also some of their predators, for instance ladybugs. This will indeed reduce to a certain extent the actual predation of trapped and non-trapped herbivorous insects on the plant's surface, but will "decorate" the plants with actual predators that certainly will deter various visually oriented insects from landing there (e.g., Sendoya et al. 2009; Dukas 2001).

In conclusion, Lev-Yadun (2014e) proposed that when arthropods are trapped on plant surfaces *via* trichomes, latex and resins, these components may not only defend the plants directly (by being toxic and sticky), or indirectly (by attracting predators), but also by visual and olfactory cues and signals of danger and death based on the trapped arthropods. This is a case of indirect aposematism, being an extended phenotype of the sticky plants, initiated by a plant character and progressing to functionality by visual and chemical characters of live and dead trapped arthropods.

Chapter 43
Gall Aposematism

Galls are abnormal plant growths induced by various parasitic organisms, mainly insects. Many herbivorous insects induce galls on various plant organs such as leaves, shoots and flowers. Gall-inducers manipulate and exploit the development, anatomy, morphology, physiology and chemistry of the host plant to their own benefit (Weis et al. 1988; Shorthouse and Rohfritsch 1992). Galls, being plant tissues, act as physiological sinks for mobilized plant resources, resulting in increased nutritional values for their inducers. They serve as "incubators" for the developing insects that gain protection from both abiotic factors (e.g., sun irradiation, wind, dessication, rain and snow) and natural enemies such as pathogens, predators and parasitoids, or even from accidental consumption by large herbivores (Price et al. 1987; Stone and Schonrogge 2003). Since the inducing insects control gall formation up to the smallest details, galls are commonly considered as their extended phenotype (Dawkins 1982; Crespi and Worobey 1998; Stone and Schonrogge 2003; Inbar et al. 2004). An earlier (but less likely) hypothesis, suggested that galls could represent adaptations of the host plants; restricting insect damages to specific organs (see Stone and Schonrogge 2003). The evolutionary and ecological contexts of many gall traits have been intensively studied (e.g. Stone et al. 2002; Raman et al. 2005). Numerous studies have examined the biochemical composition of gall tissues both from the nutritional and defensive points of view (e.g., Inbar et al. 1995; Nyman and Julkunen-Titto 2000).

Defensive gall traits against natural enemies attracted much attention from ecologists and evolutionary biologists (e.g., Cornell 1983; Abrahamson et al. 1989; Schonrogge et al. 1999). Galls are typically armed with high levels of defensive secondary metabolites of plant origin. For example, the high levels and compartmentalizing of defensive phenolics and tannins in galls are explained as an adaptive trait that protects the gall-inducing insects without compromizing the nutritive values of inner gall parts (Cornell 1983; Hartley 1998). Inbar et al. (2010a) posited that conspicuousness is a striking and common gall trait. Many galls may be conspicuous because of their size and shape, which differ from the background plant organs. Galls are often characterized by bright or contrasting (red, yellow, etc.) colors

S. Lev-Yadun, *Defensive (anti-herbivory) Coloration in Land Plants*,
DOI 10.1007/978-3-319-42096-7_43

Fig. 43.1 Colorful *Forda* sp. galls on *Pistacia terebinthus* tree growing in Makri, northern Greece

(Figs. 43.1, 43.2, 43.3 and 43.4) (Russo 2007) as a result of an accumulation of plant-derived pigments in their tissues. Since many galls are colorful, they can be clearly distinguished visually from the surrounding host plant organs. For example, the red galls of wasps (Cynipidae) induced on oaks contain high levels of carotenoids (Czeczuga 1977). Many galls, belonging to various taxa, may change color during their development, especially from green to red. Surprisingly, the potential defensive adaptability, functionality and the evolution of gall conspicuousness by aposematic signaling were practically ignored till the publication of the paper by Inbar et al. (2010a). Only a few studies casually mentioned the nature and putative function of gall coloration. For instance, Wool (2004) noted that the pigmentation in some aphid-induced galls is associated with exposure to light.

Inbar et al. (2010a) outlined a new general hypothesis (the aposematic gall hypothesis), suggesting that chemically protected galls which are also visually conspicuous are aposematic. Inbar et al. (2010a) discussed the predictions, alternative hypotheses and experimental tests of this hypothesis. Accordingly, galls that exhibit a combination of high levels of defensive compounds (e.g., Cornell 1983; Hartley 1998; Nyman and Julkunen-Titto 2000) with conspicuousness – size, shape, conspicuous coloration and possibly odor, are aposematic. The galls, which are made of host plant tissues, are manipulated by the inducing parasites to form all the components of aposematism (chemical defenses and warning coloration or odors). The components of the aposematic phenotype are expressed externally to the inducing insects, i.e., in the gall tissue, but protect the galling insects and not the host plant that produces them, as the hosts have no interest to protect their parasites. Inbar et al. (2010a) proposed that the advertisement of chemically-defended galls may reduce predation by mammalian herbivores, avian insectivores and frugivores and various arthropods. Frugivorous vertebrates (birds and mammals) are often either attracted to, or deterred by fruit coloration and odor that is stage dependent (unripe,

Fig. 43.2 A colorful *Geoica* sp. gall on *Pistacia palaestina* tree in northern Israel

ripe, or aposematic) (van der Pijl 1982; Snow and Snow 1988; Schaefer and Schmidt 2004; Hill 2006; Lev-Yadun et al. 2009a; Schaefer and Ruxton 2011; Lev-Yadun 2013a). Colorful galls may theoretically either attract, or repel frugivores, as do fruits. Nevertheless, conspicuousness (advertisement) of colorful fruits is context-dependent, based on the experience and learning of the receiver and the reward given (e.g., van der Pijl 1982; Lev-Yadun et al. 2009a; Schaefer and Ruxton 2011). Plant shape (or the gall's shape in this case), position and probably also odor in many cases (see below), may enhance the learning process of frugivores and predators and sharpen their discriminative response to colored items in the canopy arena.

Tetrachromatic avian predators that can access galls across the canopy are probably among the most important enemies involved in the evolution of defensive gall visual signaling and other defenses. Primates also efficiently use visual (coloration) cues and signaling while feeding on fruits, leaves and other plant organs, where the color of the backgrounds is critically important (e.g., Dominy and Lucas 2001; Vogel et al. 2006). Indeed, bird and mammal predation (e.g., Burstein and Wool 1992; Hill et al. 1995), may impose strong pressure on gall traits (e.g., Abrahamson et al. 1989; Zamora and Gómez 1993; Schonrogge et al. 1999). Insects, both predators and parasitoids, are also considered to be important enemies of gall formers (Price et al. 1987; Stone and Schonrogge 2003). Although some insects can see

Fig. 43.3 Colorful *Forda riccobonii* galls on *Pistacia atlantica* tree growing in the northern part of Israel. Yamazaki (2016) proposed that galls with such shapes and colors of other gall taxa mimic caterpillars for defense

wavelengths in the red part of the spectrum (Briscoe and Chittka 2001) most of them may see red galls in various shades of gray. Red galls would therefore still be much different and distinguishable from the surrounding plant coloration (e.g., green) to the arthropod's eye (see Chittka and Döring 2007). Together with size and shape (and probably their characteristic blend of volatiles), galls can present clear visual and olfactory defensive and identity signals and cues to their important natural enemies. Interestingly, it has recently been demonstrated that the coloration of leaves of lancewood (*Pseudopanax crassifolius*) from New Zealand can effectively serve as defense *via* crypsis at a young plant age and as an aposematic signal in fully grown mature plants for avian herbivores. In this case, leaf coloration changes during plant ontogeny, being cryptic when the plant is still short, changing into aposematic with bright spine coloration in leaves of taller trees as defense from moas (Fadzly et al. 2009; Burns 2010; Fadzly and Burns 2010; Kavanagh et al. 2016). Such phase changes associated with coloration seem to occur in certain galls, but this character was not studied regularly in order to address aposematism.

The predictions of the aposematic gall hypothesis as proposed by Inbar et al. (2010a) were formulated from several life history traits that are thought to promote aposematism in general (*sensu* Mallet and Joron 1999; Ruxton et al. 2004). (1) Defense level: Only chemically and possibly physically well defended galls are

Fig. 43.4 Red *Forda marginata* galls on *Pistacia palaestina* tree in the botanical garden at Oranim campus, Tivon, Israel

expected to be colorful. Galls that are less well defended (especially from avian predators) will tend to be more cryptic. Alternatively, it could be argued that advertisement of galls is a defense strategy of the host plant to attract potential enemies of the galling insects. Inbar et al. (2010a) proposed that if the plant defense by gall coloration aspect is true, less-defended galls should be more colorful and conspicuous in order to enhance the learning of their predators, but this does not seem to be the rule. (2) Aggregation: Colorful and aposematic galls will be found in species that form aggregated communities. Warning coloration in phytophagous insects (gall-formers in this case) is associated with gregariousness in various taxa (Bowers 1993; Hunter 2000). Aggregation should enhance early detection and learning by their predators, thus increasing the effectiveness of the warning signal (Edmunds 1974). (3) Longevity: Colorful galls will be more common in species with prolonged development and persistence. Long-living aposematic species can better teach predators to learn to avoid similar individuals or traits (Blest 1963). Furthermore, aggregations in long-lasting and sessile galls should increase the risk of attack if avoidance learning is not involved. (4) Size and shape: Large galls (as aggregations) or galls with irregular shapes can be more easily detected by potential enemies regardless to their color. It is therefore expected that such galls will be more often both well protected and colorful to accelerate the avoidance learning of their predators. (5) Odor: Some plants may use olfactory aposematism, since many poisonous plants emit characteristic volatiles that may deter herbivores (Atsatt and O'Dowd 1976; Eisner and Grant 1981; Rothschild 1986; Guilford et al. 1987; Provenza et al. 2000; Massei et al. 2007). Inbar et al. (2010a) therefore expected that chemically well-defended galls will tend to produce characteristic odors. (6) Ability to tolerate partial damage: The ability to overcome initial and partial damage (gall repair), and thus accelerate the enemy's learning without self-scarifying should promote the evolution of aposematism. Inbar et al. (2010a) therefore predicted that the evolution of gall aposematism should also be favored by galls with repair ability.

There are alternative (non-aposematic) explanations of gall coloration. Plant coloration in general may have multiple and even simultaneous physiological, defensive, and signaling functions (Gould et al. 2002a, b; Gould 2004; Lev-Yadun et al. 2004a; Schaefer and Wilkinson 2004; Lev-Yadun and Gould 2007; Archetti et al. 2009a; Schaefer and Ruxton 2011; Hughes and Lev-Yadun 2015). Accordingly, alternative hypotheses concerning gall coloration (both physiological and defensive) need not contrast or exclude any other functional explanation of gall coloration, as it may have more than one function. The evolution of gall coloration may reflect an adaptation to physiological pressures, defensive signaling, and defenses not involving signaling. Indeed, in many gall taxa coloration is not a fixed trait and notable polymorphism as well as ontogenetic color changes can be observed (e.g., Dias et al. 2013).

In some species, red gall pigmentation is positively associated with increased light exposure (e.g., Wool 2004), indicating a possible role in protection from the negative physiological effects of excess light (e.g., Gould et al. 2002a, b; Close and Beadle 2003), whereas anthocyanins may have accumulated as anti-oxidants not only because of excess light. If so, Inbar et al. (2010a) expected to find colorful galls only in the upper canopy or on the adaxial (upper) side of the leaves that are more exposed to lights than galls located on shaded plant parts or shaded habitats such as the understory. The aposematic hypothesis may be at least partly rejected, i.e., for certain taxa, if gall coloration is dependent only on the levels of light exposure. However, since many gall species always have their typical conspicuous coloration (e.g., Czeczuga 1977) regardless of light exposure, the aposematic hypothesis cannot be rejected in such cases.

It is theoretically possible that aposematism in galls developed as a "side effect" of multiple protective functions provided by plant pigments (i.e., anthocyanins and carotenoids). This idea originated in Fineblum and Rausher (1997) who showed that there are common biochemical pathways for flower color and defensive molecules. Anthocyanins, for instance, are derived from the phenyl-propanoid pathway, which may also produce tannins and flavonoids. The production of the pigments may therefore correlate (and reliably indicate) with higher level of chemical defenses. Elaborating on Fineblum and Rausher (1997), Schaefer and Rolshausen (2006a) formulated the "Defense Indication Hypothesis" (which will be discussed in more detail in the chapter about autumn leaf coloration), suggesting that the main reason for red color pigment accumulation in plants is physiological stresses, an explanation that cannot be true in the many cases when advertisement or other plant-animal interactions are essential (e.g., animal-pollinated flowers, animal-dispersed fruits, red abaxial leaf side). The defense indication hypothesis by Schaefer and Rolshausen (2006a) posited for autumn leaves, provides a physiological explanation for the development of aposematic galls *via* pleiotropic effects rather than direct signaling. As mentioned above, if the logical hypothesis that gall coloration indeed has a primarily physiological role (e.g., protection from photoinhibition and photo-oxidation), Inbar et al. (2010a) expected galls exposed to solar radiation to be more colorful than galls on shaded plant parts. This is partly true with certain galls that are red because of anthocyanins, but not for yellow ones and many other red ones. Inbar et al. (2010a), however, considered the probable partial contribution of physiological functions to the evolution of gall coloration. Support for the pleiotropic explanation would be an abundance of colorful but weakly-defended galls, whereas pigmentation could not be linked to signaling, but rather to biochemical cascades.

Testing the Aposematic Gall Hypothesis

Inbar et al. (2010a) proposed that several approaches can be used to test the aposematic gall hypothesis. Comparative surveys and analyses (within and between species) of gall coloration, chemical defense level and gall position (e.g., shaded *vs* direct exposure to the sun) in several systems are clearly needed. Nevertheless, only controlled experiments (field and laboratory) in which accelerated associative learning by relevant enemies, preferably naive herbivores or insectivores, mammals and especially birds is examined, can critically test the aposematic gall hypothesis. However, if the studied animals are experienced, or if the avoidance has a genetic basis, such experiments may have limited or no value. If, however, learning curves do exist, and if choice experiments between different galls and between manipulated (painted) gall coloration could be conducted, valuable results are expected. As pointed out by Chittka and Döring (2007) for other taxa, gall coloration should be examined through the eyes (visual abilities) of the potential natural enemies of a given gall former, and under the relevant natural background (see also Sumner and Mollon 2000a; Vogel et al. 2006). In cases where gall coloration is variable, manipulation of light exposure and measuring its effect on gall phenotype (color), chemical defense and predator attacks, may distinguish between the aposematism and alternative hypotheses, as was elegantly done with red leaf margins (Cooney et al. 2012). Again, like in other defensive systems, if the predators are not naive, and in nature this may be the usual situation, there might be too few attacks to allow for statistical analysis. Inbar et al. (2010a) also recommended analyses of the odors emitted from galls and their correlation with coloration, size, chemical defenses and levels of attack. This was later done concerning a small number of Near Eastern gall taxa (Rostás et al. 2013; Rand et al. 2014) and odors were indeed found to take part in the gall's defense, but much more descriptive, theoretical and experimental data is still needed.

Opposition to the Red Aposematic Gall Hypothesis

The aposematic gall hypothesis stimulated other scientists to propose alternative explanations. In the first, White (2010) argued that galls coloration is related to the senescing status of the galled tissue. White (2010) suggested that the gall inducers cause the dying gall tissues to release nutrients to the benefit of the gall inducer. White (2010) also posited that the gall inducers make the plant form the gall around the gall inducer feeding site after the rest of the plant has stopped growing. However, Inbar et al. (2010b) explained that gall coloration has nothing to do with the senescing status of the galled tissue, since red coloration can be seen in young and fast-growing galls, for instance in *Slavum wertheimae* on *Pistacia atlantica*. The tissues of young red galls are commonly meristematic and not dying, and when galls grow, the rest of the host plant grows as well. Inbar et al. (2010b) commented that although the feasibility and generality of White's (2010) physiological/nutritional hypothesis deserves a separate, thorough discussion, one should keep in mind that red and yellow coloration is not synonymous with senescence. The petals of many young and growing flowers are red or yellow (Lee 2007) and many young leaves are red (Richards 1996; Dominy et al. 2002; Lev-Yadun et al. 2012). Galls do not necessarily senesce earlier or stop the development of the galled tissue or organ. As posited in Inbar et al. (2010a), it can clearly be accepted that initially, gall coloration could have evolved along physiological routes such as stress-related coloration, or because of pleiotropic effects of defensive genes, at least in red galls. As such, White's (2010) hypothesis may theoretically be relevant for at least some gall-formers (that have not yet been identified as such), but not for all and especially not in carotenoid-colored yellow ones.

Inbar et al. (2010b) explained that there are numerous gall species, of which probably thousands are colorful. They represent many repeated but independent events of galling habit evolution. They are found in different environments and on a great variety of plant taxa, host life stages and organs. It is not reasonable to assume that all these galls have only a single and ubiquitous function for their conspicuous coloration. Indeed, Inbar et al. (2010a) discussed several alternative explanations for this phenomenon. Moreover, gall conspicuity and signaling is not dependent only on colors, but also on size, shape (Stone and Schonrogge 2003) and odor (see Rostás et al. 2013; Rand et al. 2014). A single explanation, such as the level of light exposure, also cannot explain all the variation in gall coloration, since colorful galls are also common in less illuminated microhabitats.

Whatever the reasons for the evolution of gall coloration were, it is still necessary to explain its variability and maintenance. It is widely accepted that gall formers control gall traits, their "extended phenotype". Thus, if conspicuity would have had a strong negative effect on gall survival (e.g., by attracting predators and parasitoids), Inbar et al. (2010b) expected a strong selection against it. Inbar et al. (2010b) thus posited that the common conspicuousness of galls is good evidence for its potential beneficial role, or at least for the lack of strong selection against it. An interesting example of the sophisticated ability of gall-inducers has been reported by Tooker et al. (2008). They demonstrated that gall-forming insects can "silence" the emission of conspicuous odor (volatiles) from the galled tissue, which thus reduces its detection by potential enemies. For all the above reasons, White's (2010) hypothesis is not supported by the facts of gall development and function or by host's biology.

In the second opposition, Connor et al. (2012) argued that the red color of many galls is simply a by-product of the galling process ("a fabricational noise"). Connor et al. (2012) suggested the following line of reasoning: (1) many gall-inducing bacteria (and some fungi) are known to produce cytokinins; similarly, some gall-inducing insects have been found to harbor high levels of cytokinins (Yamaguchi et al. 2012), sometimes different from the specific host plant's cytokinins; (2) insect-induced galls are mobilization sinks for photosynthates, and cytokinins are known to be involved in strengthening mobilization sinks in active plant sinks; (3) cytokinins and high sugar concentrations have been shown to be involved in production of flavonoids, including anthocyanins, and therefore, the coloration is due to accumulation of anthocyanins as a side result of the combination of high sugar and cytokinins; (4) sugars suppress carotenoid synthesis, suggesting that it is unlikely that the coloration is due to carotenoid accumulation; and (5) they refered to White (2010), accepting that reddening is associated with senescence.

In spite of these interesting suggestions, the biology of plants in general, and galls in particular, is not as simple as posited by Connor et al. (2012). Gerchman et al. (2013) in their response, argued that since galls come in many colors (red, yellow, pink, purple, green, brown, as well as in combinations of colors), and since gall coloration can change, there is more than one factor that determines gall coloration. Some galls change their color on a timely basis or when exposed to sunlight (see photos in Russo 2007; Redfern 2011), while others redden especially when induced on the shaded abaxial side of the leaf. Others however, will always remain green, both inside and outside (e.g., Inbar et al. 2010a). This variety of colors suggests that anthocyanins can be only one of multiple pigment types involved in gall coloration. For example, many insect-induced galls are yellow (e.g., Inbar et al. 2010a), and carotenoids contribute to gall coloration in *Gymnosporangium juniperi virginianae*-induced galls (Smits and Peterson 1942) as well as in other cases (Czeczuga 1977). This challenges Connor et al.'s (2012) suggestion that carotenoids are unlikely to be involved in gall pigmentation because sugars suppress its synthesis. Moreover, the fact that there are so many carotenoid-based yellow-orange sweet ripe fruits that change color from green to yellow along with a dramatic increase in sugar content (e.g., mango, peach, and certain date varieties) is an excellent demonstration that sugars do not suppress carotenoid synthesis as a rule. Moreover, since galls are composed of several tissues, the high levels of sugars are probably mostly restricted to the phloem veins and specific parenchymatic cells, posing no problem to accumulate carotenoids in other types and layers of gall tissues.

Another problem with Connor et al.'s (2012) suggestions, discussed by Gerchman et al. (2013), is that anthocyanin and cytokinin levels don't correlate in very many cases and situations. Anthocyanins are pigmented flavenoids common in many plant tissues, and more than 500 anthocyanins have been reported (Andersen and

Jordheim 2006). Anthocyanin content has been found to increase under various conditions, some of which, but definitely not all, not even the majority, are sink tissues. Elevated anthocyanins levels were demonstrated in both young expanding leaves and in senescing foliage (Richards 1996; Dominy et al. 2002; Close and Beadle 2003; Lee 2007; Archetti et al. 2009a; Lev-Yadun et al. 2012), in flowers (Grotewold 2006), and in many ripe fruits (Allan et al. 2008). While some of these (namely young expanding leaves) are sink tissues, others such as older leaves are usually source tissues and not sinks. Altogether, the connection between anthocyanin content and high cytokinins levels is very far from being deterministic. Young fruits, a definite sink tissue, contain high levels of cytokinins but are typically green. Moreover, although not as common as green young fruits, many ripe fruits tend to be red (containing high levels of anthocyanins), although they have stopped growing or being a sink (Gillaspy et al. 2003), demonstrating an inverse correlation between anthocyanin and cytokinin content. A similar trend was reported for nitrogen starvation conditions (Close and Beadle 2003), where anthocyanins accumulate but cytokinins levels do not, and in some cases even decline (Yong et al. 2000). Finally, some cytokinin-secreting organisms are known to induce "green islands" (which are, as their name implies, green) on senescing leaves (Walters et al. 2008). Altogether, there is no universal trend of cytokinins inducing reddening. In any case, the regulatory link between cytokinins and red coloration has to be thoroughly characterized in at least one model gall species in order to begin a fruitful discussion about it.

Connor et al. (2012) pointed out that in a given galling system insectivorous birds and mammals attack galls during the winter when galls have lost their red color. One could easily adopt a counter explanation to Connor et al.'s (2012) proposal, namely, that this can be regarded as an indication that the red coloration may have a defensive role, i.e., red galls are not attacked when they are still red. Such (or other) hypotheses have not yet been tested.

Gerchman et al. (2013) thus concluded that the hypothesis proposed by Connor et al. (2012), although stimulating and possibly able to explain certain cases of the gall coloration phenomenon, is an oversimplification of the involved biology in general, and is wrong in many specific cases and aspects. The gall-inducing habit has evolved independently in numerous insect (and other organisms) lineages, suggesting strong and multiple although yet only partly known adaptive values. Gerchman et al. (2013) did agree with White (2010) and with Connor et al. (2012) that theoretically, in certain cases, gall coloration may have evolved for or serves only, or mostly physiological processes and functions. However, Gerchman et al. (2013) argued that in other cases, or even when this coloration has a physiological "fabrication", gall coloration may also have been shaped by signaling selection pressures (repellent signaling). Even if the cytokinin hypothesis proposed by Connor et al. (2012) is indeed the mechanism of gall reddening in certain taxa, something that Gerchman et al. (2013) dismissed as a universal rule, it is not a negative proof for other explanations, including aposematism.

There is strong selection for gall defense from a variety of natural enemies: pathogens, parasitoids, herbivores and predators, both arthropods and vertebrates (e.g., Price et al. 1987; Schultz 1992; Zamora and Gómez 1993; van Hezewijk and Roland 2003). It is widely recognized that many gall traits are shaped by such selection pressures (Price et al. 1987; Stone and Schonrogge 2003, but see Rehill and Schultz 2012). Gall coloration may thus be part of this complicated defensive arsenal. Considering that plant-mediated signaling between herbivorous insects and their natural enemies (insect and vertebrates) has received much recent attention and recognition (e.g. Unsicker et al. 2009; Schaefer and Ruxton 2011), and the fact that galling insects can control the chemistry of their galls (e.g., Tooker et al. 2008), the probable signaling role of gall traits, including coloration, deserves much more research attention.

Chapter 44
Experimental Evidence for Plant Aposematism

The direct experimental evidence for the operation of aposematic coloration in plants is meager; however, the direct evidence that it operates is slowly accumulating. Cook et al. (1971) showed that poisonous gray seeds of *Eremocarpus setigerus* are rejected by the mountain dove. Cahn and Harper (1976) showed that sheep avoid *Trifolium repens* plants with leaf marks, but did not discuss aposematism. Lev-Yadun and Ne'eman (2004) showed that sheep, goats, camels, donkeys and cattle reject conspicuous green plants in the yellow/gray desert in the summer. Numata et al. (2004) found that leaves with delayed greening suffer lower levels of insect damage when they are still young. Hill (2006) showed that the Florida scrub jay (*Aphelocoma coerulescens*) rejects poisonous red fruits. Karageorgou and Manetas (2006) showed that red young leaves of the evergreen oak *Quercus coccifera* are attacked less by insects than green ones, but rejected the aposematic coloration hypothesis and adopted other visual defensive explanations (red masks the green color, it undermines insect camouflage). Similar results were found for other species growing in Greece (Karageorgou et al. 2008). Recently, additional data about the defensive operation of white variegation that mimics insect damage in leaves was published (Campitelli et al. 2008; Soltau et al. 2009). Gerchman et al. (2012) tested the question of whether colorful visual signaling for pollinators can simultaneously serve for aposematism. The conspicuous purple tufts of leaves ("flags"), which often terminate vertical inflorescences in the Mediterranean annual *Salvia viridis* and attract insect pollinators to the flowering patch, have an aposematic potential towards herbivores. Cooney et al. (2012) demonstrated that the width of red margins in leaves of *Pseudowintera colorata* correlated positively with concentrations of the defense compound polygodial, and negatively with the extent of leaf herbivory observed in the field. Furthermore, when these authors conducted feeding trials using larvae of the generalist herbivore *Ctenopseustis obliquana* (brown-headed leafroller moth), which are known to naturally feed on *P. colorata* leaves, a significant preference for leaves lacking red margins was observed under white light, but not under red light or darkness that cancelled the visual signaling, consistent with a visual, herbivory-deterrent function (Cooney et al. 2012). Field data from Finland

S. Lev-Yadun, *Defensive (anti-herbivory) Coloration in Land Plants*,
DOI 10.1007/978-3-319-42096-7_44

for ergot (*Claviceps purpurea*) mutualism in *Festuca rubra* plants in grazed areas, showed that mammalian grazers indeed refrain from eating infected plants (Wäli et al. 2013), thus supporting the hypothesis posited by Lev-Yadun and Halpern (2007) about ergot aposematism. Rostás et al. (2013) showed that red galls that emit various repeling volatiles are rejected by goats in choice feeding experiments. Maskato et al. (2014) studied in more detail the find by Gerchman et al. (2012) that locusts preferred green cabbage over anthocyanin-rich red cabbage. They found that female *Pieris* butterflies avoid laying eggs on anthocyanin-rich red foliage, suggesting that red color is at least a visual cue affecting oviposition behavior. They reared *P. brassicae* larvae on green and red cabbage leaves, to determine whether foliage color reliably indicates host plant quality. They did not find a difference in survival rates or maximal larval body mass in the two food treatments. However, larvae feeding on red cabbage leaves exhibited significantly lower growth rates and longer durations of larval development. Interestingly, this longer development was coupled with a higher consumption rate of dry food matter. The lower ratio of body mass gain to food consumption in larvae feeding on red cabbage leaves was coupled with significantly higher (ca. 10 %) larval metabolic rates. This suggested that insect development on red foliage may incur an increased metabolic load associated with detoxification of various secondary plant metabolites. Energy and oxygen allocation to detoxification could come at the expense of growth and thus compromise larval fitness as a result of extended development. Maskato et al. (2014) proposed that from an evolutionary perspective, red foliage color may serve as an honest defensive cue, and I say probably a signal, as it reliably indicates the plant's low quality as a substrate for larval development. The possibility of simultaneous olfactory aposematism has not been tested in any of these cases.

Altogether, there is good evidence that visual aposematism, with or without simultaneous olfactory aposematism indeed operates in plants. It is also clear that much more experimental and field evidence is needed for each of the sub-types of aposematic coloration proposed for plants in this central part of the book. While arriving at that point for many plant and herbivore species will certainly take at least several decades, these years will also allow a much better understanding of the genetic components of these characters. It is hoped that these decades will also put an end to the irrational fear of transgenic plants, a fear that parallels the medieval fears of broom-riding witches that brought storms and caused crop destruction. With the genetic and ecological data in hand, it might be possible to experiment with and apply some defenses including aposematism in agriculture using transgenes in order to reduce the use of chemical pesticides and employ ecologically greener and visually colorful agricultural practices.

Part IV
Masterpieces of Visual Defense

Chapter 45
The Complicated Enigma of Red and Yellow Autumn Leaves

My attraction to the question of autumn leaf colors started in the autumn/early-winter of 1994, in parallel to my decision to understand why *Silybum marianum* plants look like green zebras. More specifically, several field days in the autumn of 1994 at elevations of 1500–1900 m on Mount Hermon, the only place in Israel where the phenomenon of red autumn leaves is very conspicuous (Figs. 45.1, 45.2 and 45.3), triggered my attention and curiosity to understand this phenomenon. After I understood in the early spring of 1996 that visual aposematism exists in spiny plants, I became even more aware to the non-physiological aspects of leaf colors and wished to understand the biology of red autumn leaf coloration.

When I first learned in a geobotany course I took during my B.Sc. studies in 1978 about the dramatic phenomenon of red or yellow autumn leaves in the temperate region, the explanation we got was that these conspicuous colors appear when the chlorophylls are degraded in order to mobilize their amino acids and other resources for storage in the branches and trunks before leaf shed to be used for new growth in the following spring, and that these colors existed in the leaves all the time but were masked by chlorophyll. While this is true for the carotenoids that give a yellow autumn leaf color, it is not so for anthocyanins, which give rise to pink, orange, red and purple autumn leaf colors.

Before I proceed, I wish to stress that defending the free amino acids and some other resources released following the breakdown of the photosynthetic system is indeed the cornerstone of almost all the defensive, reproduction-related, and physiological hypotheses involving autumn leaf coloration.

A paper published on the physiology of autumn leaves as influenced by anthocyanins (Hoch et al. 2001) and a previous paper (Matile 2000) on the *de-novo* production of anthocyanins in leaves that will soon be shed published in a journal dealing with gerontology, intrigued me further. The fact that in Israel red autumn leaves are very conspicuous only on highest mountain (Mount Hermon), the only place in Israel with significant and regular snow and freezing temperatures in winter, caused me to think at that stage that red autumn leaves are directly and mostly related to low autumn temperatures. With the help of my colleagues Moshe

S. Lev-Yadun, *Defensive (anti-herbivory) Coloration in Land Plants*,
DOI 10.1007/978-3-319-42096-7_45

Fig. 45.1 A group of dwarf trees of *Rhus coriaria* with red autumn leaves marking the upper forest line at an elevation of 1900 m in Mount Hermon, Israel

Fig. 45.2 Red autumn leaves of *Rhus coriaria* in Mount Hermon, Israel

Fig. 45.3 Red autumn leaves of *Acer monspessulanum* subsp. microphyllum (= *Acer hermoneum*) at an elevation of 1700 m in Mount Hermon, Israel

Fig. 45.4 A red young leaf of *Ricinus communis* from the central coastal plain of Israel. Young and red and old and yellow leaves are found together on *R. communis* trees in various seasons

Flaishman and Gidi Ne'eman I formulated a partly wrong hypothesis concerning low temperatures and red autumn leaves, that luckily was rejected (for the wrong reason) by the New Phytologist. Low temperatures may indeed result in leaf reddening (Close and Beadle 2003; Hughes and Smith 2007; Hughes 2011) but red leaves are not associated only with low temperatures and low temperatures are not always associated with red leaves. For instance, young red leaves are regularly formed in the very hot summer of Israel in various taxa of wild trees and shrubs (e.g., *Ceratonia siliqua*, *Cercis siliquastrum*, *Pistacia lentiscus*, *P. palaestina*,

Ricinus communis) (Fig. 45.4) and various introduced trees [e.g., *Ailanthus altissima*, *Mangifera indica* (mango), *Eugenia uniflora* (pitango), *Eucalyptus* spp.]. Very soon, my interests in red autumn leaves turned to both red and yellow ones, and to other, more plausible explanations.

In fact, Haberlandt (1914) in his classic book "Physiological plant anatomy" reviewed several functions for leaf anthocyanins proposed earlier by various authors, and listed light screen against the excessive light that causes damage to the photosynthetic system, and heating of leaves in cold regions as possible functions of red autumn leaf coloration. I wonder if the timing (World War I) and nationality (Austrian) of Haberlandt contributed to the general ignorance of his suggestions concerning red autumn leaf coloration at the time and for almost a century. In any case, for many decades, most people believed that these colors simply appeared after the degradation of chlorophyll that masked these pigments, and that they have no function.

The change in attitude towards possible functions of autumn leaf coloration arrived simultaneously from several directions: the physiological on one hand, and the ecological-evolutionary on the other. The evolution and revolution of what scientists think about the significance and functions of autumn leaf coloration is an excellent lesson in the ways science progresses. We have the general phenomenon, the changing view concerning the basic facts of the phenomenon, the first generation of explanations, the strong opposition to the hypotheses, the emergence of additional hypotheses, the realization that the phenomenon is complicated and that it is actually made of several simultaneous different phenomena, the construction of more balanced hypotheses that take into account all the known aspects, the relaxation of the tensions and the emergence of new questions that are still only partly solved.

Several papers from Bill Hamilton's group in Oxford (Archetti 2000; Hamilton and Brown 2001; Archetti and Brown 2004) that considered red and yellow autumn leaves together under the title of "conspicuous colors" or "bright colors" as indicating the defensive (anti-herbivory) qualities of trees with bright autumn leaf colors, especially against aphids, according to Zahavi's handicap principle, triggered a wave of descriptive, experimental and theoretical studies, as well as lots of media and public interest. A special aspect of this hypothesis was that the signaling was mostly aimed at repelling aphids from laying eggs on the trees (not on the leaves), eggs that will hatch in spring, and that the color-associated defense was partly aimed towards future damage by the next insect generation. Unfortunately, Hamilton passed away in the year 2000 from a sickness he got while looking for the origin of AIDS in Congo, and we all have lost the opportunity to see his additional contributions to understanding autumn leaf coloration when the first generation of hypotheses turned out to be much too simplistic. In any case, Archetti, who was at that time a visiting student at Hamilton's lab, posited in his classic (2000) paper that bright autumn colors do not function as aposematic signals. The hypothesis formulated and published by the Hamilton school (Archetti 2000; Hamilton and Brown 2001) became known as the "co-evolutionary" colorful autumn leaf hypothesis. Several response papers that posited that contrary to the suggestion by the Hamilton group, aphids are attracted to yellow leaves rather than being repelled by them (Holopainen and Peltonen 2002; Wilkinson et al. 2002), caused me to think that defense from herbivory is more important than adaptation to low temperatures when red autumn leaves are discussed. At that time I was also interested in the operation of Zahavi's handicap principle (Zahavi 1975, 1977) in plants in general, and thought that concerning plants, it operates among other things in their biology in competitive games for capturing light in dense plant populations (Lev-Yadun 2005). The papers by Archetti (2000), Hamilton and Brown (2001), and Archetti and Brown (2004) reduced my need to fight in order to convince others that Zahavi's handicap principle, which was first presented as a hypothesis concerning animal-animal communication, may indeed operate in plants, and I was very pleased with their papers in this respect.

Twice in my 40-year-long career as a biologist I witnessed bitter and aggressive opponents concerning Zahavi's handicap principle. The first was in 1975, only several weeks after I began my university studies. Dr. Amotz Zahavi presented his handicap hypothesis (Zahavi 1975) in a lecture on the occasion of the George S. Wise Faculty of Life Sciences at Tel Aviv University Open Day. I simply understood nothing, a very funny thing since later I have studed this type of signaling in plants for about 20 years. However, I was deeply impressed by the loudness and aggressive style of the opposition to his hypothesis by several professors, which was very rude even by Eastern Mediterranean (Levantine) hot spirited standards. It was, however, a lesson concerning the difficulty to propose something very different from the mainstream, and for the need to be flexible with various theoretical aspects of a new hypothesis following constructive comments by others. The important role of the presenting personality and the type of inter-personal communication and relations in such issues was also revealed to me concerning the presentation of the handicap principle. The second time was the opposition to the application of Zahavi's handicap principle to the function of colorful autumn leaves by the Hamilton school.

Concerning the hypothesis that Zahavi's handicap principle operates in colorful autumn leaves (e.g., Archetti 2000, 2007a, b; Hamilton and Brown 2001; Hagen et al. 2003, 2004; Archetti and Brown 2004, 2006; Archetti and Leather 2005; Brown 2005), the idea was partly (Lev-Yadun 2006a; Lev-Yadun and Gould 2007, 2009; Ougham et al. 2008; Ramirez et al. 2008) or wholly (Holopainen and Peltonen 2002; Wilkinson et al. 2002; Schaefer and Wilkinson 2004; Ougham et al. 2005; Schaefer and Rolshausen 2006a, 2007a; Sinkkonen 2006a, b; Chittka and Döring 2007; Rolshausen and Schaefer 2007; Schaefer and Gould 2007; Hatier and Gould 2008; Yamazaki 2008a) discounted on various grounds. Lev-Yadun (2006a) and Lev-Yadun and Gould (2007, 2009) emphasized that the operation of aposematism in colorful autumn leaves does not exclude the possible simultaneous operation of any other types of visual or non-visual defense, as well as in providing physiological gains (see also Hatier and Gould 2008).

The opposition to the colorful autumn leaf handicap hypothesis reflected, or was the outcome of the complicated biological facts involved (which are still not yet sufficiently understood), i.e., on the simultaneous operation of various and sometimes contrasting physiological and defensive functions of yellow and red autumn leaf coloration. The various functions probably differ in their importance with time even in a single leaf, let alone in a flora or a broad geographical region (see Lev-Yadun and Gould 2007; Ougham et al. 2008). Holopainen and Peltonen (2002) suggested that leaves that have just turned yellow are a good indication to aphids of nitrogen availability in them in the form of free amino acids, an attracting cue rather than a repelling signal. Wilkinson et al. (2002) posited that rather than signaling defensive qualities to aphids, especially since these are drawn to yellow leaves, this coloration serves as a sunscreen (a physiological role), and that red colors help to warm leaves, and also function as antioxidants. Ougham et al. (2005) stressed the importance and good documentation of the physiological role of autumn leaf coloration. They argued that the signal is not costly, which, according to the most common view (but not all views, see Lachmann et al. 2001), is a basic feature of signals involved in the operation of Zahavi's handicap principle (Zahavi 1975, 1977, 1991; Zahavi and Zahavi 1997) including aposematism (Holen and Svennungsen 2012).

In the summer of 2002, I understood one possible and previously unknown defensive role of autumn leaf coloration (among other defensive and physiological functions) in undermining herbivore insects' camouflage (Lev-Yadun et al. 2004a; Lev-Yadun 2006a, 2009a; Lev-Yadun and Gould 2007) that had already been discussed in a previous chapter. Accordingly, the herbivorous invertebrates would be vulnerable to predation if they occupy leaves that don't suit their color, and in addition they may avoid inhabiting plant organs with unsuitable coloration because of their fear of predation, to the benefit of the plants. Later, Lev-Yadun (2009a) proposed that this is a type of habitat aposematism. In the year 2006, I posited that colorful autumn leaves of certain taxa may be aposematic because they are toxic for certain herbivores (Lev-Yadun 2006a) and with Kevin Gould from New Zealand as a colleague, I suggested that both yellow and red autumn leaves may signal or at least cue to herbivores that they are going to be shed soon (Lev-Yadun and Gould 2007), something that is lethal for many insects (Glinwood and Pettersson 2000; Karban 2007), and may therefore reduce insect attacks. In parallel, Holopainen (2008) proposed that defense of autumn leaves against aphids may also operate by olfactory signaling. In this case, being red or yellow matters less than in the case of defense dependant only on visual signaling.

Lev-Yadun and Gould (2007) concluded that autumn leaf colors have at least six potential visual defensive functions against herbivory in addition to the (better understood) important physiological roles or to the non-visual (chemical) defenses previously proposed. The first proposed defensive function of colors – serving as a signal that the trees are well defended (a case of Zahavi's handicap principle) (Archetti 2000; Hamilton and Brown 2001; Hagen et al. 2003, 2004; Archetti and Brown 2004; Archetti and Leather 2005; Archetti et al. 2009a), cannot be ignored, despite the important arguments of critics of this hypothesis (Holopainen and Peltonen 2002; Wilkinson et al. 2002; Schaefer and Wilkinson 2004; Ougham et al. 2005; Schaefer and Rolshausen 2006a, b, 2007a, b; Sinkkonen 2006a, b; Chittka and Döring 2007; Rolshausen and Schaefer 2007; Schaefer and Gould 2007; Hatier and Gould 2008; Yamazaki 2008a, b). The second and related hypothesis is that red and yellow autumn leaves of toxic species serve as aposematic coloration (Lev-Yadun 2006a, 2009a; Lev-Yadun and Gould 2007; Archetti et al. 2009a). The third is that it undermines herbivorous insect camouflage (Lev-Yadun et al. 2004a; Lev-Yadun 2006a, 2009a; Lev-Yadun and Gould 2007). The fourth is that they function according to the "defense indication hypothesis", stating that red leaves are chemically defended because the biosynthesis of anthocyanins correlate with the biosynthesis of various defensive compounds (Schaefer and Rolshausen 2006a, b). The fifth hypothesis proposed by Sinkkonen (2006a, b) holds that yellow foliage signals to sucking herbivores that the tree is becoming poor-quality food. The last (sixth) hypothesis is that it might deter insects from occupying the soon-to-be-shed leaves (Lev-Yadun and Gould 2007).

The issue of bright (yellow and red) autumn leaves, which was only infrequently studied for many decades, received in the years 2000–2008 lots of scientific (about 90 papers discussing it) and media attention, with several parallel but theoretically different hot debates. The first debate occurred between plant physiologists and plant ecologists (the role of various physiological functions *versus* the role of defense from herbivory in selecting for conspicuous autumn leaf colors). Some physiologists dismissed for some years the anti-herbivory role altogether. The second was among various plant physiologists (concerning the various physiological roles of anthocyanins in relation to their location within cells, tissues, and organs, as well as concerning its role in protection from UV and the proposed role of dark red color in warming the leaves in cold habitats) (see Lee 2012). The third was among ecologists (concerning the differences between yellow and red autumn leaves, in relation to various types of defensive mechanisms and attacking insects, and also whether Zahavi's handicap principle includes aposematism or not). In the beginning of the year 2008, the arena of the biology of yellow and red autumn leaves was a mess with opposition to several of the hypotheses that tried to explain the phenomenon. Surprisingly, in only about four additional years most of the theoretical conflicts were eliminated.

In general, scavenging reactive oxygen species and defense from photoinhibition under low temperatures (e.g., Matile 2000; Hoch et al. 2001, 2003; Lee 2002; Lee and Gould 2002a, b; Wilkinson et al. 2002; Schaberg et al. 2003, 2008; Gould 2004; Ougham et al. 2005; Lev-Yadun and Gould 2007, 2009) seem to be the main physiological functions of red autumn leaf coloration, but there is also a relationship to reproductive effort (Sinkkonen 2006a, b) or branch die back (Sinkkonen 2008).

Two decades earlier, Stiles (1982) proposed that strong autumn leaf colors in temperate forests may signal frugivorous birds about their ripe fruits, a hypothesis that was not accepted by all (e.g., Willson and Hoppes 1986; Burns and Dalen 2002) and has a basic inherited problem: it does not explain why dioecious species have strong autumn colors in leaves of male individuals that carry no fruit. However, if additional non-defensive signaling functions are considered, the fruit flag hypothesis may be correct although not an exclusive and full explanation.

Chapter 46
Leaf Color Variability

Another theoretical and practical issue related to autumn leaf coloration is the issue of leaf color variability. While several authors proposed that such variability in timing and intensity of autumn leaf coloration has been selected for (Archetti 2000; Hamilton and Brown 2001; Hagen et al. 2003; Ougham et al. 2005; Sinkkonen 2006a; Sinkkonen et al. 2012), theoretically, this variability may in certain cases be a non-defensive adaptation, or not adaptive at all. Variation in the timing of color change may reflect positional constraints, such as the position of leaves within the tree, the level of exposure to sunlight or wind action, the status of the tree as dominant, sub-dominant or suppressed individual within the population, or being an understory or canopy species, or else reflect physiological pressures such as fluctuations in water supply or mineral nutrition, factors that may significantly influence a tree's performance irrespective of its genotype (Kozlowski et al. 1991; Kozlowski and Pallardy 1997). The results of Hagen et al. (2003) that in *Betula pubescens* ssp. *czerepanovii* trees that turn yellow early in autumn (September) was negatively correlated with insect damage the following season, did not match the results of Isaksen and Folstad (2014) for the same subspecies but in different years. Isaksen and Folstad (2014) found no difference in the attacks of the most common caterpillar species *Epirrita autumnata* on leaves, or in the oviposition by mature females on twigs senessing early or late. Thus, it seems that at least for that taxon, the difference in succeptability to herbivory is not only related to variability in phenology, but is also year and habitat depended. Isaksen and Folstad (2014) rejected the aposematic function in that case. However, at least for various Finnish *Betula pendula* (silver birch) genotypes, Sinkkonen et al. (2012) showed that autumn leaf coloration varies significantly among genotypes, and that this influences fitness and evolution (Lev-Yadun and Keasar 2012). In that species, genotypes that expressed the strongest leaf reflectance in early autumn harbored more egg-laying females of the specialist aphid *Euceraphis betulae* than those expressing strong leaf reflectance several weeks later. Silfver et al. (2015) studied the phenology of shoot growth termination in various Finnish *Betula pendula* genotypes that differ in that character. Genotypes

S. Lev-Yadun, *Defensive (anti-herbivory) Coloration in Land Plants*,
DOI 10.1007/978-3-319-42096-7_46

that terminated shoot growth earlier suffered from higher aphid egg loads after a short, but not after a long autumn. The Sinkkonen et al. (2012) and Silfver et al. (2015) studies illuminated the need to document the variability in autumn leaf coloration and the corresponding aphid attacks in many other taxa and under various environmental conditions.

Chapter 47
What Do Red and Yellow Autumn Leaves Signal for Sure?

Autumn leaf defensive signaling of whatever type can benefit both trees and shrubs and their herbivorous insects. At a time when the phloem sap is rich with the nutrients remobilized from senescing leaves to branch and trunk tissues, defensive, and even aposematic signaling by colorful leaves (Archetti 2000; Hamilton and Brown 2001; Archetti and Brown 2004; Lev-Yadun et al. 2004a; Lev-Yadun 2006a, 2009a; Lev-Yadun and Gould 2007; Archetti et al. 2009a) could well benefit the tree by deterring phloem sap feeders, such as aphids (e.g., Dixon 1998). Lev-Yadun and Gould (2007) proposed that autumn leaf colors may also serve as a visual warning that the leaves are going to be shed soon, reliable, honest and critical information about a significant risk of insect mortality. This is vital information for insects that need leaves that will remain on the tree for longer than several days to use as a habitat, especially for those insects that cannot climb back from the ground. Leaf fall is a well known considerable agent of insect mortality (Faeth et al. 1981), and various insects may refrain from occupying leaves that are soon to abscise (Glinwood and Pettersson 2000; Karban 2007). Thus, the signaling tree will incur less herbivory, and the insects less mortality. It is already known that certain aphid species will lay eggs on trees whose leaves have just started to change color and are yellow-orange, but they usually refrain from occupying trees with red leaves (Furuta 1986). Thus the potential for such plant-insect communication by autumn leaf coloration exists. Interestingly, Holopainen et al. (2010) found that leaves of *Betula pendula* that are about to be shed, emit *Cis*-3-hexenol, an indicator of cellular disintegration, which becomes a dominant volatile just before leaf abscission. This, combined with the yellow color, may indicate to aphids that the leaves are close to shedding.

S. Lev-Yadun, *Defensive (anti-herbivory) Coloration in Land Plants*,
DOI 10.1007/978-3-319-42096-7_47

Chapter 48
The Second Generation of Hypotheses About Colorful Autumn Leaves

The strongest advocate of the defensive role of red and yellow autumn leaf coloration as a case of Zahavi's handicap principle was Dr. Marco Archetti. After the unexpected death of Professor Bill Hamilton in the year 2000, Marco Archetti was practically left by himself to carry the red/yellow autumn leaf handicap flag. As a Post Doc at Oxford University, he arranged a small conference on the autumn leaf color issue in Oxford's St. John's College in March 2008. In that conference, attended by plant physiologists, plant and animal ecologists, and several scientists that specialised in the sensory systems of herbivorous insects, the various and even contrasting views concerning the biology of autumn leaf coloration were presented. Two days of focused discussions vividly illuminated the need for a well-balanced multidisciplinary understanding, and that such a review had to be written and published by the group, and indeed, a year later it was published in Trends in Ecology and Evolution (Archetti et al. 2009a). Among other things, it became clear to all the scientists involved that yellow and red autumn leaves are in a way two different strategies (although not always and not concerning all defensive aspects) and that at least concerning the co-evolutionary hypothesis, they should be treated separately. Moreover, it also became clear that both physiology and ecolgy are involved and that the gains from autumn coloration do not stem from only one of these very different function types, and also that aposematic signaling is involved as well. This understanding was an important step to lower the flames that blazed in the arena of investigating the evolution and functions of colorful autumn leaves, but this critical understanding and even later progress did not eliminate the need for an even better understanding of the complicated scientific question of the evolution of autumn leaf colors.

After the conference I decided to make an effort to understand why yellow autumn leaves dominate the landscape of northern Europe (e.g., Holopainen and Peltonen 2002) (Figs. 48.1 and 48.2), something that dramatically differs from the dominance of red autumn leaves in landscapes of temperate South- and North America and eastern Asia (e.g., Hoch et al. 2001; Lee et al. 2003) (Fig. 48.3). I knew very well that first-hand field work is better than anything else, and contacted

S. Lev-Yadun, *Defensive (anti-herbivory) Coloration in Land Plants*,
DOI 10.1007/978-3-319-42096-7_48

Fig. 48.1 A typical autumn landscape in Finland marked by yellow *Betula* trees growing mixed with drak green conifers around a lake

Fig. 48.2 Typical gold-yellow autumn leaf coloration (Rushka) in Finland

Fig. 48.3 Typical red autumn leaf coloration in upstate New York

Professor Jarmo Holopainen from the university at Kuopio, central east Finland, in order to do some field work in Finland in the autumn of the year 2008. Jarmo responded positively and advised me concerning the right time to be in Finland.

In Finland, the phenomenon of yellow autumn leaves is called "Rushka" and it was indeed different from what I knew from the eastern USA or from Mount Hermon, Israel or even from central Europe (Hungary). When I arrived to Kuopio for the first of four long autumn visits there, some days before the onset of the dramatic yellow autumn coloration, Jarmo gave me a condensed personal workshop on the ecology of the phenomenon from a Finnish/Scandinavian insect and plant perspective. The dominance of yellow autumn leaves in Scandinavia and other parts of northern Europe was also a very good indication that trees can manage well under low temperatures without the need to turn red. This is good evidence that the physiological role of anthocyanins in withstanding low temperatures is not mandatory or even essential, and not understanding this was my major mistake in the hypothesis written in the manuscript rejected for other reasons by the New Phytologist several years earlier.

After I have digested the detailed lessons from Jarmo about the climate, botany and entomology of the Rushka phenomenon, I rented a car, bought a stock of food and drinks just to be on the safe side if I would have to sleep in the car (as I had done on a field trip in the Canadian Rockies several years earlier), and with a map of Finland, but without a mobile telephone, went for my exploration. Jarmo was worried, because I have only one leg, and I walk with the aid of crutches, but I explained to him that in this condition I had served as a Major in the army, and taken part in several wars and many other military actions, and that a map, food and drinks were sufficient and that I would manage. The only advance preparation I made was to reserve a hotel room in the town of Rovaniemi (ca. 66°30′N, about 10 km south of the Arctic Circle) for my first night, and I started to drive. The first hours of driving northward brought me to the southern part of the about 200 km wide ecological belt of the height of the yellow autumn leaf phenomenon that was gradually shifting from the north to the south of Finland in about 2–3 weeks, and lasting for several days at any specific place. When I left Kuopio it was still some days before the arrival of the Rushka there.

The next day, I drove westward in order to turn later to the northeast towards Inari in Lapland. I arrived at the Rushka belt at its height and it was incredible. I drove along the almost empty highway that had two golden walls of millions of trees with bright yellow leaves, fascinated and deeply impressed by the dramatic vegetal landscape (Fig. 48.4). Here and there I could see red spots at the side of the road, several kilometers ahead of me, and when I have approached them after several minutes of driving again and again it turned out to be an old house usually with the red-leafed type of *Sorbus aucuparia* trees planted as ornamentals in the garden (Fig. 48.5). I stopped from time to time to take pictures and to write notes about my impressions and concerning various botanical issues. After some 150 km of driving, the regularity of the autumn coloration became clear to me. All the millions of trees were yellow, but millions of low shrubs of various taxa growing under, or next to the trees, had red autumn leaves (Figs. 48.6–48.8). This was very conspicuous to me because it was very different from the shrubs growing in the Near East that are

Fig. 48.4 Two walls of gold-yellow autumn leaf coloration surrounding a road in Finland

Fig. 48.5 Red-leaved *Sorbus oucuparia* trees planted in gardens of old houses dotted the yellow tree landscape in the autumn of Finland. Again and again, they appeared from a distance or a kilometer or even more as red dots when I drove along the highway in the Rushka belt and turned to be planted trees

Fig. 48.6–48.8 Low shrubs of various taxa with red autumn leaves dominated the undergrowth of the landscape in Finland. They contrasted with the yellow autumn leaves of the trees that grew next to them

mostly evergreens, sometimes with large green winter leaves and small summer leaves (we have rain mostly in winter and no summer rain at all). After turning to the right, at a small T-junction near Kolari (67°33′N), and heading towards the heart of Lapland, I stopped at the side of the road, ate bread and sausage, drank some water, took some pictures, and the ecology and evolution of the yellow *versus* the red autumn leaves on a global scale became clear to me. I wrote various notes in order to use them in the planned manuscript, and the main goal of my trip to Finland was practically achieved.

I continued my travel to Lapland, saw the northern belt of the deciduous forest, where the Rushka and leaf shed had already ended (Figs. 48.9 and 48.10), and drove further north to see the forest of stressed and dwarfed evergreen conifers turning into tundra. Then I drove eastward and after several hours in Lapland I turned to the south in order to see the sequence of the Rushka phenomenon again, this time from north to south in Karelia. On midnight of the third day I arrived back to Kuopio after some 2000 km of driving, and the next day told Jarmo that I would give him a first draft of a manuscript in several days, so he would be able to write his share of the paper and fix my mistakes and gaps in Scandinavian ecology and in entomology (e.g., Lev-Yadun and Holopainen 2009). While all the previous detailed discussions on the evolution of colorful autumn leaves focused on the

Fig. 48.9 The fantastic phenomenon of yellow autumn leaves in Finland or elsewhere in northern Europe ends within a few days

Fig. 48.10 After the yellow leaves fall off the trees in autumn, they may accumulate in large masses on the ground

biological origin or physiological and ecological functions of autumn leaf coloration, Lev-Yadun and Holopainen (2009) focused on an unexplored aspect of its origin in time in relation to past global climatic changes, biological extinctions and migrations starting dozens of millions of years ago, not later than the early Tertiary.

For me, the conspicuous key to the enigma was that while trees and shrubs grow at the same location, they and the different specialist herbivorous insects that attack them are exposed to dramatically different conditions during winter. The facts that stand as a basis for the hypothesis are listed and discussed below. I will start with the facts as they were known in autumn 2008 and soon after. In South and North America and east Asia, autumn landscapes are dominated by many tree and shrub species with red autumn leaves (e.g., Hoch et al. 2001; Lee et al. 2003). Archetti (2009b) published a major corpus of 2368 tree and shrub species with red and yellow autumn leaves that summarized his several years of intensive field work and found that there are many more species with yellow autumn leaves than with red ones. Archetti's (2009b) finds indicated that the significantly fewer species with red autumn leaves manage much better in most temperate ecologies. However, Archetti (2009b) also found that brown autumn leaves are taxonomically more common than red ones. Lev-Yadun and Holopainen (2009) examined the natural distribution of each of the 290 tree species with red autumn leaves listed in Archetti (2009b) and found that most of them grow in North America and eastern Asia. The proportion of tree species with red autumn leaves in the flora and even more dramatically in the landscape of Northern Europe is very small; for example, there are only four indigenous tree species (*Prunus padus*, *Prunus spinosa*, *Sorbus aucuparia*, *Acer platanoides*) reaching their northernmost distribution in Northern Europe (Alanko 2001; Holopainen and Peltonen 2002), and only 24 such species exist in the whole of Europe (according to an examination of the distribution of all the tree species with red autumn leaves listed in Archetti 2009b). By contrast, in eastern North America and east Asia both the proportion in the landscape and the actual number of tree species with red autumn leaves is much greater; there are at least 89 species in a subset of the woody flora of North America (e.g., Lee et al. 2003; and examination of the distribution of the 290 tree species with red autumn leaves listed in Archetti 2009b) and at least 152 species in East Asia.

Lev-Yadun and Holopainen (2009) used the prevalence of red autumn coloration of trees in North America (Lee et al. 2003) and eastern Asia (by the examination of the geographical distribution of the 290 tree species with red autumn leaves listed in Archetti 2009b) *versus* the prevalence of yellow autumn leaves in Northern Europe (Holopainen and Peltonen 2002), along with known gross patterns of migration and extinction during the drastic climatic changes in the Tertiary and the Pleistocene (e.g., Milne and Abbott 2002), as a basis for a new hypothesis. Lev-Yadun and Holopainen (2009) proposed that the solution to the problem of the origin of red autumn leaves in general, and their limited distribution in northern Europe in particular, reflects the well known difference in the extinction histories of trees and their herbivores in eastern North America and east Asia (low extinction rate) and those in northern Europe (much higher extinction rate). If red autumn leaf coloration is the result of various ancient Tertiary adaptations of temperate floras of tropical and subtropical origins to past climates and herbivore faunas, than, while physiological adaptations are habitat and climate dependent, and a geographical shift is sufficient to allow trees to prosper when the climate changes, defense from herbivory is dependent on herbivore pressure (see below). However, the current anti-herbivory adaptations may reflect not only the current herbivore fauna and their predators and parasites, but also many extinct animal species (both vertebrates and insects). Not only did most trees with red leaves become extinct in Europe, but when many of their herbivores also became extinct, the driving selective agents for a costly red autumn coloration also declined.

The Origin of the Current Temperate Tree Flora

The current land biota, with the sharp differences between the adaptations to warm tropical and subtropical regions and cold temperate and arctic regions, is a relatively new phenomenon in geological and evolutionary time scales. After a very long warm period, with much lower thermal gradients between the tropical and polar regions than we are familiar with today (see Axelrod 1966; Tiffney 1985; Graham 1993; Manchester 1999), significant phases of cooling and glaciation alternating with warmer phases began in the mid Tertiary *c.* 35 million years ago, a process that culminated in the Pleistocene, which started about 2.6 million years ago, with its four major glaciation waves (Imbrie and Palmer-Imbrie 1979; Tiffney 1985; Zachos et al. 2001). Such dramatic climatic changes selected for various physiological adaptations, including adaptations to cold environments (e.g., Axelrod 1966; Stebbins 1974; Tiffney 1985; Takhtajan 1991; Delcourt and Delcourt 1993; Graham

1993; Wen 1999). The climatic differences between the tropical and the temperate and boreal regions as we know them today developed gradually following the breakdown of the supercontinent Pangaea and the subsequent closure of the huge ocean Panthalassa, a process that took more than 100 million years.

Several times during the Pleistocene, large areas in Asia, Europe, South- and North America were covered by thick ice sheets and could not support trees or even any plants at all (Imbrie and Palmer-Imbrie 1979; Tiffney 1985; Graham 1993; Wen 1999; Hewitt 2000; Milne and Abbott 2002). During the cold phases, trees and many other organisms survived in warmer regions of lower latitudes, termed refugia (e.g., Bennett et al. 1991; Delcourt and Delcourt 1993; Comes and Kadereit 1998; Hewitt 2000; Milne and Abbott 2002). Some refugia in Europe were relatively northern and inland (e.g., the northern Balkans) where they were exposed to very low winter temperatures that should have had a stronger influence on both plant and herbivore extinction, and some were relatively more southern and climatically milder (e.g., the Iberian Peninsula and southern Italy) (Bennett et al. 1991; Willis and van Andel 2004; Provan and Bennett 2008).

In Europe, northern refugia for broad-leaf trees during glaciations are known for a small number of deciduous species, none of which has red autumn leaves: *Alnus glutinosa* (green autumn leaves), *Betula pendula* (yellow), *Fagus sylvatica* (yellow), *Fraxinus excelsior* (green and sometimes yellow), *Salix* sp. (yellow), *Corylus avellana* (yellow) and *Frangula alnus* (yellow) (Bhagwat and Willis 2008).

Mountain Ridge Direction and the Severity of Ice-Age Extinctions

The reason for the current differences in the global distribution of red *versus* yellow autumn leaves in trees of the temperate regions lies in the geological history of the world in the Cretaceous, Tertiary and Quaternary, the outcome of plate tectonics that determined the directions of mountain ridges at the continental level and by this, indirectly determined the patterns of extinction and re-colonization throughout the Pleistocene with its repeated waves of glaciation and warming.

In eastern North America especially, but also in western North and South America and in east Asia, the direction of the mountain ridges is from north to south (Tiffney 1985; Hewitt 2000; Milne and Abbott 2002; Soltis et al. 2006), the outcome of the directions of continental *versus* ocean floor movements. By contrast, in Europe, the Alps and their eastward extensions form an east-west ridge (Tiffney 1985; Milne and Abbott 2002; Milne 2004; Soltis et al. 2006). Accordingly, in North America and east Asia, when the waves of southward-advancing ice during Pleistocenian phases of glaciation damaged the biota, tree species and their specific and non-specific insect herbivores could migrate to the warmer south in the valleys among the mountains, or along the ridges, according to their specific ecology, and *vice versa* following warming and the retreat of the ice, resulting in the preservation of many ancient Tertiary floral and faunal elements. Naturally, the directions of glaciations and glacial retreats, and of plant and animal migrations were opposite in South America, but still along a north-south trail. In Europe, however, during the repeated drastic climatic changes of the Pleistocene, the biota was trapped again and again between the advancing ice from the north on the one hand and ice from the Alps and their eastern extensions in the south on the other (Imbrie and Palmer-Imbrie 1979), and a larger proportion of the European species assembly (both plants and animals) became extinct, leaving a much smaller number of species that later spread from several refugia during warmer periods to re-occupy landscapes freed from the ice cover (Tiffney 1985; Comes and Kadereit 1998; Milne and Abbott 2002; Milne 2004; Soltis et al. 2006). The great differences in extinction rates between Europe and other continents can be seen in the much smaller number of northern European deciduous tree species compared with eastern North America and eastern Asia (e.g., Milne and Abbott 2002; Lee et al. 2003). Many more Tertiary elements are therefore found in North America and eastern Asia than in northern Europe (Tiffney 1985; Milne and Abbott 2002).

Anachronisms in Plant Adaptations as Evidence of Extinct Faunas

There are several independent sets of evidence for anachronistic adaptations in plants to extinct faunas. The first is the nature of various tropical fruits that are adapted to large mammalian frugivores (Janzen and Martin 1982; Barlow 2000; Guimarães et al. 2008). A second adaptation is the very spiny cacti that were proposed to reflect the extinct megafauna of North America (Janzen 1986), as do other defended North American plant taxa (White 1988; Barlow 2000). A similar phenomenon of spiny plants that reflect extinct large grazers such as auroches and tarpans was also proposed for northwestern Europe (Bakker et al. 2004). A third proposed anachronistic adaptation is of divaricate branching in New Zealand trees and shrubs as a defense from the extinct moas (Greenwood and Atkinson 1977; Diamond 1990; Bond et al. 2004), and in similar plants in Madagascar as defense against the extinct elephant birds (Bond and Silander 2007). Herrera (1985b), when discussing the lack of fine-scale adjustments in the coevolution of woody plants and animal seed dispersers, proposed that a slower species turnover of woody plants over geological time compared with vertebrate seed dispersers may have favored a sort of very diffuse coevolution. In all these cases, the plants may currently use the anachronistic adaptations as functional solutions in a different biological or environmental setting (Janzen and Martin 1982; Janzen 1986; Barlow 2000; Howell et al. 2002; Guimarães et al. 2008). There is no reason to assume that these cited cases of botanical anachronisms are the only ones.

While the examples of anachronistic plant adaptations to extinct faunas given above are for large herbivorous vertebrates, there is no reason to assume that the same is not true for plant-insect interactions. The fact that fossil insects are less extensively studied, and that it is harder to find a specific connection between a fossil insect and its host plant, does not rule out the probability that many insect species probably became extinct during the Pleistocene. There are solid experimental and field data concerning the sensitivity of aphid and other herbivorous insect eggs and all their life stages to very low temperatures (Niemelä 1979; Tenow and Nilssen 1990; Strathdee et al. 1995; Strathdee and Bale 1998). This sensitivity must have resulted in the extinction of many insect species during the drastic climatic changes of the Pleistocene, leaving defaunated floras in all continents, but especially in northern Europe, similarly to the defaunated but still very spiny cacti in North America (e.g., Janzen 1986).

Red Leaves in Trees

The phenomenon of red autumn leaves, spectacular as it is, is only a special case of a much broader picture of the occurance, function and evolution of red leaves. Red coloration in leaves of woody plants is common in three major situations. The first is the young red leaves that are common in the tropics (Juniper 1994; Richards 1996; Dominy et al. 2002; Lee 2007) (Fig. 20.4) as well as in subtropical and even temperate regions (Karageorgou and Manetas 2006; Lev-Yadun et al. 2012). The second is non-senescing leaves of both deciduous and evergreen species that turn red under various physiological stresses, especially those associated with low temperatures (Chalker-Scott 1999; Matile 2000; Feild et al. 2001; Hoch et al. 2001, 2003; Lee 2002; Lee and Gould 2002a, b; Close and Beadle 2003; Gould 2004; Ougham et al. 2005; Hughes and Smith 2007; Hughes 2011). The third is red autumn leaves (e.g., Matile 2000; Archetti 2000, 2009a, b; Hamilton and Brown 2001; Hoch et al. 2001; Lee 2002; Archetti et al. 2009a).

Boreal Shrubs with Red Autumn Leaves

Further support for the hypothesis by Lev-Yadun and Holopainen (2009) of an ancient Tertiary origin of red autumn coloration stems from the fact that dwarf shrubs with red autumn leaves, rather than trees, dominate the northern territories of Scandinavia (Figs. 48.6–48.8). For instance, the deciduous species *Arctostaphylos alpina* with its circumpolar distribution (Hämet-Ahti et al. 1992) is one of the most common dwarf shrub species in mountainous areas of Lapland, and has bright red autumn leaves. In lowlands and forested areas, the deciduous shrubs of *Vaccinium myrtillus* and *V. uliginosum* have reddish or darker brown autumn coloration. Several evergreen dwarf arctic shrub species of northern Europe or Alaska: *Andromeda polifolia*, *Cassiope tetragona*, *Diapensia lapponicum*, *Dryas integrifolia*, *Empetrum nigrum*, *Ledum palustre*, *Oxycoccus microcarpus*, *Pyrola grandiflora*, *Rhododendron lapponicum* and *Vaccinium vitis-idaea* (Oberbauer and Starr 2002) have red winter and spring leaves. The autumn landscape in the treeless far northern parts of both Scandinavia and Alaska is conspicuously dominated by red leaf coloration expressed only by very low shrubs.

There is a critical difference in the sensitivity of trees and shrubs to extremely cold periods and to extinction when drastic climatic changes such as glaciation and later warming occur (e.g., Milne and Abbott 2002). The trees and the insects that occupy their bark in winter (usually as eggs) are exposed to extreme low temperatures that in cold periods may be much lower than −30 °C and even as low as −50 °C, and therefore the insect eggs usually die. By contrast, the shrubs and their overwintering insects are covered by a blanket of snow, a natural igloo, and the shrubs and their insect enemies are thus insulated from most if not all of the terribly low temperatures. Therefore, Lev-Yadun and Holopainen (2009) proposed that the tree species that currently dominate northern Europe were freed from many of their insect enemies during the waves of glaciations, migration to refugia and later quick geographical expansions from small refugia when climate warmed, and the repeated waves of extinction and resulting genetic bottlenecks that occurred in the Pleistocene. All these resulted in the relaxation of the selection on northern European trees for costly chemically defended red autumn leaves. However, the insect-induced selection for red leaves as defense continued for the snow-insulated low shrubs. Trees are much larger and have a much longer life span and generation time than shrubs, which makes trees on an individual and evolutionary level less flexible and more susceptible to both local and global extinction, which leads in turn to the extinction of their specialist insect herbivores. By contrast, the boreal shrubs discussed in Lev-Yadun and Holopainen (2009) not only have much smaller size and shorter generation time, but usually have berries so their seeds can be dispersed over large distances by animals. Shrubs can also manage much better than trees in colder and less productive habitats because they manage much better with shallow, seasonally frozen soils and less fertile habitats, and their low stature allows them to enjoy an insulating snow cover in winter. Shrubs thus could find more refugia than trees in periods of glaciation. All these differences allowed shrubs with red autumn leaves to escape extinction where trees could not survive. Moreover, if red autumn leaves are at least partially an anti-herbivory adaptation (e.g., Archetti 2000; Hamilton and Brown 2001; Archetti and Brown 2004; Lev-Yadun et al. 2004a; Manetas 2006; Lev-Yadun and Gould 2007; Archetti et al. 2009a), the persistence of shrubs during periods of glaciation in the Pleistocene also allowed their insect herbivores to find refuge from extinction, thus continuing their role in selection for red and defended autumn leaves.

The Historical Origin (in Geological Era Perspective) of Red Autumn Coloration

Lev-Yadun and Holopainen (2009) proposed that, since temperate deciduous trees are of ancient (Cretaceous or Tertiary) tropical or sub-tropical origin (Axelrod 1966; Stebbins 1974; Tiffney 1985; Milne and Abbott 2002), it is possible to reconstruct a probable evolutionary route from young (e.g., Richards 1996; Lee and Collins 2001; Lee 2007) and senescing (Lee and Collins 2001) red leaves in tropical trees, through autumn- and winter-red leaves of evergreens (e.g., Chalker-Scott 2002; Hughes and Smith 2007), to red autumn leaves (e.g., Matile 2000; Hoch et al. 2001; Lee et al. 2003) in trees that acquired the deciduous habit during the Tertiary. The fact that, out of 399 tropical tree species studied, some 13.5 % expressed anthocyanin during senescence (Lee and Collins 2001), a ratio identical to the 13.5 % of dicot species with red autumn leaves found by Archetti (2009b) in his broad taxonomic review of current temperate floras, also supports the probable ancient origin of red autumn leaves.

A broad phylogenetic analysis of the origin of red autumn coloration in 2368 tree species indicated that this character evolved independently in temperate trees at least 25 times (Archetti 2009b). There are several theoretically possible periods when red autumn leaves of these deciduous trees could have evolved. First, it can be an ancient, Tertiary defensive adaptation that was further selected for because of physiological reasons during the periods of global cooling that began in the mid-Tertiary (e.g., Zachos et al. 2001). A second period that could have strongly selected for such an adaptation is the Pleistocene (last ~ 2.6 million years) with its dramatic and repeated climatic changes (e.g., Imbrie and Palmer-Imbrie 1979). Finally, it may be a recent Holocene (last 11,000 year.) adaptation. A combination of some of these is also possible.

The question is which of these scenarios is most likely. The repeated evolution of the red autumn leaf color in many tree taxa, a group of organisms that has a long generation time and therefore slow evolution, is a good indication of an ancient origin. Lev-Yadun and Holopainen (2009) posited that the conspicuous differences in distribution of red autumn coloration in eastern North America and eastern Asia (where many taxa have red autumn leaves and where they dominate the landscape) and northern Europe (which is poor in red autumn coloration) are critical for solving this puzzle. If adaptations for low autumn temperatures *per se* were the selective agent for red leaf coloration, it should be expected that the Scandinavian autumn tree landscape would have been as red as the autumn of North America or eastern Asia, but it is yellow. Alternatively, yellow autumn leaves would have dominated the autumn landscape of all continents, but it is not so. Therefore, while it is agreed that anthocyanins provide several physiological solutions under low temperatures (e.g., Matile 2000; Hoch et al. 2001, 2003; Lee 2002; Lee and Gould 2002a, b; Wilkinson et al. 2002; Schaberg et al. 2003, 2008; Ougham et al. 2005; Lev-Yadun and Gould 2007; Archetti et al. 2009a), Lev-Yadun and Holopainen (2009) argued that there is clearly no inherited physiological problem in functioning successfully with yellow autumn leaves under similar low autumn temperature, as seen in *Betula* sp., *Populus* sp. and *Salix* sp. and the many deciduous temperate tree taxa that have yellow autumn leaves in other world parts (e.g., Archetti 2009b). The possibility that the northern European tree taxa with yellow autumn leaves cannot produce anthocyanins should be dismissed because many temperate taxa with yellow autumn leaves have red coloration in various parts of their canopy, for example during spells of cold weather during leaf flush at the beginning of the growing season, or in their reproductive organs. Moreover, many of them regularly have red young leaves in spring or summer (Lev-Yadun et al. 2012), the same leaves that will be yellow in autumn!

The anti-herbivory component of the character of red autumn leaves may partly reflect anachronistic adaptations to Tertiary faunas, some of which became extinct, especially in Europe, but possibly although at a much lower rate in other continents. If this hypothesis is correct, then, this reality significantly reduces the chances of obtaining good theoretical and experimental evidence supporting anti-herbivory hypotheses concerning red autumn leaves using the current biota! Lev-Yadun and Holopainen (2009) were certain that their hypothesis may partly relax the hot debate concerning autumn leaf coloration and coevolutionary hypotheses. As with other hypotheses concerning the functions of colorful autumn leaves, this hypothesis seems to explain only part of the phenomenon, and Lev-Yadun and Holopainen continue to conduct both field and theoretical work to further understand the very complicated, intriguing and wonderful issue of colorful autumn leaves.

Chapter 49
The Shared and Separate Roles of Aposematic (Warning) Coloration and the Co-evolution Hypothesis in Defending Autumn Leaves

Originally, the co-evolutionary hypothesis addressed similarly both red and yellow autumn leaves (Archetti 2000; Hamilton and Brown 2001; Archetti and Brown 2004). However, with the general understanding that yellow leaves usually attract aphids rather than repelling them (Holopainen and Peltonen 2002; Wilkinson et al. 2002; Chittka and Döring 2007; Archetti et al. 2009a; Döring et al. 2009; Holopainen et al. 2009), the second generation of the co-evolutionary hypothesis was very restricted concerning yellow leaves, and focused mostly on red autumn leaves when aphids are concerned (Archetti et al. 2009a). The updated version of the co-evolutionary hypothesis that also includes aposematism, while not neglecting physiology (Archetti et al. 2009a), posited that with red autumn coloration, trees visually signal to all types of insects (including aphids) that migrate to the trees in autumn, about their chemical defense, lower nutritional quality, approaching leaf fall, that they have unsuitable color for insect camouflage, and possibly still unknown other characters that would result in a lower fitness in the attacking insects. In addition, yellow leaves visually signal the same defensive combination to various invertebrate herbivores, except for many aphids. When the yellow autumn leaves belong to toxic taxa, even yellow leaves may signal and even to aphids, that they are chemically defended. The visual signaling of colorful autumn leaves is probably supplemented by an olfactory one (e.g., Holopainen 2008; Blande et al. 2010; Holopainen et al. 2010), but that issue has not been studied yet on a broad scale and deserves much more research attention. The fact that there are good physiological measurements of significant volatile release from yellow autumn leaves (e.g., Keskitalo et al. 2005) supports the possibility of the simultaneous operation of olfactory, along with visual aposematism by these leaves.

The special theoretical aspect of the co-evolutionary hypothesis is that the reduced fitness of the insects is not only of those that land on the trees in autumn, those that are supposed to respond to the tree's signaling, is not only immediate, reducing insect feeding or nutrition in autumn, but it is also related to the reduced development of the next generation that will hatch in the following spring from eggs laid on the trees in the autumn (Archetti 2000; Hamilton and Brown 2001; Archetti

S. Lev-Yadun, *Defensive (anti-herbivory) Coloration in Land Plants*,
DOI 10.1007/978-3-319-42096-7_49

and Brown 2004; Archetti et al. 2009a). A second major general theoretical progress of the co-evolutionary hypothesis was the demonstration of a potentially broad operation of Zahavi's handicap principle in plants. It is important to view this definition in the light of the theoretical understanding of the time, since aposematism was not yet considered to be a handicap then (Guilford and Dawkins 1993), a situation still under debate 15 years later (Summers et al. 2015). I think that there is no theoretical difficulty to identify the overlap between Zahavi's handicap principle and aposematism in many cases of defensive signaling.

Contrasting views on the phenomenon of bright/conspicuous autumn leaf coloration have been presented concerning the hypothesis that these leaves may be aposematic. Archetti (2000), in the first paper presenting the co-evolutionary hypothesis, specifically rejected the possibility that these leaves are visually aposematic in his discussion of the defensive signaling by red and yellow autumn leaves to aphids. In other studies that favored the co-evolutionary signaling hypothesis (Hamilton and Brown 2001; Hagen et al. 2003, 2004; Archetti and Brown 2004, 2006; Archetti and Leather 2005; Brown 2005; Archetti 2007a, b), aposematism was not discussed. Interestingly, Lee and Gould (2002a, b), Lee (2002), Gould (2004), Sherratt et al. (2005), Karageorgou and Manetas (2006), Manetas (2006), Chittka and Döring (2007), Schaefer and Rolshausen (2007a) and Karageorgou et al. (2008) interpreted the co-evolutionary hypothesis of autumn coloration presented in the papers by Archetti (2000), Hamilton and Brown (2001) and Archetti and Brown (2004) as a case of visual aposematism (warning coloration), in spite of the authors' differing view.

When red-colored autumn leaves are well defended by various chemicals, even if these chemicals just happen to be there because they belong to the same biochemical pathway, as proposed by Schaefer and Rolshausen (2006a), or when yellow autumn leaves are toxic, or well defended in other ways, and when leaves of both colors are not nutritious, they should be considered aposematic (Lev-Yadun 2006a, 2009a; Lev-Yadun and Gould 2007, 2009; Archetti et al. 2009a; Archetti 2009a). Archetti (2009a) and Archetti et al. (2009a) accepted that there is an overlap between the co-evolutionary and visual aposematic hypotheses, but the fine details and degree of overlap of the two hypotheses were only partly elaborated at that time (e.g., Archetti 2009a; Archetti et al. 2009a), and a more detailed view was given by Lev-Yadun (2010).

Lev-Yadun (2010) proposed that the relationships between the co-evolutionary and aposematic hypotheses are as follows: the co-evolutionary hypothesis concerning red and yellow autumn leaves indeed equals aposematism for various herbivorous species for which it is effective, excluding the combination of non-toxic, non-defended and nutritious yellow leaves and many aphid species. Furthermore, the co-evolutionary hypothesis is actually a mixture of two types of aposematism: (1) classic aposematism towards animals that feed or refrain from feeding at the time of signaling, and (2) delayed aposematism, because the purpose of the signaling is also to deter egg laying in autumn, and the target insects are mainly the ones that would hatch in the following spring. The indications for defensive volatile signaling by colorful autumn leaves (Keskitalo et al. 2005; Holopainen 2008; Blande et al. 2010; Holopainen et al. 2010) raise the possibility that for at least certain tree species with yellow autumn leaves, the co-evolutionary hypothesis may operate even towards aphids that are attracted to yellow leaves, but in such cases mostly *via* olfactory rather than visual aposematism. A careful examination of both plant and insect species in each studied case is needed in order to make a precise classification of the signaling: simple immediate defense by aposematism/co-evolutionary, only a delayed co-evolutionary relationship, or a combination of these two strategies.

The first generation of the co-evolutionary hypothesis (Archetti 2000; Hamilton and Brown 2001; Archetti and Brown 2004) was very fruitful in stimulating many and diverse theoretical discussions and empirical studies on the defense mechanisms of red and yellow autumn leaves. The possibility that classic aposematism in plants, but in signaling systems other than red and yellow autumn leaves, has a delayed action has not been considered to the best of my knowledge, and needs further study. Studying this aspect of aposematism may be an additional important stimulatory contribution by the co-evolutionary hypothesis.

Chapter 50
Spring *Versus* Autumn or Young *Versus* Old Leaf Colors: Evidence for Different Selective Agents and Evolution in Various Species and Floras

From a global perspective, red leaf coloration in woody plants is common in three major situations. The first is the common young red leaves in the tropics (Juniper 1994; Richards 1996; Dominy et al. 2002; Lee 2007) as well as in subtropical regions (Karageorgou and Manetas 2006; Lev-Yadun et al. 2012). The second is non-senescing leaves of evergreen species that turn red under various physiological stresses, especially those associated with low temperatures (Chalker-Scott 1999; Lee and Gould 2002a, b; Close and Beadle 2003; Ougham et al. 2005; Hughes and Smith 2007; Hughes 2011). Under temperate and boreal conditions, some genotypes of both deciduous and evergreen trees and shrubs show this reddening following exposure to low temperatures (Hughes and Smith 2007; Taulavuori et al. 2011). The third is red autumn leaves (e.g., Matile 2000; Hoch et al. 2001; Archetti 2009b). However, yellow autumn leaves are particularly dominant in northern Europe (Holopainen and Peltonen 2002; Lev-Yadun and Holopainen 2009) and red ones in other temperate parts of the world (Lev-Yadun and Holopainen 2009).

Four types of functional explanations exist for the evolution of non-green spring and autumn leaf coloration: (1) advertisement by colorful autumn leaves for seed dispersers (fruit flag) (Stiles 1982), (2) physiologically oriented (Gould et al. 2002a, b; Ougham et al. 2005), (3) anti-herbivory oriented (Archetti 2000; Hamilton and Brown 2001; Archetti and Brown 2004), and (4) a combination of the last two (e.g., Gould 2004; Manetas 2006; Lev-Yadun and Gould 2007; Archetti et al. 2009a). Concerning autumn leaf coloration, enhanced recovery of foliar nitrogen and other resources by re-translocation to the branches and trunk is attributed to the protection by anthocyanins from photoinhibition and photo-oxidation (Matile 2000; Hoch et al. 2001; Feild et al. 2001; Lee and Gould 2002a, b; Gould et al. 2002a, b; Close and Beadle 2003; Ougham et al. 2005). Scavenging reactive oxygen species and defense from photoinhibition were also proposed as important physiological reason for young red leaves in the tropics and elsewhere (e.g., Lee et al. 1987; Gould 2004). Interestingly, in a comparison of mechanical defenses in 36 woody species with red young leaves with 40 woody species with green young leaves, Chen and Huang (2013) found that species with red young leaves had a much weaker mechanical

S. Lev-Yadun, *Defensive (anti-herbivory) Coloration in Land Plants*,
DOI 10.1007/978-3-319-42096-7_50

defense, indicating that the red ones are better defended chemically, finds supported by the earlier finds of (Karageorgou and Manetas 2006; Karageorgou et al. 2008).

The major hypotheses about anti-herbivory functions of non-green leaf coloration are: (1) young red leaves mimic dead or old ones or just look unlike young ones (Stone 1979; Juniper 1994; Karageorgou et al. 2008); (2) colorful young leaves attract herbivores and divert them from the more costly older ones (Lüttge 1997); (3) young leaves that are not green because of delayed greening are less attractive to herbivores since they are less nutritious (Kursar and Coley 1991, 1992, 2003; Coley and Barone 1996; Numata et al. 2004); (4) Wiens (1978) mentioned personal communication with C. Dodson that suggested that red young leaves in tropical plants may be aposematic, but did not elaborate on this. Red and yellow poisonous autumn leaves are in many cases aposematic and Müllerian and Batesian mimicry rings may occur (Lev-Yadun 2006a, 2009a; Lev-Yadun and Gould 2007, 2009; Archetti 2009a; Archetti et al. 2009a); (5) red autumn leaves (and until the year 2009 also yellow) signal to insects that the trees are well defended (Archetti 2000; Hamilton and Brown 2001; Archetti and Brown 2004; Archetti et al. 2009a) or are a nutrient-poor resource (Sinkkonen 2006a, 2008); (6) leaf coloration is implicated in tritrophic interactions that undermine herbivorous insect camouflage, allowing better visual detection of the herbivorous insects by their predators or avoidance of such leaves by the herbivores (Lev-Yadun et al. 2004a; Lev-Yadun 2006a, 2009a; Lev-Yadun and Gould 2007), or otherwise attracting species whose presence reduces the total herbivore load (Yamazaki 2008a, b); (7) autumn leaves function according to the "defense indication hypothesis" (Schaefer and Rolshausen 2006a) stating that red leaves are chemically defended because anthocyanins correlate with various defensive compounds; (8) the autumn colors signal to herbivores that the leaves are going to be shed soon (Lev-Yadun and Gould 2007). In addition to the above anti-herbivory hypotheses, anthocyanin-based red leaf coloration has been proposed to defend from fungal attacks (Coley and Aide 1989) and experimental data from red tropical leaves (Tellez et al. 2016), and from red or black fruits support this hypothesis (Schaefer et al. 2008; Schaefer 2011). Regarding the origin of autumn colors and regardless of the specific adaptive mechanism for coloration, it has been postulated that red autumn leaves are an ancient Tertiary adaptation to now partially extinct herbivore fauna (Lev-Yadun and Holopainen 2009).

Actual lower herbivory on young red leaves has indeed been shown by Karageorgou and Manetas (2006). Gerchman et al. (2012) found that the conspicuous purple tufts of leaves ("flags"), which often terminate vertical inflorescences in the Mediterranean annual *Salvia viridis* and attract insect pollinators to the flowering patch have an aposematic potential towards herbivores. Maskato et al. (2014) studied in more detail the find by Gerchman et al. (2012) that locusts preferred green cabbage over anthocyanin-rich red cabbage and found that female *Pieris* butterflies avoid laying eggs on anthocyanin-rich red foliage, and that larvae feeding on red cabbage leaves exhibited significantly lower growth rates and a longer duration of larval development. Maskato et al. (2014) thus proposed that from an evolutionary perspective, red foliage color may serve as an honest defensive cue, and I say probably a signal.

Concerning current leaf herbivore fauna and their possible role in selecting for young spring and old autumn leaf coloration, it is certain that spring and autumn faunas are not identical (e.g., Feeny 1970, Dixon and Hopkins 2010). It is highly probable that differences between spring and autumn faunas existed not later than since the emergence of strong seasonality dozens of millions of years ago. It is well-known that in the tropics, young leaves suffer most of the damage by herbivores, while mature and old ones suffer much less (Coley 1980; Coley and Barone 1996; Kursar and Coley 2003; Bixenmann et al. 2013). The basic salient difference between herbivory on young spring and old autumn leaves is that in spring, or in young leaves that emerge in any season in general, significant damage is the outcome of consumption of leaf parts or whole leaves, and in autumn leaves or in old leaves in any season, most of the damage is inflicted by aphids that suck free amino acids and other soluble resources, and leaf tissue consumption by chewing is much less important (e.g., Caldwell et al. 2016). Therefore, even if current leaf herbivory rather than that of the Tertiary was the major selective agent of leaf coloration, because of the differences between the types of damage between young and old ones, it is expected to allow disruptive evolution in various characters including leaf color.

Lev-Yadun et al. (2012) examined the possibility that different selection agents may have operated on spring *versus* autumn leaf coloration. If the same selective agents operated in spring and autumn, it was expected that when spring leaves are red, they should always be red in autumn, and if spring leaves are green, they should always be yellow or green in autumn. Alternatively, if defense rather than physiology or developmental constraints selected for the coloration, spring and autumn colors should be totally random, i.e., spring leaf color should not be directly related to autumn leaf color, and *vice versa.*

Lev-Yadun et al. (2012) examined and documented leaf color in general, but gave specific attention to the presence of red leaf color in young and old leaves of woody plants belonging to three different floras (Finland, central Japan, and Israel). This was done in order to allow for a broad ecological and evolutionary spectrum and avoid the risk of documenting convergent local adaptation or phylogenetically-based adaptation, risks that exist when a single flora is studied. Leaves were designated red when red was present even in parts of the leaves, or not in the whole population. Leaves were not designated red if red was found only next to insect damage (one species in Finland). Although it is clear that the vision of herbivores (including humans) varies among herbivore species and possibly even within species, the visual, dichotomous method selected (the separation of red from green/yellow leaves) has been used successfully in manipulative studies dealing with insect herbivore fitness (Archetti 2009a, b), mathematical modelling (Archetti 2000), and in finding large-scale differences in geographic distribution of species (Hamilton and Brown 2001; Archetti 2009b; Lev-Yadun and Holopainen 2009).

In Finland, there were five deciduous species with green young leaves; four are shrubs and one a tree and all of them have yellow autumn leaves. Lev-Yadun et al. (2012) sampled in Finland 25 species with red young leaves; 15 were shrubs and ten were trees. Seven of the shrub species were evergreens and eight were deciduous, and all ten tree species were deciduous. 14 of the 25 species with red young leaves have red coloration in autumn leaves, although 11 of the 14 species with red in autumn leaves were polymorphic with additional (green, yellow, brown) autumn colors. Notably, color polymorphism was also very common in species with red spring leaves, and the seven (out of 25) species not expressing color polymorphism were either evergreen shrubs or species for which only a small number of the individuals were examined during the field surveys. In the Finnish flora, having red in autumn leaves increases the probability that the color of spring leaves is different from the color of autumn leaves ($\chi^2=15.0$, df=1, $p<0.0005$).

In central Japan, Lev-Yadun et al. (2012) sampled 15 species with green young leaves; 11 were trees and four were shrubs. Seven species were evergreens and eight deciduous. Of these, 13 species had yellow autumn leaves, one had brown, and one had yellow, brown and very occasionally red. Lev-Yadun et al. (2012) sampled 15 species with red young leaves; nine were trees, five were shrubs and one was a climber. Seven species had red autumn leaves, and eight species had yellow autumn leaves. Several evergreen species in central Japan (e.g., *Pittosporum tobira*, *Quercus gilva*, *Rhaphiolepis indica*) had more senescing leaves in spring than in autumn. In these species new and old leaves co-existed. In *R. indica*, since it produces red young leaves and red or yellow old leaves, it was difficult to distinguish between some of the young and the old leaves. In the central Japanese flora, red-in-spring leaves increases the probability that the color of spring leaves is different from the color of autumn leaves (Fisher's exact test $p=0.014$).

In Israel, Lev-Yadun et al. (2012) sampled 14 species with green young leaves; 11 were trees and three were shrubs. Five species were evergreens, eight deciduous and one species was deciduous only in some of the years. All the species with green young leaves had yellow autumn leaves. In Israel, Lev-Yadun et al. (2012) also sampled 19 species with red young leaves; 14 were trees, four were shrubs and one was a climber. Eight were evergreens and 11 were deciduous. Five species of these 19 had red autumn leaves, two species had red and yellow, ten species had only yellow autumn leaves and two had brown autumn leaves (Table 50.1). In Israel, the probability that the color of spring leaves is different from the color of autumn leaves increased by having red spring leaf color ($\chi^2=11.3$, df=1, $p=0.001$).

Concerning the pooling of all three floras, Lev-Yadun et al. (2012) found that spring leaf color affects autumn leaf colors ($\chi^2=20.2$, df=1, $p<0.0005$). Green spring leaves were almost exclusively associated with yellow or even with green autumn leaf colors. Species with red coloration in their autumn leaves almost always had red in spring leaves. However, about half of the species with red spring leaves had yellow autumn leaves. Brown autumn leaves were not common in the species studied but they are common on a global scale (Archetti 2009b).

Anatomical evidence for different selective forces in young *versus* senescing leaf coloration emerged from a broad study of ontogenetic changes of anthocyanin and betacyanins in tropical plants (Lee and Collins 2001). Out of 399 angiosperm species, 44.9% produced anthocyanins in young leaves but only 13.5% during senescence. Of the 35 species that produced anthocyanins both in young and old leaves, several species had different cellular red pigment distribution in spring *versus*

Table 50.1 Color of young and old leaves of trees and shrubs in Israel

Species	Life form	Evergreen/deciduous	Old leaves
Green young leaves			
Amygdalus communis	T	D	Y
Calicotome villosa	S	D	Y
Calotropis procera	T	EG	Y
Ficus carica	T	D	Y
Lycium schweinfurthii	S	D	Y
Nerium oleander	S	EG	Y
Olea europaea	T	EG	Y
Phillyrea latifolia	T	EG	Y
Quercus boissieri	T	D	Y
Quercus ithaburensis	T	D	Y
Rhamnus alaternus	T	EG	Y
Salix acmophylla	T	D	Y
Styrax officinalis	T	D	Y
Ziziphus spina-christi	T	EG/D	Y
Red young leaves			
Acer monspessulanum subsp. *microphyllum*	T/S	D	R
Acer obtusifolium	T	D	Y
Arbutus andrachne	T	EG	Y
Ceratonia siliqua	T	EG	Y
Cercis siliquastrum[a]	T	D	Y
Cotoneaster racemiflorus var. *nummularia*	S	D	R
Crataegus aronia	T	D	B
Laurus nobilis	T	EG	Y
Pistacia atlantica	T	D	Y/R
Pistacia lentiscus[a]	S	EG	Y
Pistacia palaestina	T	D	R
Prunus ursina	T	D	Y/R
Quercus calliprinos	T	EG	B
Quercus libani[E]	T	D	Y
Rhus coriaria	S	D	R
Ricinus communis	T	EG	Y
Rubus sanguineus[a]	S	EG	Y
Smilax aspera	C	EG	Y
Sorbus umbellate	T	D	R

Following Lev-Yadun et al. (2012)

T tree, *S* shrub, *C* climber, *D* deciduous, *EG* evergreen, *Y* yellow, *B* brown, *R* red. [E]=red when young at high elevation, green up to 1700 m a.s.l

[a]Red young leaves also in Greece (Karageorgou et al. 2008)

autumn (Lee and Collins 2001). Similarly, in two temperate species, *Corylus avellana* and *Acer platanoides*, in spring the anthocyanins were produced in the epidermis, while in autumn in the palisade parenchyma (Merzlyak et al. 2008).

Lev-Yadun and Holopainen (2009) proposed that since temperate deciduous trees are of ancient (Cretaceous or Tertiary) tropical or sub-tropical origin, red autumn leaf coloration is an ancient adaptation. In this respect, since central Japan has more ancient Tertiary elements than Finland or Israel, it was expected that various patterns of leaf coloration that can be found in central Japan would not be found in the smaller woody floras of Finland and Israel. Indeed, a situation of plants simultaneously having a mixture of red young and red old leaves was found by Lev-Yadun et al. (2012) in central Japan, but not in the other two floras.

In Finland, most species seem to be capable of synthesizing red leaf pigments in spring, and the five species that do not express red pigmentation in spring leaves typically grow as understory shrubs or small trees in a habitat where vegetation or topography reduce fluctuation in spring and autumn temperatures. A role for successional status on autumnal leaf coloration has already been found by Koike (1990) for early *versus* late-successional trees in Japan. There, early-successional tree species change color first in the inner part of the crown, progressing to the outside. Late-successional tree species change color first in the outer parts and then progress inward. Much more descriptive and experimental research should be done before the role of successional status on leaf color at senescence will be understood.

Since leaf coloration reflects not only anti-herbivory functions but also physiological ones (e.g., Hoch et al. 2001; Lee and Gould 2002a, b; Close and Beadle 2003; Gould 2004; Ougham et al. 2005; Manetas 2006; Lev-Yadun and Gould 2007; Archetti et al. 2009a), local and seasonal physiological factors sometimes have a strong effect on spring and autumn leaf coloration. For instance, in central Finland, young leaves of certain *Betula pendula* individuals show red coloration during periods of cold weather (night frosts of −5 °C) in spring. However, day length in mid-May (spring) is 13 h, about 4 h longer than in mid-September (autumn). Therefore, spring photosynthesizing *B. pendula* leaves could be more easily exposed to photoinhibitory conditions because of low temperatures than in the shorter and darker autumn days. This may cause a stronger need to protect against photoinhibition with anthocyanins (Taulavuori et al. 2011) and to scavenge reactive oxygen in spring than in autumn (e.g., Lee and Gould 2002a, b; Ougham et al. 2005; Archetti et al. 2009a), or else red leaves may just indicate photosynthetically weaker individuals (e.g., Nikiforou et al. 2010). Whether this difference in light levels during frost periods in spring selects for red spring leaves in *B. pendula* but not in autumn has to be tested. Significantly, spring and autumn light conditions are very similar in both Israel and central Japan, which are located at much lower latitudes than Finland, indicating that spring/autumn differences in the need for photoprotection cannot explain why red spring leaf colors are more common in central Japan and in Israel than red autumn leaf colors.

While in very cold regions like Scandinavia, northern Russia and Alaska low spring or even autumn temperatures seem to explain at least part of the factors that selected for red coloration, it is not so in the tropics with their many red young leaves (e.g., Richards 1996; Lee and Collins 2001; Dominy et al. 2002) or in the Mediterranean region (Israel) where evergreen species such as *Pistacia lentiscus*, *Ceratonia siliqua* and *Ricinus communis* produce many red young leaves not only during spring, but also during summer, under night temperatures of at least 23–28 °C, and day temperatures of over 30 °C. In the tropics, red young leaves are part of a very common phenomenon of delayed greening that has been proposed to be associated with low nutritive value of these young leaves as defense from herbivory (Kursar and Coley 1992; Coley and Barone 1996). In central Japan, the evergreen tree *Elaeocarpus sylvestris* has red senescing leaves all year round (Mogi et al. 2000), indicating again that low temperatures are not the sole factor that selects for red leaves.

Dicotyledonous evergreens of the flora of Finland have red or green old leaves. However, in the floras of both central Japan and Israel there are several evergreen species with old yellow leaves. In these species, the yellow leaves are usually formed in small numbers. There are several alternative and not always mutually exclusive explanations for the patterns of yellow leaf formation in these evergreen species: (1) There is no special adaptive value to this pattern (an alternative that Lev-Yadun et al. (2012) did not consider to be true); (2) The scattered pattern results in low levels of insect attacks as the frequency of yellow senescing leaves rich in easily-available nutrients is not sufficient for building large insect populations or for their migration from other places to such leaves, similar to the defensive value of tropical tree species growing in a small proportion of individuals in each forest plot in a very diverse forest (e.g., Janzen 1970). The scattered yellow leaf pattern may thus be an escape from significant herbivory; (3) In toxic species the yellow leaves may serve as a visual aposematic signal (Lev-Yadun 2006a, 2009a; Lev-Yadun and Gould 2007) possibly along with olfactory aposematism; (4) Since yellow leaves commonly attract aphids (Holopainen and Peltonen 2002; Wilkinson et al. 2002; Chittka and Döring 2007; Archetti et al. 2009a), these small yellow islands may divert some aphid species (Holopainen et al. 2009) from other leaves including young ones, and serve for them as an indirect defense. Diversion of herbivores from more valuable mature leaves has been proposed to be the function of colorful young tropical leaves that may attract herbivores and by this divert them from the more costly older ones (Lüttge 1997); (5) The yellow color may undermine herbivorous insect camouflage and expose them to predators if they decide to occupy those, or cause them not to attack and instead seek another host because of the predation risk (Lev-Yadun et al. 2004a). The possibility that the small number of yellow leaves are visually masked by the lush green canopy and that this protects them from herbivory should also be considered.

Although most evergreen species in the data set of Lev-Yadun et al. (2012) senesce old leaves regularly so that the proportion of yellowish leaves remains small, certain central Japanese evergreens like *Gardenia jasminoides* and *Pittosporum tobira* senesce many old leaves at the same time while new ones develop in spring. As the cumulative surface area of the conspicuously colored old leaves is larger than that of the new leaves, for such cases Lev-Yadun et al. (2012) adopted and expanded the hypothesis by Lüttge (1997) and proposed that old yel-

Figs. 50.1–50.2 Young red and yellow old leaves of *Ficus religiosa* trees growing as ornamentals in Tel Aviv, photographed 1 min and 5 m apart. The tree with the red young leaves shed his yellow leaves several days before the one with the yellow leaves. This pattern of old yellow and young red leaves is common in this species, indicating that the evolution of leaf color in young versus old leaves may reflect different selective agents

low leaves may guide winged insect herbivores to land on dying leaves, regardless of the nutritive value of these. Alternatively, brown-red old leaves of certain central Japanese species (*Quercus gilva*, *Raphiolephis indica*) may create or strengthen the impression that plant foliage is a suboptimal long-term habitat for herbivorous insects because the leaves may be shed soon (e.g., Lev-Yadun and Gould 2007), or that a large part of the foliage is of low nutritive value or give the impression of dead plant tissues (e.g., Stone 1979; Juniper 1994; Chittka and Döring 2007; Karageorgou et al. 2008; Sinkkonen 2008). Even more, in species with many young leaves that are greenish-yellow in spring, the simultaneous mass senescence of brown-red old leaves could disguise the young leaves of high nutritive value from spring-migrating herbivores, as the foliage is a mixture of small patches of potentially suitable habitats in the midst of large patches of nutritionally poor dead or dying foliage. As in the case of the role of successional status on leaf color (e.g., Koike 1990), many descriptive and experimental studies must be carried out before the potential defensive role of the color of senescing evergreen leaves can be understood.

Lev-Yadun et al. (2012) concluded that on average, in about 70 % of the studied species, spring and autumn leaves have different colors, and in about 20 % autumn leaves have more than one color. It is unknown what the various historical agents selecting for young *versus* senescing leaf color were for each species. The data presented in Lev-Yadun et al. (2012) strongly support the hypothesis that in woody dicotyledons there are many cases where the selective agents on spring *versus* autumn, or on young *versus* old leaf colors are different (Figs. 50.1–50.2). In addition, their data partially support the hypotheses positing that both plant physiology and plant-herbivore interactions have had a role in the evolution of spring and autumn leaf colors.

Chapter 51
How Red Is the Red Autumn Leaf Herring and Did It Lose Its Red Color?

The term "red herring" is commonly used to oppose a proposed hypothesis considered by the critics to be a logical or factual fallacy. Pointing to and discussing problems in hypotheses is essential to science, and the use of the title red herring stimulates people to think, conduct measurements and experiments, to construct alternative hypotheses, or respond by writing. However, some hypotheses that were tagged as red herrings were later proven not only to be true, losing their red-herring status, but also to stimulate significant progress. Others remain with the red-herring tag, be it a right or a wrong tag.

A very famous and currently broadly accepted (sometimes with specific reservations) evolutionary signaling hypothesis that first faced very strong opposition was Zahavi's "handicap principle" (Zahavi 1975, 1977). One of the most significant scientists in theoretical ecology and evolution, the late John Maynard Smith strongly opposed it (Maynard Smith 1976), and only with the passing years gradually accepted it (Harper 2006). Only 15 years later, when a mathematical solution for the problem was demonstrated (Grafen 1990), its status was changed. The case of Zahavi's handicap principle is an important lesson that a herring may not always be so red, but rather pink.

The plant signaling hypothesis of the co-evolution of red and yellow autumn leaves, positing that bright autumn leaf colors signal herbivores, especially aphids, about the defensive qualities of the trees (Archetti 2000; Hamilton and Brown 2001; Archetti and Brown 2004) [also known as the "leaf signal theory" (Schaefer and Wilkinson 2004)], which has been hotly discussed and debated since its publication, was awarded two red herrings (Schaefer and Wilkinson 2004; White 2009). While the co-evolutionary hypothesis in its original form (Archetti 2000; Hamilton and Brown 2001; Archetti and Brown 2004) certainly did not explain all the aspects of the very complicated and taxonomically and geographically diverse phenomenon of red and yellow autumn leaves (e.g., Archetti 2009a, b; Archetti et al. 2009a), it seemed to explain part of this very broad, complicated and diverse phenomenon (Archetti et al. 2009a; Lev-Yadun and Holopainen 2009; Lev-Yadun 2010; Schaefer and Ruxton 2011).

S. Lev-Yadun, *Defensive (anti-herbivory) Coloration in Land Plants*,
DOI 10.1007/978-3-319-42096-7_51

Following the clear evidence that many aphids are attracted to yellow leaves rather than repelled by them (Holopainen and Peltonen 2002; Wilkinson et al. 2002; Chittka and Döring 2007; Döring et al. 2009; Holopainen et al. 2009) it became broadly recognized that yellow and red autumn leaves have a different biology including cueing/signaling to herbivorous insects, and that focusing on the differences between yellow and red autumn leaves is essential to progress in understanding the very complicated and multi-adaptive autumn leaf color phenomenon (e.g., Chittka and Döring 2007; Lev-Yadun and Gould 2007; Archetti 2009a, b; Archetti et al. 2009a; Lev-Yadun and Holopainen 2009; Lev-Yadun 2010; Schaefer and Ruxton 2011). Lev-Yadun and Holopainen (2011) discussed the question of how red the red herring tag given to the co-evolutionary hypothesis was and concluded that it was not really red, especially after the first generation of the hypothesis was amended and upgraded by Archetti et al. (2009a).

H. Martin Schaefer was the most critical figure towards the first generation of the autumn leaf co-evolution hypothesis (e.g., Schaefer and Wilkinson 2004; Schaefer and Rolshausen 2006a, b, 2007a, b; Rolshausen and Schaefer 2007; Schaefer and Gould 2007). His comments, along with comments by others, helped to formulate the upgraded second generation of the co-evolution hypothesis (i.e., Archetti et al. 2009a). Later, as a co-author in the very, if not the most, important book about plant-animal signaling (Schaefer and Ruxton 2011), the tone in the elegant review and analysis of the issue was dramatically more moderate. This welcome change, reflecting a decade of considerable scientific effort and progress in this issue, seems to remove the major red-herring tag from the co-evolutionary hypothesis.

The second red-herring tag to the co-evolutionary hypothesis (White 2009) was based on a mistake in understanding the basic facts of red or yellow autumn leaf coloration. The real facts are that in most tree and shrub species with conspicuous autumn leaf colors the color of autumn leaves changes from green to yellow (Archetti 2009b). In a smaller number of species the autumn leaves change color from green to red (Archetti 2009a; Lev-Yadun and Holopainen 2009). Very few species shed green (e.g., *Alnus glutinosa*) leaves (Archetti 2009b; Lev-Yadun et al. 2012). White (2009), however, erroneously posited that leaves turn red after they are yellow, and his logic argument (based on wrong assumptions) was that when the leaves turn red they have already lost their nitrogen in the form of free amino acids during the yellow stage, and that there is no need for signaling at that late stage of leaf life. White's (2009) statement was simply unfounded as the situation of changing autumn leaf color from green to yellow and later to red is a very rare situation, usually found in individual leaves or individual ornamental trees grown because of this attractive character, and has never been well documented, even for the rare cases when it happens. For instance, in *Sorbus aucuparia*, branches of an individual may have leaves turning from green to red, while in other branches green leaves turn to yellow and both leaf types may eventually turn brown. Sometimes, red leaves in a generally yellow tree reflect dying branches, as in the case of Norway maple *Acer plantoides* (Sinkkonen 2008). All these are exceptions rather than the rule. Again, there was no reason to tag the co-evolutionary hypothesis as a red herring on this unfounded or anecdotal basis.

Skepticism is essential for good science, but skeptics should also take their own skeptical ideas with a grain of salt. A red herring tag can be very helpful in promoting better research because it intrigues scientists, but may also sometimes unnecessarily put an end to or block a correct innovative hypothesis for years or even decades. There are cases like the phenomenon of colorful autumn leaves in which hot debates just indicate that the biological situation is so complicated that all sides involved in the argument don't fully understand it, or naively think that a specific adaptation in a species or even a group of species they know very well applies to all other species. While the red herring title sells manuscripts well, Lev-Yadun and Holopainen (2011) suggested being careful with its use, and reserving it only for extreme cases since red herrings may turn pink, as happened to the autumn leaf color co-evolutionary hypothesis.

Chapter 52
Defensive Animal and Animal Action Mimicry by Plants

It is widely accepted that quite a number of invertebrate and vertebrate animal species visually mimic or masquerade as plant parts and by this gain protection from predators or crypsis from their unaware prey (Cott 1940; Wickler 1968; Edmunds 1974; Ruxton et al. 2004; Purser 2003; Stevens 2016). Classic examples listed in Cott (1940) include: several species of fish and crabs that resemble algae; geckos and moths that look like lichens; many insects, amphibians and reptiles that masquerade as leaves; tree bark mimicked by spiders, moths, beetles, amphibians, lizards and birds; stick-insects (Phasmida) that resemble branches, and flower masquerade by thomisid spiders and bugs.

Much less attention has been paid to mimicry of animals by plants. The best-known case is of orchids of the genus *Ophrys* and several other orchid taxa that mimic female bees or wasps in size, shape, color, texture, and as was found later especially odor, and are thus pollinated by species-specific solitary males that are attracted to the flowers because they consider them to be attractive females (Kullenberg 1950, 1956, 1961; Wickler 1968; Wiens 1978; Dafni 1984; Ayasse et al. 2000; Schiestl et al. 2000; Schiestl 2005; Jersáková et al. 2006; Vereecken and Schiestl 2009). Although the *Ophrys* mimicry for the sake of pollination hypothesis is accepted by many, it may in parallel be considered to belong to a different type of deception based on the exploitation of perceptual biases in animals (Schaefer and Ruxton 2009). In either case, whether it is a signal or an exploitation of perceptual biases, the deception operates well and attracts sufficient pollinators, indicating that animal mimicry can be beneficial to plant fitness. A probable simultaneous defensive function by the visual component of this bee or wasp mimicry (Lev-Yadun and Ne'eman 2012) will be discussed later.

Wiens (1978), in the most important essay ever published on plant mimicry, estimated that about 5 % of the land plants are mimetic, and listed several types of protective (anti-herbivory) plant mimicry. Wiens (1978) also stated that defense by means of predator mimicry as found in animals is unlikely to be found in plants. In the following chapters I will show that there is sufficient evidence to posit that this latter statement by Wiens (1978) was wrong, even though the role of defensive,

S. Lev-Yadun, *Defensive (anti-herbivory) Coloration in Land Plants*,
DOI 10.1007/978-3-319-42096-7_52

anti-herbivory animal mimicry by plants has received very little attention, and even this was sometimes partly or even fully anecdotal.

Herbivores may avoid feeding on plants that have parts mimicking animals, because: (1) the newly arriving herbivores may suffer from food exhaustion or limitation if they use plants that have already been occupied by herbivores (Hilker and Meiners 2011), (2) in damaged plants various anti-herbivore defenses such as defensive chemicals, physical toughness and nectar production from extra-floral nectar glands for the attraction of body-guarding ants are commonly induced (e.g., Kessler and Baldwin 2001; Heil et al. 2001; Karban 2015), (3) large herbivores may refrain from consuming plants mimicking insects because the plants seem to be infested, especially if the "insects" seem to be distasteful, toxic, or aggressive (e.g., poisonous caterpillars, ants, bees) (Lev-Yadun and Inbar 2002; Lev-Yadun 2009d; Yamazaki 2010; Lev-Yadun and Ne'eman 2012; Aviezer and Lev-Yadun 2015; Yamazaki and Lev-Yadun 2015), (4) damaged plants may be a visual cue for insect predators (Niemelä and Tuomi 1987) as well as emit olfactory cues that attract predators (e.g., Kessler and Baldwin 2001; Karban 2015), and (5) the visual mimicry of arthropods may attract various visually oriented predators that may risk various herbivores.

It is probable that various types of defensive animal mimicry by plants may trick animals into behaving according to the plant's interests, just as they are tricked by bee mimicry of orchid flowers during pollination (e.g., Dafni 1984; Schiestl 2005; Jersáková et al. 2006; Jürgens and Shuttleworth 2016). The earliest proposed cases of defensive animal mimicry by plants are more than 100 years old, but they were practically forgotten. They include bee mimicry by *Ophrys* species that was proposed to scare off insects (Brown 1831; mentioned in Wickler 1968 without giving the reference), insect mimicry by dark flowers in umbels of *Daucus carota* (Detto 1905; cited in Yeo 1972) to discourage grazing animals, and Robinson (1910) that proposed that visual bee-mimicry by orchid flowers has not evolved for pollinator attraction, but rather to deter grazing cows and other animals (cited in Rolfe 1910). Defensive animal mimicry by plants exists in several forms: (1) feeding damage, (2) tunneling damage, (3) butterfly eggs, (4) caterpillars, (5) aphids, (6) ants, (7) beetles, (8) spider web, (9) bee and wasp, (10) gall midges, (11) arthropod wing movement, (12) snake, (13) eye spots, and (14) feces and carrion. A related issue is of extended phenotypes, including either live or dead insects, an issue that will be discussed later.

Chapter 53
Caterpillar and Other Herbivore Feeding Damage Mimicry as Defense

I will begin the detailed discussion of defensive animal mimicry by plants with the mimicry of animal feeding damage, because it was my first hypothesis on visual anti-herbivory defense in plants. I also do it in order to demonstrate the common situation of several simultaneous probable functions for a single trait. I was still an M.Sc. student, when during my regular weekly reading of new biology journal issues, I came across a paper by Heinrich and Collins (1983) about the feeding and hiding habits of defended and visually conspicuous aposematic, *versus* non-defended and non-conspicuous caterpillars. Since I regularly examined the reference lists in many of the papers I read, I found an earlier relevant paper by Heinrich (1979) on this issue. From those two papers I understood that many caterpillars do not eat leaves just as they reach them, and that birds are a fierce enemy of many herbivorous caterpillar species. Twenty years later, I used these facts along with other facts to propose another (but related) hypothesis of plants undermining herbivorous insect camouflage (Lev-Yadun et al. 2004a), that was described and discussed in detail in a previous chapter. However, at that specific moment, while I was still sitting in the library, it caused me to think that this was probably the reason for the existence of the conspicuous holes found as a normal development in the huge leaves of the tropical climber *Monstera deliciosa* (see their description in Melville and Wrigley 1969) commonly grown as an ornamental (Fig. 53.1). I thought that these leaves mimic caterpillar damage and by this attract insectivorous birds because the holes cue such birds about potential food availability, and while the attracted birds patroll there, they will find and consume various herbivorous invertebrates that occupy the leaves.

Since at that time I was also greatly interested in plant biomechanics and functional plant architecture, I figured that the holes may also allow the illumination of lower leaves for better photosynthesis, as do lobed leaves in dense temperate forests (e.g., Horn 1971), influence boundary layer thickness and leaf water economy (e.g., Gates 1980) and lower the mechanical impact of strong wind and especially of rain that can drain through the holes, partly functioning like drip tips that are common in tropical leaves (e.g., Richards 1996). Thirty years later I realized that allowing water

S. Lev-Yadun, *Defensive (anti-herbivory) Coloration in Land Plants*,
DOI 10.1007/978-3-319-42096-7_53

Fig. 53.1 A leaf of *Monstera deliciosa* (Swiss cheese plant) with its typical holes. Mimicking of caterpillar feeding damage by this plant was my first visual defensive hypothesis, formulated when I was a M.Sc. student in the year 1983

to drip through the holes in *Monstera* leaves had been proposed already in the nineteenth century, and that reducing the thickness of the boundary layer was proposed in 1977 (see Muir 2013 and citations therein). Because of the complexity of the issue of leaf biophysics (see Vogel 2012), because I had more urgent papers to write, and because I was simply not yet ripe as a scientist to bring the theoretical idea and discussion to the needed level in a text written in decent English, I opened a file for the reprints of relevant papers and for my notes and waited for the appropriate time.

This time came, but to other people who published the same idea but on other taxa (Moraceae) with much smaller lobed leaves (Fig. 53.2). Niemelä and Tuomi (1987), without discussing the biophysical aspects that are indeed much less important in smaller leaves, proposed that if visually searching herbivores avoid previously colonized host plants, or if such leaves attract predators and parasitoids that prey on the herbivores of these plants, then such mimicry would be beneficial to the plant. They published it while I was already a Ph.D. student, and in spite of the disappointment, it gave me further confidence that sometimes my theoretical ideas might not be too daring. More than 20 years later I had the pleasure of meeting Professor Pekka Niemelä several times in his office in Turku, southern Finland, and we spent some wonderful hours in his university's tropical botanical garden greenhouses, discussing defensive plant morphology and coloration to our mutual pleasure. Currently we are collaborating on visual defenses from herbivory in various types of leaves.

In general, sun leaves of many species with lobing growing in the vicinity of Denver (Colorado) have more pronounced lobing than the shade leaves of the same plants (Talbert and Holch 1957). While these authors did not propose an explanation,

Fig. 53.2 A leaf of *Morus alba* (white mulberry) showing its special juvenile or wounding related morphology. Niemelä and Tuomi (1987) proposed that the lobes mimic caterpillar damage for defense from herbivory

Horn (1971) gave a convincing physiological-ecological explanation for leaf lobing (illumination of deeper leaf layers of the canopy). However, this non-random distribution of lobed leaves in the exterior of plant canopies seems also to agree with the Niemelä and Tuomi (1987) hypothesis of attracting insectivorous birds to the plants, since they will first see these external leaves and may think that the lobing cues about caterpillar activity and investigate the plants. The hypothesis of feeding damage mimicry by leaves proposed by Niemelä and Tuomi (1987) was accepted by Brown and Lawton (1991) and it seems to characterize various plants in many floras, although not a single flora was investigated in depth from the morphological point of view for this caracter.

A related morphology, of leaves in certain palms with distal ends that appear to be partly eaten was also proposed as defense from herbivores by Dirzo (2002). Hansen et al. (2003) while studying heterophyllous woody species of the Mascarene Islands (Réunion, Mauritius and Rodrigues) in relation to herbivory by the recently extinct giant tortoises and large flightless birds, proposed that red leaf margins, midribs and petioles may mimic damage by browsing and may be considered by such herbivores not worth further consumption.

Chapter 54
Tunneling Damage Mimicry

Insect damage mimicry, especially of tunneling (Fig. 26.2), was the first specific ecological hypothesis that tried to explain the function, evolution and ecology of white leaf variegation. The first to discuss in detail the function of white leaf variegation in the context of possible aposematism was Smith (1986), and while he rejected the aposematic hypothesis for the species (*Byttneria aculeata*) he studied, he gave a clear and detailed formulation of the aposematic hypothesis for poisonous plants: "The benefits to the plant of chemical defense against herbivores would be greater if herbivores avoided such plants altogether, rather than testing leaves for palatability, and so causing some damage. A distinct leaf color pattern linked with chemical defense might function in this way. Polymorphism for leaf color should then coincide with polymorphisms for chemical defense. Müllerian and Batesian mimicry could result in evolution of similar patterns of variegation, with or without associated toxicity, among other species which have herbivore species in common with the model species". Smith (1986) found that the variegated morph of *B. aculeata* was attacked less by herbivorous insects than the non-variegated one.

Mimicry of tunneling insect damage was also one of the several types of proposed defensive functions of white leaf variegation of *Silybum marianum* (Lev-Yadun 2003a), making it look as if it were already infested to prevent further insect attacks (Figs. 1, 19.1 and 26.2). Originally, this was the first potential defensive function of the variegation of this species that I thought about, but later, aposematism took priority because the three annual plant species of the Near Eastern flora that express this morphology to the highest extent (*Silybum marianum*, *Notobasis syriaca* and *Scolymus maculatus*) are all very spiny (Lev-Yadun 2003a, 2006a, 2009a, b, c, 2011, 2014a). Lee (2007) also proposed that certain types of white variegation may serve as a defense from herbivory by mimicking leaf infestation. Soltau et al. (2009) experimentally manipulated the visual appearance of *Caladium steudneriifolium* (Araceae), an understory plant from the *Podocarpus* National Park in southeast Ecuador, by painting artificial white variegation on non-variegated green leaves. The leaves of *C. steudneriifolium* are either plain green or patterned with whitish variegation. In nature, about a third of the leaves are variegated and

S. Lev-Yadun, *Defensive (anti-herbivory) Coloration in Land Plants*,
DOI 10.1007/978-3-319-42096-7_54

both morphs are frequently attacked by mining moth caterpillars. The variegated zones of the leaves strongly resemble recent mining damage and was hypothesized to mimic recent mining attacks. Infestation was found to be 4–12 times higher in plain green leaves than for variegated ones. They studied the level of herbivore damage and showed that painting artificial white variegation on plain green leaves resulted in a reduction in attacks from 7.88 to 0.41 %, leading them to propose that the variegation is probably the mimicry of mining damage to deter ovipositing moths (Soltau et al. 2009). Yamazaki (2010) also proposed that actual tunneling damage in the form of white variegation may deter herbivores.

There is solid evidence that plants infested by insects are avoided by various other insects because of several risks: cannibalism or interspecific predation, competition, induction of host defensive mechanisms that reduce its palatability and increase its toxicity, and in the case of leaves, flowers, fruits and young branches, also the risk of organ shed (habitat destruction) (e.g., Addicott 1982; Lev-Yadun and Gould 2007; Karban 2007; Yamazaki 2010; Schaefer and Ruxton 2011). The common phenomenon of wounded leaves signaling *via* odor to attract predators and parasitoids (e.g., Kessler and Baldwin 2001; De Moraes et al. 2001; Kappers et al. 2005) should also be considered in plants with real infestation damage. For instance, Finch and Jones (1989) reported that large colonies of the cabbage aphid *Brevicoryne brassicae* and the peach aphid *Myzus presicae* deter ovipositing by the root fly *Delia radicum*. Inbar et al. (1999) demonstrated that homopterans (whiteflies) not only alter adult cabbage looper (*Trichoplusia ni*) host selection, but also actually reduce the feeding efficiency of their offspring. Thus, variegation that mimics herbivore damage might serve as mimicry of an already infected or infested leaf and deter female insects from laying eggs (see Smith 1986; Soltau et al. 2009).

Chapter 55
Butterfly Egg Mimicry

For many years, the best known case of defensive animal mimicry by plants that was also tested experimentally was of butterfly egg mimicry. It has been proposed to reduce egg laying by *Heliconius* butterflies on *Passiflora* species, but to also operate for other plant and butterfly taxa (Benson et al. 1975; Gilbert 1980, 1982; Shapiro 1981a, b; Williams and Gilbert 1981). It was a development of the emerging understanding that butterflies can assess the presence of butterfly eggs on plants (e.g., Prokopy 1972; Rothschild and Schoonhoven 1977; Rausher 1979). Benson et al. (1975) studied interactions between Heliconiini butterflies and many plant species belonging to several genera of the Passifloraceae, showing several types of egg mimicry by tendril meristems, bulbous yellow petiolar glands, yellow deciduous flower buds, and yellow nectar glands on the abaxial leaf side (Fig. 55.1), and suggested that all these are aimed at defending from butterfly egg laying. Williams and Gilbert (1981) experimentally showed that *Heliconius* butterflies are less likely to oviposit on *Passiflora cyanea* and *P. oerstedii* host plants that have mimetic egglike structures. Shapiro (1981a) showed that egg mimics of *Streptanthus breweri* (Brassicaceae) in the shape of pigmented callosities on its upper leaves deter oviposition by *Pieris sisymbrii*. Removal of these egg mimics in the field significantly increased the probability of oviposition. Shapiro (1981b) showed that the same is true in another member of the genus *S. glandulosus*. Gilbert (1982) in his well-illustrated review of the issue in the genus *Passiflora* explained the multitudes of defense from *Heliconius* caterpillars, which include bodyguards in the shape of ants attracted by extrafloral nectar secration, changing leaf shape with time, mimicking non-host plants when the plants are juvenile, and having hooked trichomes that capture caterpillars. These tested cases of visual animal mimicry by plants as defense from herbivory provide a strong support for other phenomena of visual defensive animal mimicry by plants. I think that some plant taxa by colorful spots along leaf margins (Fig. 55.2) may mimic aposematic spines towards vertebrate herbivores and insect eggs towards arthropods. In this case, they may by this egg mimicry also attract predators and parasitoids.

S. Lev-Yadun, *Defensive (anti-herbivory) Coloration in Land Plants*,
DOI 10.1007/978-3-319-42096-7_55

Fig. 55.1 Butterfly egg mimicry in the underside of a *Passiflora* sp. leaf growing in a botanical garden in Konstanz, south Germany

Fig. 55.2 *Red spots* along leaf margins of the Israeli coastal plant *Aegialophila pumilio*. They may mimic aposematic spines towards vertebrate herbivores and insect eggs towards arthropods

Chapter 56
Caterpillar Mimicry

Defensive caterpillar mimicry by plants was first suggested by Rothschild (1974, 1984) concerning the shape of stipules found along the branches of *Passiflora caerulae* that look like caterpillars, slugs or snails climbing along the stems. Soon after, Benson et al. (1975) proposed that the yellow-orange serrulate stipules of *Passiflora platyloba* and *P. maliformi* seem to mimic young heliconian caterpillars, but all this was not known to me till the year 2001.

The moment that I understood that plants may mimic colorful aposematic caterpillars is related both to the study of the origin of agriculture (legume domestication) and to defensive plant coloration. In the late spring of 1999, while I was a scientist in the Department of Natural Resources, the Agricultural Research Organization (ARO), I conducted a field survey of wild legumes as part of the study of the origin of agriculture in the Near East about 10,600 years ago (Lev-Yadun et al. 2000), including a broad study of wild genetic resources of peas for the study led by my colleague Shahal Abbo (e.g., Abbo et al. 2008, 2013a, b). At that time I had already been paying regular attention to color patterns of plants as a way of camouflage on one hand and as aposematic (warning) coloration on the other. On that special day I sampled plants at the bottom of a dormant volcano (Mount Odem) in the Golan Heights, Israel, not far from the Syrian border. The diverse plant species growing there expressed many color patterns in leaves, stems, flowers, fruits and seeds, and I was looking for both the variation, and if possible also for some types of regularity of color patterns. Suddenly I saw a nice large caterpillar among the plants. It was yellowish with several reddish-purple spots along its body, a typical colorful poisonous and aposematic caterpillar (Fig. 56.1). A second look indicated that it was a mature pod of the wild legume *Pisum humile*, a species that was recognized by a classic genetic study to be the immediate wild progenitor of the domesticated pea (Ben-Ze'ev and Zohary 1973). Interestingly, the defensive pod coloration was selected against in the domesticated pea (Aviezer and Lev-Yadun 2015). I had examined that population of *P. humile* plants for several consecutive years, always some weeks after the flowering season when its flowers had disappeared, and therefore, I considered them to be specimens of *P. fulvum*, a species in

S. Lev-Yadun, *Defensive (anti-herbivory) Coloration in Land Plants*,
DOI 10.1007/978-3-319-42096-7_56

Fig. 56.1 A mature pod of *Pisum humile*, the wild projenitor of *Pisum sativum* that was domesticated some 10,500 years ago. When I saw it I thought for several seconds that it is a caterpillar. The photo-lab owner who printed the pictures I took asked me "What are these worms?" and a professor of entomology from Padova, Italy asked me in a lecture I gave there "What are these caterpillars?" Two very good indications that it is a very good caterpillar mimic

which the pods use defensive coloration for camouflage (see Aviezer and Lev-Yadun 2015; and the chapter about camouflage) rather than conspicuous caterpillar mimicry. Therefore, the paper by Lev-Yadun and Inbar (2002) included a taxonomic mistake.

Plants that visually and olfactorilly mimic insects were especially known as an adaptation for pollinator attraction (e.g., Wickler 1968; Dafni 1984), and here was a wonderful case of caterpillar mimicry that fooled me for a second, and clearly did not serve for either pollinator attraction or seed dispersal. Members of the genus *Pisum* disperse their seeds by quickly twisting their dry pods till it ends with an explosion (Fahn and Zohary 1955) (Fig. 56.2) and so mimicry to attract seed dispersers is not relevant. After the photo-lab owner who printed the pictures I took asked me "What are these worms?" I knew that it was a good mimicry type, and when 15 years later, a professor of entomology from Padova, Italy asked me in a lecture I gave there "What are these caterpillars?" I was further convinced that these are very good mimics.

In the next two springs, I documented in various habitats in Israel several additional wild legume species with pods that putatively mimic caterpillars. I found that the immature pods of several wild annual Near Eastern legumes (*Lathyrus ochrus*, *Pisum humile*, *Vicia peregrina*) have conspicuous red to purple spots, arranged

Fig. 56.2 A twisted pod of *Pisum elatius* after it dispersed its seeds by quick explosion

Figs. 56.3–56.4 Two caterpillar mimicking *Vicia peregrina* pod morphs, southern Mount Carmel, Israel

along the pods, which appears to mimic lepidopteran caterpillars. In one of the species (*V. peregrina*) two distinct morphs were found. Red spots characterize the first, resembling those of *L. ochrus* and *P. humile*, and red circles with green centers characterize the second morph (Figs. 56.3–56.4). Other wild legume species har a solid red-purple line along the pods (Fig. 56.5). Since it was an entomological-related issue, in the year 2001 I presented my ideas and finds to Moshe (Moshik) Inbar, our enthusiastic departmental entomologist, and asked him to join me in this study that progressed very well to proposing several types of defensive animal mimicry by plants (Lev-Yadun and Inbar 2002; and see John Whitfield, "Plants feign infestation" in Nature science update, November 14, 2002).

Following the initial suggestion that the pods of several wild legume species of the flora of Israel mimic aposematic caterpillars (Lev-Yadun and Inbar 2002) (Figs. 56.1, 56.3–56.4) and previous suggestions for defensive caterpillar mimicry by plants (Rothschild 1974, 1984; Benson et al. 1975), Aviezer and Lev-Yadun

Fig. 56.5 A pod of *Vicia* sp. with a *red-purple* straight line along it from the coastal plain of Israel, about 10 km north of Tel Aviv

decided to further pursue this issue and examined the pods of all three wild *Pisum* taxa growing in Israel. Aviezer and Lev-Yadun (2015) found that all the very young pods and some fully grown ones of all *Pisum* taxa are green, a color that provides a basic camouflage within green canopies. In general, the defensive pod coloration other than green does not appear at once and its ontogeny takes some days. In all *Pisum* taxa and individuals that expressed red/purple/black pod pigmentation, there was a stronger dark pigmentation on the side exposed to the sun, i.e., facing the open area. Several seasons of field work and some common garden experiments showed that many of the pods of *P. elatius* probably mimic several animal types and not only aposematic caterpillars, all these as a probable defense from herbivory (Aviezer and Lev-Yadun 2015). Two of the color patterns found in the pods of *P. elatius* were similar to those of *P. humile* (plain green and a line of red/purple spots) (Figs. 56.6, 56.7 and 56.8). Four other red-purple pod coloration types were unique to *P. elatius*: (1) An intermediate situation between spots and a line (Fig. 56.9), (2) A straight line (Fig. 56.10), (3) A zigzag line (Figs. 56.11 and 56.12), and (4) A pod purple all over (Fig. 56.13).

The common spot coloration along the pods of *P. humile* and *P. elatius* causes them to bear a strong resemblance to aposematic caterpillars. Such pods mimic the general shape, size and color of lepidopteran caterpillars (Fig. 56.14) ornamented with spiracles or other spots on their sides, such as those of a pieride moth (Pieridae) (Lev-Yadun and Inbar 2002; Aviezer and Lev-Yadun 2015). The straight line coloration on some pods of *P. elatius* may also mimic the coloration of an aposematic caterpillar, although it can simultaneously mimic the dorsal coloration of certain individuals of the common local Israeli viper (this will be discussed below). The common morph of pods with a line of spots of red/purple coloration has been suggested to be a case of aposematic caterpillar mimicry, a type of Batesian mimicry,

Fig. 56.6 A young pod of *Pisum elatius* showing the beginning of the expression of a line of *red* spots

Fig. 56.7 Two still young pods of *Pisum elatius* showing the progress of the expression of a line of *red/purple* spots

and to serve as a defense mechanism of the pods against herbivory by various mammalian species that feed on the plant in general and on the pods in particular (Lev-Yadun and Inbar 2002; Aviezer and Lev-Yadun 2015).

Caterpillars employ a large array of defenses that reduce predation, such as camouflage, mimicry of their enemies' predators, exploiting various physical shelters and having chemical defenses (Bowers 1993). Unpalatable caterpillars with stinging and irritating hairs, functional osmeteria or body-fluid toxins often advertise their presence by aposematic coloration and aggregation (Cott 1940; Bowers 1993;

Fig. 56.8 A mature, almost dry pod of *Pisum elatius* showing the line of *red/purple* spots, several minutes or hours before its explosive seed dispersal

Fig. 56.9 A young pod of *Pisum elatius* showing an intermediate situation between *spots* and a *line*

Ruxton et al. 2004; Eisner et al. 2005; Lichter-Marck et al. 2015). The usual warning colors of caterpillars are yellow, orange, red, black and white with stripes along the body and/or arranged in spots, especially around the abdominal spiracles. Aposematism in caterpillars should reduce predation for several reasons, including an innate tendency of vertebrates to avoid warning colors and faster avoidance learning of bright as opposed to cryptic colors (Coppinger 1970; Gibson 1980; Lindström et al. 2001; Ruxton et al. 2004; Lichter-Marck et al. 2015). Interestingly, aposematic caterpillar mimicry is also performed by nestlings of the Amazonian

Fig. 56.10 A young pod of *Pisum elatius* showing a straight line

Fig. 56.11 A mature, but still *green* pod of *Pisum elatius* showing a zigzag line

birds *Laniocera hypopyrra* and *Laniisoma elegans* in order to reduce predation (D'Horta et al. 2012; Londoño et al. 2015). Therefore, Lev-Yadun and Inbar (2002) suggested that by mimicking aposematic caterpillars with red "spiracle spots", wild legumes may reduce immature pod predation.

It has been shown that ungulates may actively select leaves in the field by shape and color and avoid spotted ones (e.g. Cahn and Harper 1976) but Lev-Yadun and Inbar (2002) have failed to find good data on the response of vertebrate herbivores to aposematic (or cryptic) caterpillars. This proposed visual defense (spots and straight line) also has the potential of defending the pods and seeds from herbivory

Fig. 56.12 A mature, almost dry pod of *Pisum elatius* showing a zigzag line

Fig. 56.13 A pod of a rare *Pisum elatius* morph purple all over

of various insects by causing them not to lay eggs, since the plants seem to be already occupied and therefore the insects may choose another, non-occupied plant, a phenomenon similar to the defense by butterfly egg mimicry in *Passiflora* and other taxa (e.g., Benson et al. 1975; Williams and Gilbert 1981; Shapiro 1981a, b). In addition, visual caterpillar mimicry may attract visually oriented predators such as birds to the plant and they may attack real insects when they are found there.

Because the fruiting habit of all *Pisum* taxa is indeterminate (e.g., Abbo et al. 2009, 2011), both young and mature pods are usually found on the same plant. Thus, in the beginning of the fruiting season the *P. humile* and *P. elatius* plants will

Fig. 56.14 A lepidopteran caterpillar from southern Mount Carmel, Israel ornamented with spots from a habitat in northern Israel that has several wild legume species with pods that resembling it (Figs. 56.3–56.4)

have only green and cryptic pods; later, the defensive coloration of the more mature pods may defend not only the more mature ones, but also the younger and more susceptible ones. Towards the end of the season, only pods that mimic aposematic caterpillars (or other animals) by coloration are found on the plants (e.g., Lev-Yadun and Inbar 2002; Aviezer and Lev-Yadun 2015). A similar scenario will be described below concerning caterpillar mimicry by transparent young legume pods.

Partly Transparent Young Legume Pods: Do They Mimic Caterpillars for Defense?

Being transparent as defense from predation or as crypsis towards prey is mostly known in various groups of aquatic animals (Purcell 1980; Johnsen 2001; Ruxton et al. 2004; Cronin et al. 2014). Theoretically, all transparent organisms can be seen, although not at all parts of the spectrum and not under all lighted conditions. Therefore, in spite of being imperfect, being transparent is a practical solution for crypsis in many animals. In terrestrial habitats, and even in shallow aquatic habitats, being transparent for defense from predation or as crypsis from prey increases the risks of UV damage (Cronin et al. 2014). The number of terrestrial animals that use whole body transparency for crypsis is low, and this character is found mostly in terrestrial arthropod wings. The great differences in reflection of transparent organisms in very different media such as air *versus* water, as well as the generally much lower visibility in water, may also explain why transparency for crypsis is common in aquatic animals but is a much rarer defensive solution in terrestrial animals

(Ruxton et al. 2004; Cronin et al. 2014). The use of transparency for crypsis from prey in combination with animal mimicry is found in some transparent pelagic siphonophores that perform aggressive mimicry of copepods or fish larva in order to lure prey that does not fear them and so approaches a deadly but cryptic predator because most of its body is transparent (Purcell 1980).

Plants, being photosynthetic, cannot be highly transparent in general (Vogelmann 1993). Moreover, the fact that plant cell walls are made of various polymers (Fahn 1990; Albersheim et al. 2011) further reduces the ability of plants to be highly transparent. In spite of these inherent limitations, many succulent plant species of hot arid zones have partially transparent "windows" to allow below-ground photosynthesis as defense from both herbivores and from harsh environmental conditions, or have such "windows" in thick, aerial organs (Krulik 1980; Moore and Langenkamp 1991; Christensen-Dean and Moore 1993; Vogelmann 1993). Recently, Lev-Yadun (2013a) proposed that similarly to the succulent leaves of window-plants, in many unripe fleshy fruits there are partly transparent whitish spots that among their several proposed defensive functions, serve as windows that enable light to penetrate deeper into the photosynthetic layers in the thick developing fruit, this being a solution to overcome the limitations of light harvest because of the high volume-to-surface ratio of developing fleshy fruits.

Soon after I began my field work with the Eastern Mediterranean wild legumes as described above, it became obvious that the young pods of many of them, i.e., various species belonging to the genera *Lathyrus*, *Pisum* and *Vicia*, are temporarily partially transparent when viewed with back illumination. Of the genus *Lathyrus*, the species *L. ochrus* and several other un-identified *Lathyrus* species have transparent young pods. In all three wild *Pisum* species of Israel, *P. elatius*, *P. fulvum* and *P. humile*, the young pods are also transparent. Similarly, the young pods of several unidentified *Vicia* species are transparent too. With their elongated shape and showing their row of dark developing seeds that are conspicuous at that transparent stage when back-illuminated, the pods look like caterpillars, very different from their classic pod look when they are not transparent (Figs. 56.15, 56.16, 56.17 and 56.18). Lev-Yadun (2015a) thus proposed that being partly transparent probably simultaneously serves two functions: (1) acting as defensive caterpillar mimicry to reduce their consumption by various herbivores, and (2) allowing better photosynthesis in the rapidly growing seeds and pods.

Plant-animal communication *via* transparent tissues was considered in two plant types: (1) For pitcher carnivorous plants (Moran et al. 2012; Schaefer and Ruxton 2014), although *via* two different mechanisms, demonstrating that illumination through transparent plant tissues can be used in more than one way. Moran et al. (2012) found that in the pitcher plant species *Nepenthes aristolochioides*, the light coming through the translucent pitcher windows causes a visual illusion of false exits, helping to keep the prey inside the trap. However, Schaefer and Ruxton (2014) found that the windows in the species *Sarracenia minor* do not serve as false exits but rather function in long-range visual attraction of prey. (2) "Windows" in traps aimed for pollination that cause trapped pollinating insects to stay longer within the trap (Faegri and van der Pijl 1979; Oelschlägel et al. 2009). There is therefore no

Fig. 56.15 Fully developped *Lathyrus ochrus* pods in a plant growing in the central coastal plain of Israel. The combination of shape, size and coloration give the pods the look of a caterpillar

Fig. 56.16 Young green *Lathyrus ochrus* pods in a plant growing in the campus of Oranim. When the sunlight falls on it they look like regular *green* pods

theoretical reason to dismiss the possibility that partly transparent young legume pods may visually defend themselves by mimicking caterpillars in order to deter herbivores, possibly by using their perceptual exploitation *sensu* Schaefer and Ruxton (2009) concerning fear of various animals. The animal mimicry by the partly transparent young legume pods described and discussed in Lev-Yadun (2015a) is also a kind of aposematism (Batesian mimicry) for herbivores that do not like to consume caterpillars; in addition, it may give insects searching for unoccupied

Fig. 56.17 The same young green *Lathyrus ochrus* pods are partly transparent with back sun illuminationls and look like caterpillars

Fig. 56.18 A transparent young pod of an unidentified *Lathyrus* or *Vicia* species that look like a caterpillar

plants to lay their eggs on the impression that the plant is already occupied, or else it can serve as a type of masquerade *sensu* Lev-Yadun (2014b), by looking like an uninteresting food item to herbivores searching for pods or for other plant parts. At the same time, looking like a caterpillar may visually attract predaceous birds and invertebrates (*sensu* Niemelä and Tuomi 1987), and when patrolling there they may attack invertebrate herbivores that have already occupied these plants. Contrary to the use of being transparent in animals for crypsis, in the case of young legume pods it is used for enhancing apparency and by this they visually mimic caterpillars for defense. Because flowering and the formation of pods in legumes is indeterminate

(Abbo et al. 2009), in the beginning of the fruiting season the plants will have only transparent pods that look like caterpillars when back lighted. Some weeks later, when the pods start to express red/purple coloration, some of the plants will simultaneously have pods that mimic aposematic caterpillars (e.g., Lev-Yadun and Inbar 2002; Aviezer and Lev-Yadun 2015). Towards the end of the season, only pods that mimic aposematic caterpillars are found on the plants. Since real aposematic caterpillars are found in the fields where these plants grow (e.g., Fig. 10 in Lev-Yadun and Inbar 2002), such aposematic caterpillars seem to be the models for the mimicry. The lesser resemblance of the partly transparent pods to aposematic caterpillars compared to pigmented ones seems to be compensated by the principle introduced by Miriam Rothschild (1984) "aide mémoire mimicry", according to which the herbivore is made to recall previous unpleasant experiences or risks, and so there is no need for very good mimicry.

Lev-Yadun (2015a) concluded that being partly transparent may be used by young legume pods of several species and genera to establish a visual illusion that the pods are caterpillars. This may decrease their consumption by large mammalian herbivores, and in turn may attract predaceous birds or arthropods to the plants, which then might hunt real herbivorous insects infesting them. This may also cause butterflies to refrain from laying eggs on these plants, as occurs in *Passiflora* plants that mimic butterfly eggs (e.g., Williams and Gilbert 1981). Being transparent may simultaneously allow fast-growing young legume pods to perform higher levels of photosynthesis and possibly to heat cold tissues on cold spring days by serving as a greenhouse. It seems that while in animals being transparent serves both defense and attack *via* crypsis, in plants it serves defensive mimicry/aposematism and masquerade by being more conspicuous. Of the thousands of legume species worldwide there are probably many that express this character but were not described. The possibility that it may also be expressed in young fruits of various other taxa should also be considered and studied. Such a database is essential for evaluating the physiological, ecological and evolutionary aspects of this phenomenon.

Visual Caterpillar Mimicry by Certain Galls as Potential Defense from Herbivory

While gall aposematism was discussed above, Yamazaki (2016) proposed that various galls from Japan, the Middle East, or of Palearctic distribution visually mimic caterpillars and by this gain protection from various herbivores that will refrain from consuming or colonizing occupied plants. Moreover, Yamazaki (2016) also proposed that predators and parasitoids of caterpillars may be attracted to the plants because of such visual mimicry and then attack caterpilars that harbor them and by this lower competition with the galls.

Yamazaki's (2016) hypothesis seems to fit the features of several gall taxa proposed in one of the previous chapters to be aposematic, e.g., *Forda* sp. galls on

Pistacia terebinthus tree growing in Makri, northern Greece (Fig. 43.1), *Forda riccobonii* galls on *Pistacia atlantica* tree growing in the northern part of Israel (Fig. 43.3), and *Forda marginata* galls on *Pistacia palaestina* tree from the botanical garden of Oranim campus, Tivon, Israel. There is no need to assume that this proposed defense by visual caterpillar mimicry is exclusive, and that morphology and coloration may serve other physiological or defensive functions.

Chapter 57
Aphid Mimicry

My understanding that plants may mimic aphid infestation arrived in a flash when Moshik Inbar and I were teaching field ecology. The class was large, so each of us had about 25 students and we sat about 20 meters apart along a stream that emerged and flowed slowly from a spring in cave. At the same moment, we spotted the same plant species (*Paspalum paspaloides* = *P. distichum*; Poaceae), a wind-pollinated plant, with numerous dark anthers that look like aphids (Fig. 57.1). We each picked a flowering branch, and without saying a word lifted it up to show the other. We immediately understood each other and discussed it at the end of the field day.

Aphid mimicry, not only by dark anthers of wind-pollinated grasses, was proposed to operate by Lev-Yadun and Inbar (2002) and discussed later by Lev-Yadun (2006a, 2009a) and Lev-Yadun and Gould (2009). Lev-Yadun and Inbar (2002) described aphid mimicry in *P. paspaloides*, where the dark and about 2–3 mm long anthers are the size, shape and color of aphids, and they sway in the wind like swivelling aphids (Fig. 57.2). Similarly, the stems of *Alcea setosa* are also covered with dark flecks that look like aphids. The outer cover of young and still not open inflorescences of the aroid geophyte *Eminium spiculatum* (= *Helicophyllum crassipes*) found in grazed areas in Israel has typical dark spots that give it the look of being covered with aphids (Fig. 57.3). I also found such anther morphologies in several wild grasses growing in North Carolina (USA) (Lev-Yadun unpublished).

Scale insects are commonly white. They may attack plants in large numbers and cause them to look as if they are variegated in white (Fig. 57.4). It is possible that certain cases of white variegation (Fig. 57.5) may mimic such aphids.

It has been argued by Lev-Yadun and Inbar (2002) that plants that look infested may be left untouched by both grazers and other aphids or insects. Several studies have shown that early infestation by aphids and other homopterans has a negative impact on host plant preferences and on larval performance of other insect

S. Lev-Yadun, *Defensive (anti-herbivory) Coloration in Land Plants*,
DOI 10.1007/978-3-319-42096-7_57

Fig. 57.1 A branch of *Paspalum paspaloides* (= *P. distichum*; Poaceae), a wind-pollinated plant, with numerous dark anthers that look like aphids

Fig. 57.2 An aphid infested plant from Israel

Fig. 57.3 The outer cover of young and still not open inflorescence of the aroid geophyte *Eminium spiculatum* (= *Helicophyllum crassipes*) found in grazed areas in Israel with its typical dark spots that give it the look of being infested with aphids

Fig. 57.4 White scale aphids on a *Hedera helix* leaf of a plant growing in Padova, Italy

Fig. 57.5 Potential scale insect mimicry by *white* variegation

herbivores (Finch and Jones 1989; Inbar et al. 1999). Aphids respond to crowding by enhanced dispersal (Dixon 1998) and it is thus also probable that they may avoid previously infested or infestation-mimicking hosts. This clear zoological data supports the hypotheses about the potential defensive value of aphid mimicry, but experimental data is needed to fully accept this hypothesis. Again, the possible simultaneous involvement of olfactory cues should not be ruled out.

Chapter 58
Ant Mimicry

Ants are common in many and varying ecologies from the tropics to the arctic; they bite and sting and are aggressive, and many animals, including large mammalian herbivores, will avoid them whenever they can. Thus, ants have become models for a variety of arthropods that have evolved to visually mimic ants as defense from predation (Wickler 1968; Edmunds 1974). The aggressiveness of ants and their danger brought even to innate aversion to ants and ant mimics (e.g., Nelson et al. 2006).

Already in the year 1995, I became aware of the numerous conspicuous dark spots and flecks 2–10 mm in size, found on the epidermis of stems, branches and many petioles of *Xanthium strumarium* (Asteraceae), a common summer plant in Israel (Fig. 58.1). Dots predominate in some individual plants, fleacks in others. However, I could not figure out any physiological function for these scattered dark spots and flecks. It intrigued me, and I examined and photographed *X. strumarium* plants in many habitats. At the moment that I understood with Moshik Inbar the probable mimetic function of the dark anthers of *Paspalum paspaloides* described above, I thought that aphid mimicry is also involved in this case. On the next morning, I cut a large *X. strumarium* plant and as soon as I entered Moshik Inbar's room, before I could say anything, he said "ants". The reason for this was that many of the spots and flecks in that plant are arranged in lines and look like ant columns. Similarly to *X. strumarium*, the petioles and inflorescence stems of *Arisarum vulgare* and *Eminium spiculatum* (= *Heliophyllum crassipes*) (Araceae) are covered by many dark flecks (Figs. 58.2 and 58.3). Thus, to the human eye at least, the shoots of such species appear to be covered by a swarm of ants (Formicidae). Ant swarms are typically made of many moving dark flecks, varying in size from 2–3 mm to over 1 cm. The swaying of leaves, stems, or branches in the wind in combination with the dark spots and flecks, some of which are arranged in lines, may give the illusion that the "ants" move (Lev-Yadun 2009a).

Plants employ various predators and parasitoids as bodyguards that act as an indirect defense against herbivores. Many plant species are already known to emit herbivore-induced volatiles to attract predatory mites (e.g., Dicke and Sabelis 1988; Dicke et al. 1990; Takabayashi and Dicke 1996; Heil 2008; Karban 2015) and

S. Lev-Yadun, *Defensive (anti-herbivory) Coloration in Land Plants*,
DOI 10.1007/978-3-319-42096-7_58

Fig. 58.1 The stems, branches and petioles of *Xanthium strumarium*, a common summer plant in Israel look as if infested by ants

parasitoids (e.g., van Poecke et al. 2001; Shiojiri et al. 2010). Well-known invertebrates allies of plants are ants, predatory mites and parasitoids. Many myrmecophytes (ant-plants) invest resources in attracting ants by making living space (i.e., domatia) and nourish their ants *via* extrafloral or floral nectaries and by various types of solid food bodies (Huxley and Cutler 1991). It is well known that ants defend plants from insect or mammalian herbivory and in certain cases their relations with their hosts have been recognized as mutualistic (e.g., Madden and Young 1992; Jolivet 1998; González-Teuber and Heil 2015). The potential benefit from ant-attendance mimicry is obvious. Mutualistic ants use the domatia and/or foods and in turn defend their host plants from various natural enemies including herbivorous arthropods, mammals and even from other plant competitors (Janzen 1966; Yano 1994; Jolivet 1998; Frederickson et al. 2005; Heil 2008; González-Teuber and Heil 2015). Although aphids and other hemipterans are sometimes important plant enemies, they secrete honeydew that in turn attracts ants, resulting in the protection of the plants from various herbivores (Bristow 1991; Dixon 1998; Styrsky and Eubanks 2007; Yamazaki 2008a, b).

The importance of ants in defending plants was demonstrated in a field experiment in which ant and aphid removal resulted in a 76 % increase in the abundance of other herbivores on narrow-leaf cottonwoods (Wimp and Whitham 2001). Mappes and Kaitala (1995) found that female parent bugs (*Elasmucha grisea*)

Fig. 58.2 The stems and petioles of *Arisarum vulgare* also look as is infested by ants

choose to oviposit their eggs on *Betula* trees with no ants rather than trees with ants. Thus, it is not surprising that like in various arthropods, visual ant mimicry is found in plants. Plants may benefit significantly from defensive ant mimicry if herbivores consider them to be occupied by real ants and thus refrain from occupying and consuming them, especially their reproductive organs (Lev-Yadun and Inbar 2002). When plants can effectively mimic the actual signals of their bodyguards, or mimic reliable cues for bodyguard activity, they may benefit from the protection from herbivores without investing in volatiles, nectar, and cavities for housing. Lev-Yadun and Inbar (2002) proposed that the aggressive and efficient anti-herbivore activities of ants makes it highly beneficial for plants such as *X. trumarium* and *Arisarum vulgare* to mimic ant attendance in order to deter herbivores (both insects and vertebrates) without paying the cost of feeding or housing ants. Again, the possibility of the involvement of olfactory mimicry of ants has not been studied yet.

Fig. 58.3 The petioles and the underside of inflorescences of *Eminium spiculatum* (= *Heliophyllum crassipes*), a species that grows in grazed areas look as is infested by ants

Ant Mimicry by *Passiflora* Flowers?

The predominantly American genus *Passiflora* comprises more than 500 species, some of which commonly attract ants by means of extrafloral nectaries found on their bracts or on the back of the sepals (Ulmer and MacDougal 2004). After finding dark dots and short stripes that appear to the human eye to resemble ants, on the anthers and stigmas in many *Passiflora* spp. plants growing as ornamentals in Israel (Figs. 58.4–58.5), I studied the plates in Ulmer and MacDougal (2004), which describe several dozens of *Passiflora* species, to examine whether this putative ant mimicry is a common phenomenon in this genus. This examination indicated that probable visual ant mimicry exists in at least 22 *Passiflora* species. In 17 species (*P. amethystina*, *P. cincinnata*, *P. deltoifolia*, *P. eichleriana*, *P. exura*, *P. foetida*, *P. garckei*, *P. gibertii*, *P. incarnata*, *P. mayarum*, *P. menispermifolia*, *P. mooreana*, *P. oerstedii*, *P. speciosa*, *P. subrotunda*, *P. tenuifila* and *P. trisulca*) the anthers and stigmas were mottled, and in five other species (*P. candida*, *P. gracilis*, *P. lancearia*, *P. rhamnifolia* and *P. sclerophylla*) the petals were mottled. It is likely that additional such species exist among the many *Passiflora* species that were not documented in that book (Lev-Yadun 2009d). Ant mimicry is not necessarily the sole

Figs. 58.4–58.5 The flowers of dozens of *Passiflora* species look as if infested (and protected) by ants

option of defensive animal mimicry by dark spots on flowers of *Passiflora*. Such dark spots may also mimic aphids. As described above, aphid mimicry has already been suggested to defend plants from herbivory because aphids refrain from colonizing plants already occupied by other aphids (Lev-Yadun and Inbar 2002). The suggestion of ant mimicry by coloration in *Passiflora* flowers is supported by the activity of actual ants that were found to protect flowers of *P. coccinea* from both herbivores and nectar thieves and by this significantly increase seed set (Leal et al. 2006).

The American genus *Passiflora* is the best-known plant genus in regard to several morphological/coloration adaptations suggested to have evolved to reduce herbivory *via* animal mimicry. The best-studied case is of butterfly egg mimicry by tendril meristems, bulbous yellow petiolar glands, yellow deciduous flower buds, and yellow nectar glands on the abaxial leaf side of several *Passiflora* species, which has been suggested to reduce egg-laying by *Heliconius* butterflies, but seems to operate also for other plant and butterfly taxa (Benson et al. 1975; Shapiro 1981b; Schaefer and Ruxton 2009). The second type of defensive animal mimicry in this genus was noted by Rothschild (1974, 1984) for the stipules along the branches of *Passiflora caerulae* that resemble caterpillars, slugs or snails crawling along the stems, and young heliconian caterpillar mimicry by the yellow-orange serrulate stipules of *Passiflora platyloba* and *P. maliformi* (Benson et al. 1975). Lev-Yadun (2009d) showed that many *Passiflora* species have dark dots and short stripes on their flowers, and suggested that these visually mimic ants, adding to the two previously described types of animal mimicry that may protect *Passiflora* plants from herbivory.

The hypothesis that visual ant (or aphid) mimicry, in the shape of dark dots and short stripes, occurs in flowers is not exclusive. Such coloration may also attract pollinators and lead them towards the nectar and the pollen within the flowers (Dafni and Giurfa 1999; Biesmeijer et al. 2005). A dual purpose of flower characters for both pollination and defense has been suggested many times for various visual and chemical flower characters (e.g., Hinton 1973; Strauss and Whittall 2006; Hansen et al. 2007; Lev-Yadun 2009a).

Lev-Yadun (2009d) concluded that in the flowers of many *Passiflora* species the coloration pattern raises the possibility of a visual defensive ant or aphid mimicry. While the evolution of such coloration patterns in flowers could probably be attributed primarily or solely to pollinator attraction (Dafni and Giurfa 1999; Biesmeijer et al. 2005), the existence of such patterns of coloration on stems and leaves (e.g., Lev-Yadun and Inbar 2002) seems to be purely defensive. Thus, there is a possibility that visual ant mimicry by plants is a broad phenomenon, the extent of which should be studied globally. Accordingly, when studying the complicated issue of plant-ant relationships, visual defensive ant mimicry by plants, a type of Batesian mimicry and masquerade should also be taken into account.

Chapter 59
Beetle Mimicry

A large number of arthropod taxa including spiders, cockroaches and longicorn beetles mimic poisonous beetles such as ladybirds, fireflies and leaf beetles for defense (Wickler 1968). Till recently, visual beetle mimicry by plants has been proposed and studied only concerning pollinator attraction. Dark spots in ray flowers of *Hesperantha vaginata* or *Gorteria diffusa* (Asteraceae) and similar markings in other plant taxa attract pollinating beetles and other insects (Johnson and Midgley 1997, 2001; van Kleunen et al. 2007). Therefore, since beetles are mimicked for defense by various animals and for pollination by plants, and because there are so many chemically defended beetle species (Eisner et al. 2005), beetle mimicry for defense by plants should be expected.

Yamazaki and Lev-Yadun (2014) proposed a new type of visual beetle and beetle feces defensive plant mimicry, and suggested following their field experience that this type of putative defensive mimicry against mammalian and insect herbivores is probably a widespread but overlooked phenomenon. In addition to direct herbivore repelling, such mimicry may also attract visually oriented arthropod predators and parasitoids that can attack various invertebrate herbivores actually occupying these plants. For instance, the dark axils of *Lycium chinense* plants growing in Japan visually mimic poisonous feces or feces-covered larvae of the leaf beetle *Lema decempunctata*, which may result in reduced herbivory by various mammalian and insect herbivores. Both adults and larvae of *L. decempunctata* feed on *L. chinense* leaves and the larvae carry slimy feces on their backs. As they crawl on the leaves and stems, the feces fall and thus scatter on the plants surface. In addition, feeding traces of larvae and adults of *L. decempunctata* and also mite galls sometimes give a dirty-looking appearance to the plants. To the human eye, the dark axils of *L. chinense* resembled the feces or feces-covered *L. decempunctata* larvae, and even undamaged *L. chinense* plants looked as if they were colonized by *L. decempunctata* larvae. The field survey for this coloration pattern in the very different wild flora of Israel revealed that the same morphology/coloration exists in various wild plant species, both monocotyledons and dicotyledons, including both annual and perennial species: *Avena sterilis*, *Lolium temulentum*, *Hyparrhenia hirta*, *Aegilops*

S. Lev-Yadun, *Defensive (anti-herbivory) Coloration in Land Plants*,
DOI 10.1007/978-3-319-42096-7_59

Fig. 59.1 Dark internodes are common along the stems of *Phalaris paradox*a and many other non-woody gramineous taxa. They resumble beetles

peregrina, *Phalaris paradox*a, *Piptatherum blancheanum*, *Sorghum halepense* and *Hordeum bulbosum*, belonging to the monocotyledons, and in two dicotyledon species the perennial *Polygonum palaestinum* and in the annual ruderal herb *Malva nicaeensis*. In all the monocotyledonous species the nodal zone or in the *Polygonum* the bract leaves covering it were brown, sometimes with a yellow part, resembling small beetles (Fig. 59.1). In *M. nicaeensis*, a small area where the leaf blade connects to the petiole is brown. In all these species there was variability in the extent of brown coloration and not all plants or all stems of an individual plant expressed it.

This proposed defensive beetle mimicry may be effective against both large herbivores and herbivorous insects. Mammalian herbivores learn to avoid poisonous or distasteful plants by odor and sight (Pfeister 1999). Thus, large herbivores may learn to avoid feeding on feces-scattered plants and prefer feces-free plants over them, and thus may avoid plants with dark axils or with similar types of coloration due to the confusion with the feces. Herbivorous insects in general detect previous damage by conspecific or other herbivore taxa using visual and chemical cues, and commonly refrain from colonizing of or ovipositing on the damaged plants (Prokopy and Owens 1983; Niemelä and Tuomi 1987; Wold and Marquis 1997; Papaj and Newsom 2005). There is increasing evidence that demonstrates the importance of vision in finding and selecting host plants by phytophagous insects (Reeves 2011). Therefore, herbivorous insects and possibly *L. decempunctata* adults as well might refrain from approaching *L. chinense* with dark axils when other plants are available (Yamazaki and Lev-Yadun 2014).

Interestingly, many leaf-beetle larvae use their poisonous slimy feces as defense and scatter them on the plants they occupy (Morton and Vencl 1998; Vencl et al. 1999; Furth 2004; Weiss 2006; Chaboo et al. 2007; Chaboo 2011). Criocerinae leaf beetles (e.g., *Lema* spp., *Lilioceris* spp., *Oulema* spp., etc.) have diversified, using as habitat and food various monocotyledons such as grasses, lilies and yams (Kimoto and Takizawa 1994), which often have dark axils or propagules, and leave their defensive feces on their host plants. Several beetle taxa in Israel, e.g., *Chrysolina* spp., *Omophlus* spp. and *Chilocorus bipustulatus* have a similar size and morphology to the brown spots. Thus, the common dark colored nodal zones of various monocotyledons and the brown axils of *L. chinense*, the bract leaves of *P. palaestinum*, and the brown spots on *M. nicaeensis* leaves may have a similar defensive effect against herbivores, because they resemble poisonous beetle feces in addition to the aposematic, distasteful leaf beetle and even adult ladybirds. Altogether, such putative defensive beetle mimicry by plants may be widespread. The fact that brown axil coloration may have still unknown physiological functions along with defensive ones as do many reddish-brown color patterns (see Gould 2004; Lev-Yadun and Gould 2009) causes no theoretical difficulty.

Chapter 60
Spider Web Mimicry

Spiders are very common voracious predators of various herbivorous arthropods on and around terrestrial plants, they play an important role in regulating arthropod populations (Wise 1993; Nyffeler et al. 1994; Nyffeler 2000) and their activity results in positive effects on plant growth in various ecologies (e.g., Louda 1982; Wise 1993; Nyffeler et al. 1994). In the various geographical regions occupied by dangerous spiders such as black widows (e.g., Ori and Ikeda 1998), which can significantly damage and even kill mammals, there is a possibility that even herbivorous mammals do not approach spider-web-like structures, as suggested by human arachnophobia (e.g., Peira et al. 2010; Vetter 2013). Accordingly, spiders are considered to satisfy the requirements for being the models of mimicry, and certain arthropods belonging to diverse taxa indeed mimic spiders in order to prevent predation (Eisner 1985; Mather and Roitberg 1987; Zolnerowich 1992; Rota and Wagner 2006). Hence, there is no theoretical reason to dismiss the possibility that plants may also mimic spiders or their webs.

Decades of progress in studies of predator-prey interactions and community ecology has revealed that predators negatively affect prey not only by direct consumption but also by changing prey's behavior, including reduced feeding and emigration (Hlivko and Rypstra 2003; Werner and Peacor 2003; Rypstra and Buddle 2013). These non-consumptive effects of predators on prey can also positively affect plant populations and communities, since herbivores exposed to cues of danger may avoid feeding on the plants and look for other, less dangerous habitats (Hlivko and Rypstra 2003; Ripple and Beschta 2004; Rypstra and Buddle 2013). Hence, plants probably deter herbivors if they possess predators' cues, as suggested by Lev-Yadun et al. (2009b), and by Yamazaki (2010).

Nyffeler (2000) calculated the average spider density across various habitats in England to be 152 individuals/m^2. With so many spiders in the system, various plant species actually form some facultative partnerships with spiders. For example, *Chamaecrista nictitans* (Fabaceae) herbs in the USA attract jumping spiders in addition to attracting ants *via* extrafloral nectar, and the resulting reduced herbivory allows for an increased seed set (Ruhren and Handel 1999). Similarly, *Acacia*

S. Lev-Yadun, *Defensive (anti-herbivory) Coloration in Land Plants*,
DOI 10.1007/978-3-319-42096-7_60

lingulata (Fabaceae) shrubs in Australia bear extrafloral nectaries and attract and harbor a subsocial amaurobiid spider species that make dense webs on the plants, thus reducing seed predation by bugs, wasps and weevils (Whitney 2004). A different strategy is employed by *Trichogoniopsis adenantha* (Asteraceae) and *Rhynchanthera dichotoma* (Melastomataceae) shrubs in South America. These plants are equipped with glandular trichomes that capture tiny arthropods. Lynx spiders prey on free-ranging herbivorous arthropods found on these plants, and also feed on the prey trapped by the trichomes, resulting in a reduction in herbivory (Romero et al. 2008; Morais-Filho and Romero 2010; Krimmel and Pearse 2013).

As already discussed concerning ant mimicry, when plants can effectively mimic the actual signals of their bodyguards (spiders in this case), or mimic reliable cues for their activity, they may benefit from the protection from herbivores without investing in attracting, feeding and/or housing the bodyguards. Taking into account the fact that spiders are abundant and ubiquitous predators on plants, plants are expected to mimic spiders, or at least spider activity cues for their defense. Also, because plants already attacked by herbivorous arthropods are inferior food sources (Karban and Baldwin 1997) or turn into dangerous habitats that commonly attract natural enemies (Niemelä and Tuomi 1987; Kessler and Baldwin 2001; Karban 2015), plants may also mimic for defense the cues of the herbivores. Yamazaki and Lev-Yadun (2015) have thus explored the issue of defensive spider cue mimicry by plants and proposed that certain plants indeed mimic spider webs for defense, as well as mimicking lepidopteran and spider-mite web nests.

During field work in the environs of Makri in northern Greece in the summer of the year 2003, I came across some plants of *Carthamus* sp. (Asteraceae) that had very long whitish trichomes (Fig. 60.1). It took me a long time to decide that they were trichomes and not a spider web. I examined these plants several times during the days I spent there in the field and decided to see if the same phenomenon occurs in other taxa and ecologies. Three years later, in the summer of the year 2006, I realized that two other species that I examined in the Baltic coastal region of Estonia, i.e., inflorescences of *Arctium tomentosum*, and very young leaves of *Tussilago farfara* (Asteraceae) have similar masses of white trichomes (Figs. 60.2 and 60.3). Several years later, Dr. Kazuo Yamazaki from Japan consulted me on the same phenomenon in plants from Central Japan, and we decided to collaborate. As a result of our field surveys, possible spider web mimicry by various plant organs was recognized in additional taxa. Dense white villose trichomes on newly-extending stems and expanding new leaves seem to visually and structurally mimic spider webs. In Israel, many very young inflorescences of the hemicryptophytes *Gundelia tournefortii* and several members of the genus *Onopordum* (Figs. 60.4–60.5) and annuals such as *Carthamus lanatus* and *Centaurea melitensis* (Asteraceae) found in the European part of the Mediterranean region also bear silk-like trichome masses, especially on their inflorescences but also on young shoots. In Japan, new shoots of the fern *Osmunda japonica* (Osmundaceae) all have dense masses of wooly trichomes and thus appear to be covered with spider webs. None of the abovementioned plant species are very common, but they usually grow as individuals or in small groups of several plants among large numbers of other species that do not

Fig. 60.1 Spider web mimicry by *Carthamus* sp. from northern Greece. It took me a long time to realize that the web is of plant origin and not of arthropod origin

Fig. 60.2 Inflorescences of *Arctium tomentosum* from Estonia that look as if harboring a spider web

Fig. 60.3 The very young leaves of *Tussilago farfara* in Estonia and elsewhere have masses of white trichomes giving the impression of a dense spider web

Figs. 60.4–60.5 The young stems and buds of *Onopordum* sp. from the Golan Heights, Israel look as if covered by a dense spider web

express such white trichome masses. In Japan, *O. japonica* grows along montane streams in small soil patches, although farmers often cultivate such plots.

Since a spider web (Fig. 60.6) is reliable evidence of spider activity, many herbivorous insects or other arthropods usually avoid colonizing plants with conspicuous spiders or spider webs if they manage to spot them. In fact, it was experimentally demonstrated that actual spider silk greatly reduced herbivore damage on plants

Fig. 60.6 An actual spider web, one out of very many in the Oranim campus, Tivon, Israel

(Hlivko and Rypstra 2003; Rypstra and Buddle 2013). Thus, if herbivorous arthropods visually confuse the spider-web-like trichome masses produced by various plants with actual spider webs, they may refrain from colonizing plants expressing such trichome masses and thus not consume them. Herbivorous insects use visual cues as well as chemical cues for finding host plants (Prokopy and Owens 1983; Reeves 2011). Reeves (2011) pointed out that all seven orders of herbivorous insects more or less use visual cues for host detection, that some herbivores find hosts primarily using visual cues, and that there are some insect species that can find hosts using only visual cues. These facts suggest that visual Batesian mimicry of spider webs by plants is likely. The requirements for mimicry evolution are that mimics must be rare in general relative to their models (Ruxton et al. 2004; Kikuchi and Pfennig 2010), and that the mimicked cue (spider webs) should be a reliable indicator of a threat (spider presence). According to Higginson et al. (2007), although crab spiders close lavender bracts with silk that seems an important cue of spiders for honeybees, the bees do not avoid artificially-closed bracts, possibly because the spiders frequently move between closed bracts, and thus the closed bracts are not highly reliable cues of spider presence. However, spiders and their webs are much more abundant (Nyffeler 2000) relative to the few plant species bearing silk-like trichomes, and even if the spiders are absent, unlike closed bracts, the webs may still trap and kill prey, further enabling this mimicry system.

Although vertical, symmetric orb webs by Araneidae, Tetragnathidae and Uloboridae are typical and well-known, very diverse spider taxa produce various irregular webs on and between plants (Ono 2009; Blackledge et al. 2011) and thus, such web types are apparently visually similar to the white trichome masses described by Yamazaki and Lev-Yadun (2015). Moreover, these white trichome masses may in many cases also look like sheets of webs of Coelotidae spiders and Agelenidae spiders, bowl-shaped webs of Linyphidae spiders or irregular three-dimensional webs of Theridiidae spiders, as well as lepidopteran and spider-mite web nests.

Yamazaki and Lev-Yadun (2015) proposed that the apparent spider-web mimicry by masses of thread-like trichomes exists in the above-mentioned plant taxa found in various ecologies of at least two continents. This may be a direct mimicry, or operate *via* perceptual exploitation of herbivores *sensu* Schaefer and Ruxton (2009), in which potential herbivores have perceptual biases of fear of conspicuous silk-like structures and some of the plants discussed here develop even more exaggerated traits than many actual spider webs.

In addition to direct defense from herbivory and various physiological gains provided by the trichome masses, the spider-web-like or the lepidopteran- and spider-mite web nests like trichome masses may attract visually-hunting predators because spiders, caterpillars and spider-mites are suitable prey for predators such as birds and lizards (Schoener and Spiller 1987; Gunnarsson 2007) and also for wasps (e.g., Field 1992) and for other predators and parasitoids (e.g., Dicke and Sabelis 1988; Hawkins 1994). All these result in concomitant predation on, or escape of the herbivorous arthropods from the plants, although those predators may reduce actual spider density on the plants, which in turn may potentially increase herbivory. Furthermore, spider-web-like trichomes may attract other predators that potentially prey on invertebrate herbivores that occupy the plants and on those that may be trapped in spider webs (Thornhill 1975; Sivinski and Stowe 1980; Whitehouse 1997) or on sticky trichomes (Krimmel and Pearse 2013).

One negative point for spider-web-like trichome masses is that the trichomes on inflorescences in species such as *Arctium tomentosum*, *Carthamus* sp. and *Centaurea melitensis* may deter some pollinators, resulting in reduced fertilization and seed set. However, as tubular flowers of the Asteraceae usually extend and cover the top of the inflorescences at flowering (Shimizu et al. 2001) and most bees and hoverflies can alight on the flowers without full contact with the inflorescences (Yokoi and Fujisaki 2009). Yamazaki and Lev-Yadun (2015) proposed that repellence of pollinators by spider web mimicry in the cases when it is associated with flowers is not too damaging.

Various arthropods other than spiders also spin silk to construct their nests on plants (Craig 1997), and the networks made of dense villose trichomes of the web-mimicking plants described by Yamazaki and Lev-Yadun (2015) are also likely to resemble the web nests constructed by many lepidopteran caterpillars belonging to taxa such as Arctiidae, Choreutidae, Crambidae, Lasiocampidae, Lymantriidae, Nymphalidae, Pieridae, Pylaridae, Scythrididae and Yponomeutidae (cf., Ishiwata et al. 2005) and some spider mites (Clotuche et al. 2011). When plants have already been used by lepidopteran caterpillars and spider mites, what remains of them is usually of low nutritive quality because the more nutritious parts have already been consumed, and because of the activation of various induced defenses (e.g., Karban and Baldwin 1997), such infected plants may also harbor or attract predators and parasitoids (van Poecke et al. 2001; Shiojiri et al. 2010; Yamazaki 2010). Moreover, various invertebrate predators such as spiders live as secondary inhabitants in lepidopteran web nests (Kunimi 1983; Fukui 2001), and herbivorous insects avoid lepidopteran silk, although to a lesser extent than spider silk (Rypstra and Buddle 2013). Thus, if herbivorous arthropods perceive the trichome webs as caterpillar or mite silk, they may not be attracted to or colonize the mimicking plants. However, since the web nests of caterpillars and spider mites are profitable for some herbivorous arthropods looking for concealment habitats, or for physiological gains from the mild microclimate inside shelters, and because there are no web-building costs (Fukui 2001; Yano 2012), such herbivores may prefer using or at least probing the trichome-covered plants.

Yamazaki and Lev-Yadun (2015) also proposed an alternative hypothesis, which may operate simultaneously with the above-mentioned one about spider-web mimicry. Accordingly, these white trichome masses may mimic masses of fungal hyphae. In humid climates such as in Estonia, Canada and Finland, where plant-pathogenic fungi including gray mold (*Botryotinia*) and powdery mildew (Erysiphaceae) are common and the infected plants frequently contain harmful substances and are less nutritious, the trichomes may act as defensive mimicry of such fungal hyphae. Visual fungal hyphae mimicry by plants as defense from herbivory has been proposed for various plants that have white or whitish leaves (Lev-Yadun 2006a, b, 2014a; Yamazaki 2010) or white variegation (Lev-Yadun 2014a) (Fig. 16.1) because fungal-infested plants may be toxic and of low nutritive value. However, there is no theoretical difficulty to assume that various physiological and defensive functions of trichome masses operate simultaneously, and that the relative importance of these functions would vary in various biotic and abiotic environments.

Testing of the Hypothesis

To test the spider-web mimicry hypothesis, conspecific plant variants regarding mass trichome production may be used for the evaluation of this plant trait for deterring herbivory. Addition and removal of web-like trichomes are also candidates for experimentation if wound-induced volatiles that may influence insect activity can be blocked or avoided. Predator effects on lower trophic levels are sometimes complex. One example is that top predators reduce the number of intermediate predators, resulting in an increase of herbivore numbers and a subsequent enhancement of herbivory (Schmidt-Entling and Siegenthaler 2009). Similarly, intraguild interactions (e.g., ants *vs* spiders) on plants may also affect herbivore density, herbivory and plant fitness (Nahas et al. 2012). Thus, carefully designed field experiments that consider such contrasting effects are required to test the defensive plant spider-web mimicry hypothesis. Moreover, if artificial optical models of what different plants would look like through the eyes of typical herbivores can be created (e.g., Döring et al. 2009), the visual aspects of the spider-web-mimicry hypothesis may be tested with better resolution. When visual perception by the herbivores of plants with dense trichomes is similar to or stronger than that of plants with spider webs, the hypothesis of spider-web mimicry or perceptual exploitation may be supported, respectively. Though indirectly, if villose trichomes of plants visually resemble webs of the local spider fauna in various parts of the world, the mimicry hypothesis is favored.

Chapter 61
Defensive Bee and Wasp Mimicry by Orchid Flowers

A variety of arthropods (e.g., flies, spiders, butterflies) are known to mimic bees or wasps as defense from predators (Cott 1940; Wickler 1968; Edmunds 1974; Plowright and Owen 1980; Rothschild 1984; Howarth and Edmunds 2000; Bain et al. 2007; Chittka and Osorio 2007; Penney et al. 2012) (Fig. 61.1). Bees and wasps may be aggressive and sting, and by this deter many animals including herbivores (Breed et al. 2004).

General visual bee-mimicry and specific volatile mix chemical mimicry by flowers in order to imitate solitary female bees or wasps for the sake of pollination is well known in several orchid genera, e.g., the Mediterranean genus *Ophrys* (Fig. 61.2), the Australian genera *Cryptostylis* and *Chiloglottis*, and the south African *Disa*. This mimicry has been shown to attract solitary male bees or wasps, which are their species-specific pollinators. The visual and chemical signals are considered to be a type of deceptive pollination mechanism based on mimicry for the exploitation of perceptual biases of animals (Dafni 1984; Schiestl 2005). Accumulating evidence indicated that the chemical-olfactory mimicry of female bee pheromone mix, first described by B. Kullenberg (Vereecken et al. 2009), and not the visual one (as originally proposed) is the dominant species-specific deceiving attractant of the solitary male bees to flowers of *Ophrys* and other orchid taxa (Kullenberg 1950, 1956, 1961; van der Pijl and Dodson 1966; Dafni 1984; Ayasse et al. 2000; Schiestl et al. 2000; Schiestl 2005). The flower fragrance mimics exactly the structure and composition of the specific sexual pheromone exerted by the females of the pollinating male bees (Schiestl 2005; Jürgens and Shuttleworth 2016).

Following the various hypotheses that plants employ defensive animal mimicry described above and below, Lev-Yadun and Ne'eman (2012) proposed reconsidering and experimentally testing the almost forgotten, more than a century-old anti-herbivory role of visual bee-mimicry discussed by Rolfe (1910) and briefly mentioned by van der Pijl and Dodson (1966). Mr. E. Kay Robinson, in a letter to the newspaper "Daily News", was the second to propose that visual bee-mimicry by orchid flowers has not evolved for pollinator attraction, but rather to deter grazing

S. Lev-Yadun, *Defensive (anti-herbivory) Coloration in Land Plants*,
DOI 10.1007/978-3-319-42096-7_61

Fig. 61.1 A vasp-mimicking spider from southern Germany

cows and other animals (cited in Rolfe 1910). In this connotation, Wickler (1968) mentioned, without giving the reference, that Brown expressed in 1831 an opinion that *Ophrys* species scare off insects with their bee-mimicking flowers. While Rolfe (1910) dismissed the defensive mimicry hypothesis, which was later practically forgotten, Lev-Yadun and Ne'eman (2012) argued that it may partly explain the visual bee- or wasp-mimicry by orchid flowers, especially in the view that olfactory female bee or wasp mimicry is the dominant species-specific pollinator attractor. Experiments showed that the specific chemical mimicry dominates the species-specific pollinator attraction and not the general color polymorphism of *Ophrys* flowers (Vereecken and Schiestl 2009), and that the variation in the visual stimulus prevents recognition of the deceptive flowers by the male bees that could lead to the lack of pollination (Schiestl 2005).

Moreover, in the orchid genus *Cryptostylis* there is even a visually contrasting difference between the deceptive orchid flower, which reflects UV seen by hymenoptera, and the red, green and white orchid flower to which the hymenoptera eye is less sensitive (Gaskett and Herberstein 2010). Moreover, the orchid flower's visual details, including their hairy surfaces are too delicate for the limited spatial resolution of the bee eye (e.g., Spaethe and Chittka 2003), but observable for sharper sighted herbivorous mammals. Thus, concerning pollination, the general visual

Fig. 61.2 The bee-mimicking flowers of the Mediterranean orchid *Ophrys carmeli* (= *Ophrys dinsmorei*)

bee-mimicry of *Ophrys* flowers remains partly unexplained, and Lev-Yadun and Ne'eman (2012) thought that it deserved further consideration. In the genus *Ophrys*, the combination of non-species-specific visual bee-mimicry with the species-specific role of chemical-fragrance mimicry of the female bee pheromone for attracting pollinating solitary male bees may indicate an additional, non-reproductive cause for the evolution of such special floral bee-mimicking morphology (but see Schiestl 2005).

Lev-Yadun and Ne'eman (2012) proposed that various large herbivores and some herbivorous insects may be deterred by visual bee- or wasp-mimicry, and that defensive mimicry of bee pheromone in *Ophrys* flowers should also be considered in this context. This proposed defensive Batesian mimicry is not exclusive and probably plays a secondary role to pollination. This mimicry may deter large mammalian herbivores, and possibly also insects from the plants and especially from their flowers. In addition, Lev-Yadun and Ne'eman (2012) proposed that the visual and chemical bee-mimicking signals can also be considered to belong to a different type of deception based on exploitation of perceptual biases of animals (*sensu* Schaefer and Ruxton 2009). They presented the question of whether the visual bee- and wasp mimicry of orchid flowers or the exploitation of perceptual biases of animals towards aggressive bees and wasps can serve also as a sophisticated anti-herbivory defense. The emerging evidence on various types of odor mimicry by orchids

(Schiestl and Ayasse 2001; Brodmann et al. 2008, 2009) and the plethora of types of deception by plants (Wickler 1968; Wiens 1978; Schaefer and Ruxton 2009, 2011) strongly indicate that a very sophisticated anti-herbivory visual bee- or wasp-mimicry can be expected.

The potential benefit from bee- or wasp mimicry is obvious. Plants may benefit from defensive bee- or wasp-mimicry if herbivores consider them to be occupied by bees or wasps and refrain from eating them, especially their flowers. While *Ophrys* and other bee- or wasp-pollinated taxa mimic solitary bees and wasps (Wickler 1968; Dafni 1984; Schiestl 2005; Jersáková et al. 2006; Gaskett and Herberstein 2010), and the worst bee attacks are of social bees such as the honeybee that form large colonies (Breed et al. 2004), there is no need for perfect mimicry to deter enemies (e.g., Rothschild 1984; Howarth and Edmunds 2000; Chittka and Osorio 2007; Janzen et al. 2010; Penney et al. 2012) and imperfect bee and wasp mimicry should not be different (see Kauppinen and Mappes 2003; Penney et al. 2012). From other systems of bee mimicry it is known that the mimicry does not have to be perfect for human eyes, but if at, all rather for the eyes and perception of the target animal (Bain et al. 2007), even though the human eye was found to be a good judge (Penney et al. 2012). While male bees are stingless, wasps do have stings and it is not always possible or easy, and of course not without risk, to distinguish between stingless bees and sting-owning bees or wasps. Moreover, the fact that bees and wasps are actually found on orchid flowers, adds a Müllerian effect (extended phenotype) to this visual defense, an issue that will be discussed later.

There are four potential components of bee- or wasp mimicry: (1) visual (shape, texture and color), (2) chemical (odor), (3) movement, and (4) sound production. Out of these, plants are known to mimic or exploit the first three. The pollination-related visual aspect of bee- or wasp mimicry by orchid flowers has been discussed above and there is not sufficient specific, target-oriented data about defensive bee- or wasp mimicry by plants to discuss it. Therefore, Lev-Yadun and Ne'eman (2012) discussed the potential role of plant movement in combination with visual bee- and wasp mimicry in defense from herbivory.

The visual role of plant movement as a way of communication (and deception) between flowers and pollinators has been studied in two cases of attraction mentioned in previous chapters but will be repeated here to help the reader. In the first case, some orchid species belonging to the genus *Oncidium* induce attacks by territorial male *Centris* bees when the inflorescences move in the wind because the male bees mistake them for rival males. During the attacks of the pseudo-rivals, the male bees transfer the pollen (Dodson and Frymire 1961; Dodson 1962 cited in Wiens 1978). In the second case, Warren and James (2008) showed that movement ("waving") of the inflorescences of *Silene maritima* increases pollination success. Movement of plant parts that visually mimic aphids and ants (Lev-Yadun and Inbar 2002; Lev-Yadun 2006a, 2009a, d) was discussed in previous chapters. Thus, following such tested examples from pollination biology, there is no theoretical reason to dismiss the possibility that plant movements also take part in visual defense from herbivory (see also Yamazaki 2011).

In order to specifically propose that bee mimicry may defend plants, it is essential to demonstrate that bees indeed deter herbivores and thus defend plants. Vollrath and Douglas-Hamilton (2002) and King et al. (2007, 2009, 2011) showed in a series of field experiments that even the largest living terrestrial herbivore, the African elephant (*Loxodonta africana*), is deterred by (1) honeybee attacks, (2) just the recorded sound of bee buzzing, and (3) the sight of empty beehives. Elephants refrained from attacking local *Acacia* trees and crops growing on local farms in the proximity of beehives. Moreover, at the opposite end of the herbivore size scale, caterpillars are also deterred by wasp buzzing (Tautz and Markl 1978) and exposure to honeybee buzzing reduced caterpillar damage in soybean plants (Tautz and Rostás 2008). Therefore, Lev-Yadun and Ne'eman (2012) proposed that visual and olfactory bee- or wasp mimicry by orchid flowers may not only serve for pollination, but also defend them from herbivory by both large herbivores and insects.

The common occurrence of eight bee-mimicking *Ophrys* species in heavily grazed areas in Israel (Feinbrun-Dothan and Danin 1991; and my vast field experience) is another indirect support for the hypothesis. Similarly; it has been found that a chemically well-defended and possibly aposematic Mediterranean geophyte, *Anemone coronaria*, which is poisonous and has red flowers, benefits from heavy grazing that also reduces competition with grasses in the same habitats (Perevolotsky et al. 2011). Orchids in heavily grazed areas in the Near East also benefit from selective grazing and increase in numbers.

Additional Testing of the Defensive (Anti-herbivory) Role of Bee and Wasp Mimicry

As discussed above, defensive bee and wasp mimicry is well known in the animal kingdom and therefore there is no need to test this principle. It should, however, be tested specifically for deceptive orchids, and in general when bees and wasps are abundant on various other taxa of flowering plants. (1) Grazers and relevant herbivorous insects have to be inspected to see if they regularly avoid such orchids, (2) visually mimicking plastic models should be used to study grazer responses on seeing them, (3) olfactory signals, when known, should be examined with and without the visual aspect, (4) the innate and learning components of deterrence should be tested in mature versus very young grazers, (5) increasing and decreasing signal strength by changing plant densities combined with or without external application of relevant volatile blends should also be done, (6) the biochemical and molecular aspects of herbivore responses to these signals should be studied, (7) the fitness of the orchids as expressed in seed set and plant demography should be studied over many years under various grazing pressures, and (8) all these should also be done with non-mimicking flowers that according to the hypothesis posited by Lev-Yadun and Ne'eman (2012) are potentially protected by an abundance of pollinating bees and wasps.

Chapter 62
Gall Midge Mimicry

Many of the white umbels of wild carrot (*Daucus carota*, Apiaceae), also known as Queen Anne's Lace, are decorated by a group of modified flowers in their center. The modified flowers are very dark, ranging when mature from very dark purple to black (Fig. 62.1). Another member of the family, *Artedia squamata* also has such dark central flowers (Fig. 62.2). The proposed possible functions of these dark central flowers, and the differences in their function between habitats, as well as the proposed lack of function, is an excellent demonstration of the level of ecological/evolutionary complexity of such visual and colorful characters.

Discussions concerning these dark flowers began with Darwin (1877), who could not figure out a function for this character. Detto (1905) cited in Yeo (1972) proposed that the dark flowers in umbels of *D. carota* are a type of insect mimicry evolved in order to discourage grazing animals. Rothschild (1972) thought that these dark flowers are crude insect mimics that may serve to attract pollinators. Eisikowitch (1980) proposed that these dark flowers mimic insects and by this attract pollinators in *D. carota* plants growing in Israel. Westmoreland and Muntan (1996) found that this may operate in pollinator attraction in certain *D. carota* populations studied in West Virginia, Georgia, Iowa, Kentucky and New York, but not in many others, depending on the composition of the pollinating fauna. Lamborn and Ollerton (2000) did not find any influence of these dark flowers on pollination in *D. carota* plants growing in central England, but Goulson et al. (2009) found that it attracted pollinators in plants growing in Portugal. Polte and Reinhold (2013) showed that these flowers may provide some defense against the parasitic gall midge *Kiefferia pericarpiicola* in *D. carota* plants growing in Germany. Altogether, the various functions of these visually conspicuous dark modified flowers are not universal, but are rather site and situation dependent. This is an excellent demonstration of the multiple functions of insect mimicry by plants including defense from herbivory and attraction of pollinators, and sheds light on the spatial and temporal differences in the function of such characters.

S. Lev-Yadun, *Defensive (anti-herbivory) Coloration in Land Plants*,
DOI 10.1007/978-3-319-42096-7_62

Fig. 62.1 The umbels of the wild carrot *Daucus carota*, also known as Queen Anne's Lace, are decorated by a group of modified dark flowers in their center

Fig. 62.2 Like *Daucus carota*, *Artedia squamata* also has modified dark flowers in their umbels, enhanced in the picture by actual beetles

Chapter 63
Arthropod Wing Movement Mimicry

Adaptive active plant organ movement has attracted considerable scientific attention since Darwin's time (Darwin 1881). The best known examples are growth towards the light (Darwin 1881), repositioning of trunks and branches of trees in relation to the vector of gravity (Timell 1986; Groover 2016), sun tracking (Koller 2011), prey-driven movements, i.e., of *Dionaea* (Simons 1992) and movements during self-seed-dispersal (autochory) (Koller 2011; Abraham and Elbaum 2013). Quick anti-herbivory plant movements are best known in *Mimosa pudica* and several other related taxa that down-fold their leaflets to reduce visibility and accesibility when mechanically disturbed (Simons 1992), and Braam (2005) proposed that the quick *Mimosa* leaflet movement may also deter herbivores. Leaflets of *Schrankia micropylla* down-fold when touched, and by doing this expose their thorns (Eisner 1981), a character later proposed to be a type of visual aposematism (Lev-Yadun 2009a). In the herb *Cardamine scutata*, the very fast silique bursting expels and even kills chewing caterpillars, thus defending its seeds (Yano 1997). In addition, Yamazaki (2011), in a broad and intriguing theoretical treatment of potentially defensive plant movements, advocated the hypothesis of considering the possible roles of passive leaf movement induced by wind and rain as a common anti-herbivory defense. The hypothesis of this mode of passive defensive leaf movement was tested experimentally and was strongly supported (Warren 2015; Leonard et al. 2016). Interestingly, all plant taxa expressing quick active leaf or leaflet movements (*Dionaea*, *Mimosa pudica*, *Schrankia micropylla* and *Desmodium motorium*) are found in warm parts of the globe, probably because high temperatures allow quick movements in plants, a group that is usually and even mostly ectothermal.

One of the most dramatic and enigmatic quick active plant movements, for which till recently there was no proposed adaptive hypothesis, is that of the Indian telegraph (semaphore) plant *Desmodium motorium* (= *D. gyrans* = *Hedysarum gyrans* = *Codariocalyx motorius*). The plant has large leaves that perform typical slow sleep (nyctinastic movement) movements at night as well as smaller stipules that move during the day in conspicuous quick elliptical circles every few minutes, and under high temperatures (35 °C) may execute a round every 90 s (Simons 1992).

S. Lev-Yadun, *Defensive (anti-herbivory) Coloration in Land Plants*,
DOI 10.1007/978-3-319-42096-7_63

Since the plants have many stipular leaves, there is always some leaf movement in their canopy. The structural-physiological mechanism allowing this quick leaflet rotation is based on pulvinar motor cells that induce movement by volume-change, mediated *via* membrane depolarization (Antkowiak et al. 1991; Antkowiak and Engelmann 1995). The enigmatic function of this conspicuous movement has intrigued scientists for centuries. Darwin (1881:364) wrote: "No one supposes that the rapid movements of the lateral leaflets of *D. gyrans* are of any use to the plant; and why they should behave in this manner is quite unknown", a statement that was still valid till recently.

Lev-Yadun (2013b) proposed that the enigmatic quick diurnal leaflet movements in elliptical circles every few minutes is a type of defensive butterfly or general winged arthropod (herbivorous or predaceous) mimicry by this plant. Resting butterflies and various other winged arthropods commonly move their wings to control exposure to the sun (Clench 1966). Since the plant has many stipules, and each pair moves once in a few minutes (see *Desmodium* in youtube.com), to a passing butterfly searching for an unoccupied site suitable for depositing its eggs, the plant may look as if it is already occupied. Moreover, stipule movement in *D. gyrans* may attract predacious birds, lizards or arthropods, deceived by the movements concerning the existence of potential winged arthropod prey, and while carefully examining the plant at close range, may catch and consume or parasitize insects and other invertebrates that occupy it. The visual predator attraction hypothesis was proposed previously irrespective of plant movement for the lobed shape of various leaves that mimics caterpillar feeding damage (Niemelä and Tuomi 1987; Brown and Lawton 1991). The possibility that diurnal mammalian or avian herbivores may also be deterred by these movements should not be dismissed. The defensive role for plants of predators of insects is well documented (e.g., Holmes et al. 1979; Price et al. 1980). Predator attraction to leaves by producing extrafloral nectaries that feed bodyguards (Bentley 1977), by emitting volatiles (Kessler and Baldwin 2001; Kappers et al. 2005; Li et al. 2012), and following the sticking of small insects to leaves by sticky trichomes (Krimmel and Pearse 2013) was found to increase plant fitness in many cases.

Concerning experimentation, the best experiment is to compare herbivore attacks and attraction of predators to mutants of *D. motorium* that do not move their leaflets (such mutants have not been described yet) as compared with the wild-type. The alternative is to do so with mechanical models in which it is possible to control "leaflet" movements, or to block leaflet movements in *D. motorium* by various inhibitors if they do not change its olfactory signature.

Chapter 64
"Eye Spot" Mimicry

The pattern of colored spots on the pods of *P. humile* and *P. elatius* proposed to function as aposematic caterpillar mimicry (Lev-Yadun and Inbar 2002; Aviezer and Lev-Yadun 2015) could also mimic eyes, a character that was only recently suggested to operate as defense in plants (Aviezer and Lev-Yadun 2015) (Figs. 56.1, 56.3, 56.7, 56.8 and 56.15) but is well-known in animals (Cott 1940; Edmunds 1974; Ruxton et al. 2004; Stevens et al. 2007, 2008b; De Bona et al. 2015). In animals, the "eyespots" are often not more than circular features (Stevens et al. 2008b), but the resemblance to an eye, to a face with two eyes, or even to more than two "eyes" or more than a single face, can intimidate or startle potential predators (herbivores), since the "eyespots" indicate the probable existence of a bigger and maybe even extremely dangerous creature (see Wickler 1968; Edmunds 1974; Ruxton et al. 2004; Stevens et al. 2007; Janzen et al. 2010). However, it seems that two "eyespots" intimidate better than other numbers (Mukherjee and Kodandaramaiah 2015). It seems that concerning "eyespots", and concerning the possible snake mimicry by certain *P. elatius* pods discussed below, both proposed mimicry types do not necessarily have to be exact (see Stevens et al. 2008b; Penney et al. 2012). There is no need to mimic eyes exactly, since the price of a mistake in identifying a very dangerous predator can be lethal (see Janzen et al. 2010) and it is better not to spend more time in close proximity to such a potential danger. The so-called "eyespots" that certainly defend in many cases from attacks, may operate not only as eye mimicry, but also by just being a salient stimulus, through neophobia (Stevens and Ruxton 2014).

Fully grown green and mature brown seed cones of *Pinus halepensis* and of various other pine species have conspicuous spots that look like eyespots (Fig. 14.3). In these cases the number of "eyes" depends on cone size, its cover by leaves or by other cones and with deterioration of the coloration with age. There are various other types of plant coloration that potentially mimic eyes, but never described.

Certainly, more descriptive, field and experimental work must be done in order to understand the scope of the phenomenon of "eyespots" in plants in general and in *Pisum* and other legume pods that show similar patterns of pod coloration (e.g., Lev-Yadun and Inbar 2002; Aviezer and Lev-Yadun 2015) in particular.

S. Lev-Yadun, *Defensive (anti-herbivory) Coloration in Land Plants*,
DOI 10.1007/978-3-319-42096-7_64

Chapter 65
Snake Mimicry

Dorsal zigzag lines in viper snakes are a well-known visual deterrent towards predators or towards mammalian herbivores that may just trample snakes (Wüster et al. 2004; Mappes et al. 2005; Valkonen et al. 2011). Mimicking venomous snakes as an anti-herbivory defense mechanism by organisms other than snakes was studied especially in the insect world (snake head mimicry) (Cott 1940; Morrell 1969; Ruxton et al. 2004; Janzen et al. 2010; Stevens 2016), and as expected is also found in various non-venomous snakes inhabiting several continents (Wickler 1968; Edmunds 1974). Field experiments in the UK, Finland and Spain with plastic snake models having dorsal European viper markings showed that bearing the characteristic dorsal zigzag band significantly reduced attacks compared to plain models (Wüster et al. 2004; Niskanen and Mappes 2005).

The fully developped pods of some individuals of *P. elatius* have a zigzag or straight red/purple marking along their sides (Figs. 56.10, 56.11 and 56.12). Aviezer and Lev-Yadun (2015) proposed that they resemble the conspicuous dorsal markings of the local common venomous viper, potentially mimicking it. This was the first time that potential defensive snake mimicry was proposed to occur as defense from herbivory in the plant world. This coloration pattern on such *P. elatius* pods, if it is indeed mimicking the dorsal color pattern of a viper snake (something that can potentially be determined by complicated experiments of herbivore behavior), has a great theoretical potential of intimidating large mammalian herbivores and herbivorous birds that approach the plants. The probable model snake for *P. elatius* defensive pod coloration in the apparent shape of snake mimicry found in Israel seems to be the common local viper species (*Vipera palaestinae*) and when this morphology/coloration exists in other parts of the very broad distribution of *P. elatius*, of other, allopatric viper species. In Israel and adjacent countries there are several non-venomous snakes (e.g., *Coluber ravergieri*, *C. nummifer*) that mimic the dorsal zigzag markings of vipers for defense from predation (Bouskila and Amitai 2003) indicating that the model was sufficiently frequent and intimidating to allow for its mimics to have evolved. Even the *Pisum* pods with a straight red line (Fig. 56.10) and other legume taxa with similar pod coloration (Fig. 56.5) look similar to some

S. Lev-Yadun, *Defensive (anti-herbivory) Coloration in Land Plants*,
DOI 10.1007/978-3-319-42096-7_65

local viper specimens since the common local viper is polymorphic concerning its dorsal color pattern. While having the zigzag line is the dominant local viper morph, there are viper individuals in which the line is straight in their proximal part and the zigzag pattern is formed only in their distal part (Mendelssohn 1963).

Although the pods are not as long as vipers, even compared to young and small vipers, in many cases our local vipers as well as many other snakes are seldom seen in their whole length as I know very well from the years of snake watching and catching when I still wished to be a zoologist and study reptiles. Moreover, in probably the best known defensive snake mimicry by non snake taxa, that of the caterpillar *Panacra mydon*, which mimics the head of a pit viper (Morrell 1969), the caterpillar is in the size of *Pisum* pods. The putative snake mimics by smaller-sized organisms of various taxa may also use the principle introduced by Miriam Rothschild (1984) "aide mémoire mimicry", according to which the repelled herbivore is made to recall previous unpleasant experiences or risks and so there is no need for very good or actual size mimicry.

Chapter 66
Visual and Olfactory Feces and Carrion Mimicry

Wiens (1978) was the first to describe visual feces mimicry by plants as defense from herbivory, but said that this is a doubtful category. Examples of possible visual feces mimicry listed in Wiens (1978) are *Anacampseros papyracea* (Portulacaceae), which has some resemblance to the fecal droppings of large birds, and other, smaller species, e.g., *Anacampseros crinita* and *Crassula alstonii*, which look like ungulate feces.

Lev-Yadun et al. (2009b) proposed that carrion and dung odors of various flowers belonging to different taxa that have traditionally been considered an adaptation for attracting flies and beetles for pollination (e.g., Faegri and van der Pijl 1979; Jürgens and Shuttleworth 2016) also have another, overlooked, anti-herbivore defensive function. They suggested that such odors may also deter mammalian herbivores, especially during the critical period of flowering (Fig. 66.1). It was based on the fact that carrion odor is a good predictor of the proximity of carnivores. Similarly, dung odor predicts feces-contaminated habitats that may present high risks of parasitism and pathogens. These were two new types of repulsive olfactory aposematic mimicry by plants: (1) olfactory feigning of carcass (thanatosis), a well-known behavioral defensive strategy in animals (see Edmunds 1974; Ruxton et al. 2004), and (2) olfactory mimicry of feces, which also has a defensive visual parallel in animals (see Hinton 1973; Eisner and Eisner 2000; Eisner et al. 2005; Weiss 2006; Forbes 2009).

The hypothesis that such plant odors may also deter mammalian herbivores, especially during the critical period of flowering (Lev-Yadun et al. 2009b) was supported by field data demonstrating cattle's fear of carrion that resulted in avoidance of very productive grassy plots contaminated by carrion (Lev-Yadun and Gutman 2013). Byers (2015) studied the possibility that mimicry of carrion odors by the insect earwig (*Labidura riparia*), which spits a rotting-flesh odor when bitten by anole lizards (*Anolis carolinenesus*) is an anti-predator defense. Lizards that attacked an earwig and rejected it following the exposure to stinky spit refrained from attacking another one for weeks.

The use of feces as defense against natural enemies is widespread among insects (Weiss 2006). Yamazaki and Lev-Yadun (2014) described plants of *Lycium chinense*

S. Lev-Yadun, *Defensive (anti-herbivory) Coloration in Land Plants*,
DOI 10.1007/978-3-319-42096-7_66

Fig. 66.1 *Arum dioscoridis* has a strong foetid odor proposed not only to attract pollinators, but also to repel herbivores

that have dark axils that look like feces-covered *Lema decempunctata* larvae. The leaf beetle *L. decempunctata*, found in China, Japan, Korea, Siberia and Mongolia, commonly uses this shrub as a host plant, forming on it four generations per year (Kimoto and Takizawa 1994; Ishiwata et al. 2005). Many leaf beetle larvae comprising the five subfamilies Cassidinae, Criocerinae, Cryptocephalinae, Lamprosomatinae and Galerucinae use their feces as defense from predators (Chaboo et al. 2007). The feces contain sequestered secondary substances of their host plants and deter natural enemies such as ants (Morton and Vencl 1998; Vencl et al. 1999; Weiss 2006; Chaboo 2011). Although the chemicals of *L. decempunctata* feces have not been analyzed and the deterrent effects have not been studied, it is likely that the feces have an adverse effect on natural enemies because *L. chinense* contains various secondary chemicals including the alkaloid betaine (Potterat 2010), and *L. decempunctata* larvae would sequester such chemicals in their feces, similar to other congeneric species (Morton and Vencl 1998). Yamazaki and Lev-Yadun (2014) proposed that these observations imply that the dark axils of *L. chinense* may also function as defensive mimicry of beetle feces.

Altogether, there are good indications for a potential defensive role of feces and carrion odor mimicry not only for pollination (e.g., Faegri and van der Pijl 1979) or seed dispersal (Midgley et al. 2015), but also for defense from herbivory (Lev-Yadun et al. 2009b). Experiments to test this hypothesis are certainly needed.

Chapter 67
Extended Phenotype

Rothschild (1972, 1986) proposed that aposematic insects seem to defend certain plants from large herbivores. Accordingly, when various poisonous aposematic insects aggregate on poisonous plants, they add to the plant's aposematic odor and possibly also to its coloration. This is a type of plant-insect mutualism, a case of an extended phenotype, which deserves much more descriptive, theoretical and experimental studies.

This line of thinking was continued by Lev-Yadun and Ne'eman (2012) when they extended their hypothesis of defensive visual bee- and wasp flower mimicry to many other plants that produce highly rewarding bee-pollinated flowers that attract many bees, and proposed that the sight and sound of massive pollinating bee activity may deter herbivore insects and vertebrates from such plants, especially from their flowers, in the critical time of peak flowering season. The well-known deterrence of caterpillar feeding and movement by wasp and bee buzz (Tautz and Markl 1978; Tautz and Rostás 2008), and the great fear of bees by African elephants (Vollrath and Douglas-Hamilton 2002; King et al. 2007, 2009, 2011) are strong indications of the potential of such a defense. Lev-Yadun and Ne'eman (2012) also proposed that if experiments show that pollinating bees deter herbivores, some of the costs of attracting bees by flowers should be regarded as the cost of defense from herbivory rather than pure costs of reproduction. However, this hypothesis still deserves significant field observations and experimental work.

S. Lev-Yadun, *Defensive (anti-herbivory) Coloration in Land Plants*,
DOI 10.1007/978-3-319-42096-7_67

Chapter 68
A General Perspective of Defensive Animal Mimicry by Plants

The proposed cases of defensive mimicry of insects and other animals, and animal action mimicry by plants, commonly by colors, include a wealth of types. Some of them were tested experimentally and found to defend plants from herbivory, e.g., butterfly egg mimicry (Williams and Gilbert 1981; Shapiro 1981a, b) and tunnelling mimicry (Soltau et al. 2009). I am certain that additional animal mimicry types exist, and that many more taxa express the mimicry types already proposed and those that are not yet identified. I estimate the number of plant species that visually mimic animals or their action for defense in the thousands. Similarly, defensive olfactory mimicry of animals or their action by plants is probably not smaller although almost unknown. The combination of the two mimicry types for defense from herbivory should also occur. Although only a few of these hypotheses have been tested directly, these few cases and the much better known indirect data indicate that such defensive animal and animal action mimicry may indeed defend plants from herbivory.

S. Lev-Yadun, *Defensive (anti-herbivory) Coloration in Land Plants*,
DOI 10.1007/978-3-319-42096-7_68

Chapter 18
A General Perspective of Defensive Animal Mimicry by Plants

Chapter 69
Currently Temporary Final Words

This book, the outcome of 35 years of thinking, 22 years of targeted field work and 15 years of publishing, is not the end and not the final word in anything, but rather the beginning of many things. Writing it was an opportunity for a realistic evaluation and presentation of the current status of defensive (anti-herbivory) plant coloration and some related issues. Seeing the progress in the study and understanding of defensive plant coloration and morphology since the year 2000, and knowing that the many studies I currently conduct (for instance on defensive bark coloration) will need some years of progress before ripening, and the other still unpublished studies that I either reviewed or was asked for advice concerning them, I think that within the next decade considerable progress can be expected in this area. Much more so than in zoology, descriptive studies of patterns of plant coloration (including their genetics), are greatly needed. This is the outcome of the fact identified by Harper (1977), who posited almost 40 years ago that "botanists have been reluctant to accept precisions of adaptations that are commonplace to zoologists". When no function was attributed to the non-reproductive or non-physiological color patterns, botanists did not bother to describe and study these patterns, and when they did, certainly not on a regular basis.

Hypotheses about defensive animal coloration that were posited in the nineteenth century such as Batesian (Bates 1862) and Müllerian mimicry (Bates 1862; Müller 1879), aposematism (Bates 1862; Müller 1879; Poulton 1890), and in the first half of the twentieth century such as counter shading (Thayer 1918), dazzle coloration in 1917 (see Wilkinson 1969) and many other hypotheses (e.g., Cott 1940) are still under intensive study. I have no illusions that understanding defensive plant coloration and related visual aspects will be fully understood even at the end of the twenty-first century. Many of the above cited authors who discussed defensive animal and military coloration, based their hypotheses on intuition, the result of reading, thinking, and vast field experience. Many later studies showed that the intuition of the above authors turned out to be a good and realistic guide in many cases, even when the original hypotheses were crude. Concerning defensive plant coloration, the role of intuition is still much too large, but the proposed hypotheses, crude as

S. Lev-Yadun, *Defensive (anti-herbivory) Coloration in Land Plants*,
DOI 10.1007/978-3-319-42096-7_69

they are, are a must in order to allow further detailed and delicate experimentation and modelling that will lead to progress.

Colors play an important role in plant-animal communication. The best-known such communication system is color serving as an advertisement to attract various animals to flowers for the sake of pollination, hence as gene-dispersing agents when they distribute pollen among flowers (Darwin 1877; Faegri and van der Pijl 1979; Clegg and Durbin 2003; Schaefer and Ruxton 2011). The second best-known system is fruit colors, which alert seed-dispersing frugivores to the ripening stage of the fruits (van der Pijl 1982; Willson and Whelan 1990; Schaefer and Ruxton 2011). Although these plant-animal signaling systems are very far from being sufficiently understood, it will take a very long time and enormous efforts (and considerable funding) before defensive plant coloration is understood like these two systems. A good collaboration between botanists (plant scientists) and experts in animal perception and behavior is critical in studying defensive plant coloration and not enough was done concerning this. Moreover, since plant coloration plays a significant role in plant physiology, it is not always easy to distinguish between the physiological and defensive roles of the patterns of coloration. In addition to potential anti-herbivory functions, anthocyanin-based red leaf coloration seems also to defend from fungal attacks (Coley and Aide 1989; Schaefer et al. 2008; Tellez et al. 2016). This antifungal function may further complicate the study of defensive plant coloration. The possibility that defense from herbivory by visuality (coloration, shape, movement) is enhanced by odors, a combination common in aposematic insects (Rowe and Halpin 2013), has received very little attention in plants and must be documented and tested in many more taxa and ecologies.

Even when defensive animal coloration was experimentally tested, this was usually done with a single predator species. However, in the wild, several if not many predators attack or consider an attack on each visually defended species, and the defensive coloration was selected to operate simultaneously against such a mixture of visual systems. This was discussed theoretically by Endler and Mappes (2004) and experimentally by Mappes et al. (2014) and by Turini et al. (2016). In plants, this aspect was never addressed except for yellow *versus* red autumn coloration (Archetti et al. 2009a).

A serious problem in studying defensive plant coloration that is expected to pose even more difficulties in studying defensive plant coloration in the future, is the influence of what is called the Anthropocene, with the loss of natural biomes to agriculture, urbanization, roads and other man-made systems (see Waters et al. 2016). The disappearance of large mammalian and avian herbivores and carnivores, the break-down of trophic cascades and decline of herding as a lifestyle in traditional societies all make defensive plant coloration and other anti-vertebrate defenses unnessesary in many ecosystems. It is known that plant defenses are selected against because of their cost in island biota (e.g., Bowen and Van Vuren 1997; Kavanagh 2015). I can see how it happens in the wild flora of the central coastal plain of Israel where I live. Many thorny, spiny and prickly species that grew there when I was young disappeared after the seasonal grazing by nomadic Bedouins ceased around the year 1980. In Tel Aviv and other cities, the weeds found in gardens and various unbuilt plots are also non-spiny. The same problematic change is also true concerning other systems such as insect herbivores and their major enemies, i.e., birds. Therefore, in many ecosystems the Anthropocene is probably causing evolution towards new optima that include the loss of certain characters and the gain of others. Fieldwork in this quickly changing world may be misleading concerning past adaptations.

In spite of all these problems, the defensive role of plant coloration was tested and demonstrated by various authors, e.g., Numata et al. (2004), Hill (2006), Karageorgou and Manetas (2006), Campitelli et al. (2008), Klooster et al. (2009), Soltau et al. (2009), Wong and Srivastava (2010), Cooney et al. (2012), Strauss and Cacho (2013), Niu et al. (2014), Menzies et al. (2016). I see a colorful future for defensive plant coloration in the coming decades. I hope that this book will contribute to the expected progress by stimulating others to pursue this issue. If after a decade some of the chapters of this book are partly outdated, I will consider it a success.

References

Abbo S, Zezak I, Schwartz E, Lev-Yadun S, Gopher A (2008) Experimental harvesting of wild peas in Israel: implications for the origins of Near East farming. J Archaeol Sci 35:922–929
Abbo S, Saranga Y, Peleg Z, Kerem Z, Lev-Yadun S, Gopher A (2009) Reconsidering legumes *vs.* cereals domestication. Q Rev Biol 84:29–50
Abbo S, Rachamim E, Zehavi Y, Zezak I, Lev-Yadun S, Gopher A (2011) Experimental growing of wild pea in Israel and its bearing on Near Eastern plant domestication. Ann Bot 107:1399–1404
Abbo S, Lev-Yadun S, Heun M, Gopher A (2013a) On the 'lost' crops of the Neolithic Near East. J Exp Bot 64:815–822
Abbo S, Zezak I, Zehavi Y, Schwarts E, Lev-Yadun S, Gopher A (2013b) Six seasons of wild pea harvest in Israel: bearing on Near Eastern plant domestication. J Archaeol Sci 40:2095–2100
Abeles FB, Morgan PW, Saltveit ME Jr (1992) Ethylene in plant biology, 2nd edn. Academic Press, San Diego
Abraham Y, Elbaum R (2013) Hygroscopic movements in Geraniaceae: the structural variations that are responsible for coiling or bending. New Phytol 199:584–594
Abrahamson WG, Sattler JF, McCrea KD, Weis AE (1989) Variation in selection pressures on the goldenrod gall fly and the competitive interactions of its natural enemies. Oecologia 79:15–22
Acherar M, Lepart J, Debussche M (1984) La colonisation des friches par le pin d'Alep (*Pinus halepensis* Miller) en Languedoc méditerranéen. Acta Oecol Oecol Plant 5:179–189
Addicott FT (1982) Abscission. University of California Press, Berkeley
Agrawal AA (1999) Induced responses to herbivory in wild radish: effects on several herbivores and plant fitness. Ecology 80:1713–1723
Agrawal AA, Karban R (2000) Specificity of constitutive and induced resistance: pigment glands influence mites and caterpillars on cotton plants. Entomol Exp Appl 96:39–49
Agrawal AA, Konno K (2009) Latex: a model for understanding mechanisms, ecology, and evolution of plant defense against herbivory. Annu Rev Ecol Evol Syst 40:311–331
Agrawal AA, Spiller DA (2004) Polymorphic buttonwood: effects of disturbance on resistance to herbivores in green and silver morphs of a Bahamian shrub. Am J Bot 91:1990–1997
Agrawal AA, Petschenka G, Bingham RA, Weber MG, Rasmann S (2012) Toxic cardenolides: chemical ecology and coevolution of specialized plant-herbivore interactions. New Phytol 194:28–45
Alanko P (2001) Finnish indegenous trees. Sorbifolia 32:125–133 (in Finnish with English abstract)
Albersheim P, Darvill A, Roberts K, Sederoff R, Staehelin A (2011) Plant cell walls. From chemistry to biology. Garland Science, New York

S. Lev-Yadun, *Defensive (anti-herbivory) Coloration in Land Plants*,
DOI 10.1007/978-3-319-42096-7

Allan AC, Hellens RP, Laing WA (2008) MYB transcription factors that colour our fruit. Trends Plant Sci 13:99–102

Allen JA, Cooper JM (1985) Crypsis and masquerade. J Biol Educ 19:268–270

Allen JA, Knill R (1991) Do grazers leave mottled leaves in the shade? Trends Ecol Evol 6:109–110

Allen LJS, Allen EJ, Kunst CRG, Sosebee RE (1991) A diffusion model for dispersal of *Opuntia imbricata* (cholla) on rangeland. J Ecol 79:1123–1135

Allen WL, Cuthill IC, Scott-Samuel NE, Baddeley R (2011) Why the leopard got its spots: relating pattern development to ecology in felids. Proc R Soc B 278:1373–1380

Alm T (2003) The witch trials of Finnmark, northern Norway, during the 17th century: evidence for ergotism as a contributing factor. Econ Bot 57:403–416

Amador-Vargas S, Dominguez M, León G, Maldonado B, Murillo J, Vides GL (2014) Leaf-folding response of a sensitive plant shows context-dependent behavioral plasticity. Plant Ecol 215:1445–1454

Ambrose MJ, Ellis THN (2008) Ballistic seed dispersal and associated seed shadow in wild *Pisum* germplasm. Pisum Genet 40:5–10

Andersen ØM, Jordheim M (2006) The anthocyanins. In: Andersen ØM, Markham KR (eds) Flavonoids, chemistry, biochemistry and applications, 2nd edn. CRC Press, Boca Raton, pp 452–471

Antkowiak B, Engelmann W (1995) Oscillations of apoplasmic K+ and H+ activities in *Desmodium motorium* (Houtt.) Merril. pulvini in relation to the membrane potential of motor cells and leaflet movements. Planta 196:350–356

Antkowiak B, Mayer W-E, Engelmann W (1991) Oscillations of the membrane potential of pulvinar motor cells *in situ* in relation to leaflet movements of *Desmodium motorium*. J Exp Bot 42:901–910

Archetti M (2000) The origin of autumn colours by coevolution. J Theor Biol 205:625–630

Archetti M (2007a) Autumn colours and the nutritional translocation hypothesis: a theoretical assessment. J Theor Biol 244:714–721

Archetti M (2007b) Colour preference as evidence for the theories on the evolution of autumn colours. J Theor Biol 245:595–596

Archetti M (2009a) Classification of hypotheses for the evolution of autumn colours. Oikos 118:328–333

Archetti M (2009b) Phylogenetic analysis reveals a scattered distribution of autumn colours. Ann Bot 103:703–713

Archetti M (2009c) Evidence from the domestication of apple for the maintenance of autumn colours by coevolution. Proc R Soc B 276:2575–2580

Archetti M, Brown SP (2004) The coevolution theory of autumn colours. Proc R Soc Lond B 271:1219–1223

Archetti M, Brown SP (2006) Putting 'red alerts' in an ecological and evolutionary context. BioEssays 28:959

Archetti M, Leather SR (2005) A test of the coevolution theory of autumn colours: colour preference of *Rhopalosiphum padi* on *Prunus padus*. Oikos 110:339–343

Archetti M, Döring TF, Hagen SB, Hughes NM, Leather SR, Lee DW, Lev-Yadun S, Manetas Y, Ougham HJ, Schaberg PG, Thomas H (2009a) Unravelling the evolution of autumn colours: an interdisciplinary approach. Trends Ecol Evol 24:166–173

Archetti M, Döring TF, Hagen SB, Hughes NM, Leather SR, Lee DW, Lev-Yadun S, Manetas Y, Ougham HJ, Schaberg PG, Thomas H (2009b) Ultraviolet reflectance in autumn leaves and the un-naming of colours. Trends Ecol Evol 24:237–238

Archibold OW (1995) Ecology of world vegetation. Chapman & Hall, London

Arditti J, Rodriguez E (1982) *Dieffenbachia*: uses, abuses and toxic constituents: a review. J Ethnopharmacol 5:293–302

Armbruster WS (1997) Exaptations link evolution of plant-herbivore and plant-pollinator interactions: a phylogenetic inquiry. Ecology 78:1661–1672

Armbruster WS, Howard JJ, Clausen TP, Debevec EM, Loquvam JC, Matsuki M, Cerendolo B, Andel F (1997) Do biochemical exaptations link evolution of plant defense and pollination systems? Historical hypotheses and experimental tests with *Dalechampia* vines. Am Nat 149:461–484

Arnold GW (1966) The special senses in grazing animals. I. Sight and dietary habits in sheep. Aust J Agric Res 17:521–529

Atsatt PR, O'Dowd DJ (1976) Plant defense guilds. Science 193:24–29

Augner M (1994) Should a plant always signal its defence against herbivores? Oikos 70:322–332

Augner M (1995) Low nutritive quality as a plant defence: effects of herbivore-mediated interactions. Evol Ecol 9:605–616

Augner M, Bernays EA (1998) Plant defence signals and Batesian mimicry. Evol Ecol 12:667–679

Aviezer I, Lev-Yadun S (2015) Pod and seed defensive coloration (camouflage and mimicry) in the genus *Pisum*. Isr J Plant Sci 62:39–51

Axelrod DI (1966) Origin of deciduous and evergreen habits in temperate forests. Evolution 20:1–15

Ayasse M, Schiestl FP, Paulus HF, Löfstedt C, Hansson B, Ibarra F, Francke W (2000) Evolution of reproductive strategies in the sexually deceptive orchid *Ophrys sphegodes*: how does flower-specific variation of odor signals influence reproductive success? Evolution 54:1995–2006

Bain RS, Rashed A, Cowper VJ, Gilbert FS, Sherratt TN (2007) The key mimetic features of hoverflies through avian eyes. Proc R Soc B 274:1949–1954

Baker G, Jones LHP, Wardrop ID (1959) Cause of wear in sheeps' teeth. Nature 184:1583–1584

Bakker ES, Olff H, Vandenberghe C, De Maeyer K, Smit R, Gleichman JM, Vera FWM (2004) Ecological anachronisms in the recruitment of temperate light-demanding tree species in wooded pastures. J Appl Ecol 41:571–582

Ballaré CL (2003) Stress under the sun: spotlight on ultraviolet-B responses. Plant Physiol 132:1725–1727

Bannister P (1989) Nitrogen concentration and mimicry in some New Zealand mistletoes. Oecologia 79:128–132

Barbosa P, Hines J, Kaplan I, Martinson H, Szczepaniec A, Szendrei Z (2009) Associational resistance and associational susceptibility: having right or wrong neighbors. Annu Rev Ecol Evol Syst 40:1–20

Bar-Joseph M, Frenkel H (1983) Spraying citrus plants with kaolin suspensions reduces colonization by the spiraea aphid (*Aphis citricola* van der Goot). Crop Prot 2:371–374

Barlow C (2000) The ghosts of evolution. Nonsensical fruits, missing partmers, and other ecological anachronisms. Basic Books, New York

Barlow BA, Wiens D (1977) Host-parasite resemblance in Australian mistletoes: the case for cryptic mimicry. Evolution 31:69–84

Barnett JB, Cuthill IC (2014) Distance-dependent defensive coloration. Curr Biol 24:R1157–R1158

Bar-Oz G, Lev-Yadun S (2012) Paleolithic cave rock art, animal coloration and specific animal habitats. Proc Natl Acad Sci USA 109:E1212

Barton KE (2016) Tougher and thornier: general patterns in the induction of physical defence traits. Funct Ecol 30:181–187

Barton KE, Koricheva J (2010) The ontogeny of plant defense and herbivory: characterizing general patterns using meta-analysis. Am Nat 175:481–493

Bas JM, Gómez C, Pons P (2005) Fruit production and predispersal seed fall and predation in *Rhamnus alaternus* (Rhamnaceae). Acta Oecol 27:115–123

Baskin JM, Baskin CC, Li X (2000) Taxonomy, anatomy and evolution of physical dormancy in seeds. Plant Species Biol 15:139–152

Bates HW (1862) Contributions to an insect fauna of the Amazon valley. Lepidoptera: Heliconidae. Trans Linn Soc Lond XXIII:495–566

Bayly M, Kellow A (2006) An illustrated guide to New Zealand hebes. Te Papa Press, Wellington

Bedford GO (1978) Biology and ecology of the Phasmatodea. Annu Rev Entomol 23:125–149

Bennett JW, Bentley R (1999) Pride and prejudice: the story of ergot. Perspect Biol Med 42:333–355

Bennett KD, Tzedakis PC, Willis KJ (1991) Quaternary refugia of north European trees. J Biogeogr 18:103–115

Benson L (1982) The cacti of the United States and Canada. Stanford University Press, Stanford

Benson WW, Brown KS, Gilbert LE (1975) Coevolution of plants and herbivores: passion flower butterflies. Evolution 29:659–680

Bentley BL (1977) Extrafloral nectaries and protection by pugnacious bodyguards. Annu Rev Ecol Syst 8:407–427

Ben-Ze'ev N, Zohary D (1973) Species relationships in the genus *Pisum* L. Isr J Bot 22:73–91

Benzing DH (2000) Bromeliaceae: profile of an adaptive radiation. Cambridge University Press, Cambridge

Bergvall UA, Rautio P, Kesti K, Tuomi J, Leimar O (2006) Associational effects of plant defences in relation to within- and between-patch food choice by a mammalian herbivore: neighbour contrast susceptibility and defence. Oecologia 147:253–260

Bernays EA, Chapman RF (1994) Host-plant selection by phytophagous insects. Chapman & Hall, New York

Bernays EA, Driver GC, Bilgener M (1989) Herbivores and plant tannins. Adv Ecol Res 19:263–302

Bernier B, Brazeau M (1988) Nutrient deficiency symptoms associated with sugar male dieback and decline in the Quebec Appalachians. Can J For Res 18:762–769

Bhagwat SA, Willis KJ (2008) Species persistence in northerly glacial refugia of Europe: a matter of chance or biogeographical traits? J Biogeogr 35:464–482

Biesmeijer JC, Giurfa M, Koedam D, Potts SG, Joel DM, Dafni A (2005) Convergent evolution: floral guides, stingless bee nest entrances, and insectivorous pitchers. Naturwiss 92:444–450

Birkett MA, Campbell CAM, Chamberlain K, Guerrieri E, Hick AJ et al (2000) New roles for *cis*-jasmone as an insect semiochemical and in plant defense. Proc Natl Acad Sci USA 97:9329–9334

Bixenmann RJ, Coley PD, Kursar TA (2013) Developmental changes in direct and indirect defenses in the young leaves of the Neotropical tree genus *Inga* (Fabaceae). Biotropica 45:175–184

Blackledge TA, Kuntner M, Agnarsson I (2011) The form and function of spider orb webs: evolution from silk to ecosystems. Adv Insect Physiol 41:175–262

Blande JD, Korjus M, Holopainen JK (2010) Foliar methyl salicylate emissions indicate prolonged aphid infestation on silver birch and black alder. Tree Physiol 30:404–416

Blechman H (2004) Disruptive pattern material. An encyclopedia of camouflage. Firefly Books, Buffalo

Blest AD (1963) Longevity, palatability and natural selection in five species of new world saturnid moth. Nature 197:1183–1186

Blick RAJ, Burns KC, Moles AT (2012) Predicting network topology of mistletoe-host interactions: do mistletoes really mimic their hosts? Oikos 121:761–771

Bobich EG, Nobel PS (2001) Vegetative reproduction as related to biomechanics, morphology and anatomy of four cholla cactus species in the Sonoran desert. Ann Bot 87:485–493

Boege K, Marquis RJ (2005) Facing herbivory as you grow up: the ontogeny of resistance in plants. Trends Ecol Evol 20:441–448

Bond AB (2007) The evolution of color polymorphism: crypticity, searching images, and apostatic selection. Annu Rev Ecol Evol Syst 38:489–514

Bond AB, Kamil AC (1998) Apostatic selection by blue jays produces balanced polymorphism in virtual prey. Nature 395:594–596

Bond WJ, Silander JA (2007) Springs and wire plants: anachronistic defences against Madagascar's extinct elephant birds. Proc R Soc B 274:1985–1992

Bond WJ, Lee WG, Craine JM (2004) Plant structural defences against browsing birds: a legacy of New Zealand's extinct moas. Oikos 104:500–508

Borowicz VA (1988) Do vertebrates reject decaying fruit? An experimental test with *Cornus amomum* fruits. Oikos 53:74–78

Bosch J, Retana J, Cerd X (1997) Flowering phenology, floral traits and pollinator composition in a herbaceous Mediterranean plant community. Oecologia 109:583–591

Bottalico A (1998) *Fusarium* diseases of cereals: species complex and related mycotoxin profiles, in Europe. J Plant Pathol 80:85–103

Bouskila A, Amitai P (2003) Handbook of amphibians & reptiles of Israel, 2nd edn. Keter, Jerusalem (in Hebrew)

Bowen L, Van Vuren D (1997) Insular endemic plants lack defenses against herbivores. Conserv Biol 11:1249–1254

Bowers DM (1993) Aposematic caterpillars: life-styles of the warningly colored and unpalatable. In: Stamp NE, Casey TM (eds) Caterpillars: ecological and evolutionary constraints on foraging. Chapman & Hall, New York, pp 331–371

Braam J (2005) In touch: plant responses to mechanical stimuli. New Phytol 165:373–389

Bradbury JH, Nixon RW (1998) The acridity of raphides from the edible aroids. J Sci Food Agric 76:608–616

Bradshaw HD Jr, Schemske DW (2003) Allele substitution at a flower colour locus produces a pollinator shift in monkeyflowers. Nature 426:176–178

Brady J, Shereni W (1988) Landing responses of the tsetse fly *Glossina morsitans morsitans* Westwood and the stable fly *Stomoxys calcitrans* (L.) (Diptera: Glossinidae & Muscidae) to black-and-white patterns: a laboratory study. Bull Entomol Res 78:301–311

Brandt M, Mahsberg D (2002) Bugs with a backpack: the function of nymphal camouflage in the West African assassin bugs *Paredocla* and *Acanthaspis* spp. Anim Behav 63:277–284

Brayton RD, Capon B (1980) Productivity, depletion, and natural selection of *Salvia columbariae* seeds. Aliso 9:581–587

Breed MD, Guzmán-Novoa E, Hunt GJ (2004) Defensive behavior of honey bees: organization, genetics, and comparisons with other bees. Annu Rev Entomol 49:271–298

Breen JP (1994) *Acremonium* endophyte interactions with enhanced plant resistance to insects. Annu Rev Entomol 39:401–423

Briscoe AD, Chittka L (2001) The evolution of color vision in insects. Annu Rev Entomol 46:471–510

Bristow CM (1991) Why are so few aphids ant attended? In: Huxley CR, Cutler DF (eds) Ant-plant interactions. Oxford University Press, London, pp 104–119

Brodmann J, Twele R, Francke W, Hölzler G, Zhang Q-H, Ayasse M (2008) Orchids mimic green-leaf volatiles to attract prey-hunting wasps for pollination. Curr Biol 18:740–744

Brodmann J, Twele R, Francke W, Yi-bo L, Xi-qiang S, Ayasse M (2009) Orchids mimics honey bee alarm pheromone in order to attract hornets for pollination. Curr Biol 19:1368–1372

Brooks R, Owen-Smith N (1994) Plant defences against mammalian herbivores: are juvenile *Acacia* more heavily defended than mature trees? Bothalia 24:211–215

Brown JS (1999) Vigilance, patch use and habitat selection: foraging under predation risk. Evol Ecol Res 1:49–71

Brown SP (2005) A view from Mars. In: Ridley M (ed) Narrow roads of gene land – the collected papers of W. D. Hamilton volume 3 – last words. Oxford University Press, Oxford, pp 350–356

Brown JS, Kotler BP (2004) Hazardous duty pay and the foraging cost of predation. Ecol Lett 7:999–1014

Brown VK, Lawton JH (1991) Herbivory and the evolution of leaf size and shape. Philos Trans R Soc Lond B 333:265–272

Brown JH, Lieberman GA, Dengler WF (1972) Woodrats and cholla: dependence of a small mammal population on the density of cacti. Ecology 53:310–313

Bryan J, Plum F (1972) Persistent vegetative state after brain damage. Lancet 299:734–737

Buchholz R, Levey DJ (1990) The evolutionary triad of microbes, fruits, and seed dispersers: an experiment in fruit choice by cedar waxwings, *Bombycilla cedrorum*. Oikos 59:200–204

Burns KC (2010) Is crypsis a common defensive strategy in plants? Speculation on signal deception in the New Zealand flora. Plant Signal Behav 5:9–13

Burns KC, Dalen JL (2002) Foliage color contrasts and adaptive fruit color variation in a bird-dispersed plant community. Oikos 96:463–469

Burstein M, Wool D (1992) Great tits exploit aphid galls as a source of food. Ornis Scand 23:107–109

Burtt EH Jr, Ichida JM (2004) Gloger's rule, feather-degrading bacteria, and color variation among song sparrows. Condor 106:681–686

Bush LP, Wilkinson HH, Schardl CL (1997) Bioprotective alkaloids of grass-fungal endophyte symbioses. Plant Physiol 114:1–7

Byers JA (2015) Earwigs (*Labidura riparia*) mimic rotting-flesh odor to deceive vertebrate predators. Sci Nat 102:38

Cahn MG, Harper JL (1976) The biology of the leaf mark polymorphism in *Trifolium repens* L. 2. Evidence for the selection of leaf marks by rumen fistulated sheep. Heredity 37:327–333

Caldwell E, Read J, Sanson GD (2016) Which leaf mechanical traits correlate with insect herbivory among feeding guilds? Ann Bot 117:349–361

Camazine S (1983) Mushroom chemical defense: food aversion learning induced by hallucinogenic toxin, muscimol. J Chem Ecol 9:1473–1481

Camazine S (1985) Olfactory aposematism: association of food toxicity with naturally occurring odor. J Chem Ecol 11:1289–1295

Campitelli BE, Steglik I, Stinchcombe JR (2008) Leaf variegation is associated with reduced herbivore damage in *Hydrophyllum virginianum*. Botany 86:306–313

Canyon DV, Hill CJ (1997) Mistletoe host-resemblance: a study of herbivory, nitrogen and moisture in two Australian mistletoes and their host trees. Aust J Ecol 22:395–403

Capon B, Brecht PE (1970) Variations in seed germination and morphology among populations of *Salvia columbariae* Benth. in southern California. Aliso 7:207–216

Carlquist S (1974) Island biology. Columbia University Press, New York

Caro T (2005) Antipredator defenses in birds and mammals. University of Chicago Press, Chicago

Caro T (2009) Contrasting coloration in terrestrial mammals. Philos Trans R Soc B 364:537–548

Carpenter KL, Keidel TS, Pihl MC, Hughes NM (2014) Support for a photoprotective function of winter leaf reddening in nitrogen-deficient individuals of *Lonicera japonica*. Molecules 19:17810–17828

Carrascal LM, Díaz JA, Huertas DL, Mozetich I (2001) Behavioral thermoregulation by treecreepers: trade-off between saving energy and reducing crypsis. Ecology 82:1642–1654

Caveney S, Charlet DA, Freitag H, Maier-Stolte M, Starratt AN (2001) New observations on the secondary chemistry of world *Ephedra* (Ephedraceae). Am J Bot 88:1199–1208

Cebrian J, Lartigue J (2004) Patterns of herbivory and decomposition in aquatic and terrestrial ecosystems. Ecol Monogr 74:237–259

Chaboo C (2011) Defensive behaviors in leaf beetles: from the unusual to the weird. In: Vivanco JM, Weir T (eds) Chemical biology of the tropics. Springer-Verlag, Berlin, pp 59–69

Chaboo C, Grobbelaar E, Larsen A (2007) Fecal ecology in leaf beetles: novel records in the African arrow-poison beetles, *Diamphidia* Gerstaecker and *Polyclada* Chevrolat (Chrysomelidae: Galerucinae). Coleopt Bull 61:297–309

Chalker-Scott L (1999) Environmental significance of anthocyanins in plant stress responses. Photochem Photobiol 70:1–9

Chalker-Scott L (2002) Do anthocyanins function as osmoregulators in leaf tissues? Adv Bot Res 37:103–127

Chang P-K, Bennett JW, Cotty PJ (2001) Association of aflatoxin biosynthesis and sclerotial development in *Aspergillus parasiticus*. Mycopathologia 153:41–48

Chauser-Volfson E, Shen Z, Hu Z, Gutterman Y (2002) Anatomical structure and distribution of secondary metabolites as a peripheral defence strategy in *Aloe hereroensis* leaves. Bot J Linn Soc 138:107–116

Chen Y-Z, Huang S-Q (2013) Red young leaves have less mechanical defence than green young leaves. Oikos 122:1035–1041

Chittka L, Döring TF (2007) Are autumn foliage colors red signals to aphids? PLoS Biol 5:e187

Chittka L, Osorio D (2007) Cognitive dimensions of predator responses to imperfect mimicry? PLoS Biol 5:e339

Christensen-Dean GA, Moore R (1993) Development of chlorenchyma and window tissues in leaves of *Peperomia columella*. Ann Bot 71:141–146

Cieslak M, Reissmann M, Hofreiter M, Ludwig A (2011) Colours of domestication. Biol Rev 86:885–899
Cipollini ML, Levey DJ (1997) Secondary metabolites of fleshy vertebrate-dispersed fruits: adaptive hypotheses and implications for seed dispersal. Am Nat 150:346–372
Cipollini ML, Stiles EW (1993) Fruit rot, antifungal defense, and palatability of fleshy fruits for frugivorous birds. Ecology 74:751–762
Clay K (1988) Clavicipitaceous fungal endophytes of grasses: coevolution and the change from parasitism to mutualism. In: Pirozynski KA, Hawksworth DL (eds) Coevolution of fungi with plants and animals. Academic Press, London, pp 79–105
Clay K (1990) Fungal endophytes of grasses. Annu Rev Ecol Syst 21:275–297
Clay K, Schardl C (2002) Evolutionary origins and ecological consequences of endophyte symbiosis with grasses. Am Nat 160(Suppl):S99–S127
Clegg MT, Durbin ML (2003) Tracing floral adaptations from ecology to molecules. Nat Rev Genet 4:206–215
Clement CR, Manshardt RM (2000) A review of the importance of spines for pejibaye heart-of-palm production. Sci Hortic 83:11–23
Clench HK (1966) Behavioral thermoregulation in butterflies. Ecology 47:1021–1034
Close DC, Beadle CL (2003) The ecophysiology of foliar anthocyanin. Bot Rev 69:149–161
Clotuche G, Mailleux A-C, Astudillo-Fernández A, Deneubourg J-L, Detrain C, Hance T (2011) The formation of collective silk balls in the spider mite *Tetranychus urticae* Koch. PLoS One 6:e18854
Cloudsley-Thompson JL (1979) Adaptive functions of the colours of desert animals. J Arid Environ 2:95–104
Cloudsley-Thompson JL (1999) Multiple factors in the evolution of animal coloration. Naturwiss 86:123–132
Cole DT (1970) Lithops in habitat. In: Sprechman DL (ed) Lithops. Fairleigh Dickinson University Press, Cranbury, pp 21–32
Cole DT, Cole NA (2005) Lithops – flowering stones. Cactus & Co., Venegono
Coley PD (1980) Effects of leaf age and plant life history patterns on herbivory. Nature 284:545–546
Coley PD, Aide TM (1989) Red coloration of tropical young leaves: a possible antifungal defence? J Trop Ecol 5:293–300
Coley PD, Barone JA (1996) Herbivory and plant defenses in tropical forests. Annu Rev Ecol Syst 27:305–335
Coley PD, Bryant JP, Chapin FS III (1985) Resource availability and plant antiherbivore defense. Science 230:895–899
Comes HP, Kadereit JW (1998) The effect of Quaternary climatic changes on plant distribution and evolution. Trends Plant Sci 3:432–438
Connor EF, Bartlett L, O'Toole S, Byrd S, Biskar K, Orozco J (2012) The mechanism of gall induction makes galls red. Arthropod Plant Interact 6:489–495
Cook LM (2000) Changing views on melanic moths. Biol J Linn Soc 69:431–441
Cook LM (2003) The rise and fall of the *Carbonaria* form of the peppered moth. Q Rev Biol 78:399–417
Cook AD, Atsatt PR, Simon CA (1971) Doves and dove weed: multiple defenses against avian predation. Bioscience 21:277–281
Cook RJ, Barron JC, Papendick RI, Williams GJ III (1980) Impact on agriculture of the Mount St. Helens eruptions. Science 211:16–22
Cooney LJ, van Klink JW, Hughes NM, Perry NB, Schaefer HM, Menzies IJ, Gould KS (2012) Red leaf margins indicate increased polygodial content and function as visual signals to reduce herbivory in *Pseudowintera colorata*. New Phytol 194:488–497
Cooper SM, Owen-Smith N (1986) Effects of plant spinescence on large mammalian herbivores. Oecologia 68:446–455
Coppinger RP (1970) The effect of experience and novelty on avian feeding behavior with reference to the evolution of warning coloration in butterflies. II. Reactions of naïve birds to novel insects. Am Nat 104:323–335

Cornell HV (1983) The secondary chemistry and complex morphology of galls formed by the Cynipinae (Hymenoptera): why and how? Am Midl Nat 110:225–234
Cornett JW (1986) Spineless petioles in *Washingtonia filifera* (Arecaceae). Madrono 33:76–78
Cott HB (1940) Adaptive coloration in animals. Methuen & Co. Ltd., London
Craig CL (1997) Evolution of arthropod silks. Annu Rev Entomol 42:231–267
Cramer JM, Cloud ML, Muchhala NC, Ware AE, Smith BH, Williamson GB (2003) A test of the bicolored fruit display hypothesis: berry removal with artificial fruit flags. J Torrey Bot Soc 130:30–33
Cranshaw W (1998) Pests of the west, 2nd edn. Fulcrum Publishing, Golden
Crawford-Sidebotham TJ (1972) The role of slugs and snails in the maintenance of the cyanogenesis polymorphisms of *Lotus corniculatus* and *Trifolium repens*. Heredity 28:405–411
Crawley MJ (1983) Herbivory. The dynamics of animal-plant interactions. Blackwell Scientific Publications, Oxford
Crespi B, Worobey M (1998) Comparative analysis of gall morphology in Australian gall thrips: the evolution of extended phenotypes. Evolution 52:1686–1696
Cronin TW, Johnsen S, Marshall NJ, Warrant EJ (2014) Visual ecology. Princeton University Press, Princeton
Cyr H, Pace ML (1993) Magnitude and patterns of herbivory in aquatic and terrestrial ecosystems. Nature 361:148–150
Czeczuga B (1977) Carotenoids in leaves and their galls. Marcellia 40:177–180
Dafni A (1984) Mimicry and deception in pollination. Annu Rev Ecol Syst 15:259–278
Dafni A, Giurfa M (1999) The functional ecology of floral guides in relation to insects behaviour and vision. In: Wasser SP (ed) Evolutionary theory and processes: modern perspectives, papers in honour of Eviatar Nevo. Kluwer Academic Publishers, Dordrecht, pp 363–383
Dafni A, Lehrer M, Kevan PG (1997) Spatial flower parameters and insect spatial vision. Biol Rev 72:239–282
Danin A (1977) The vegetation of the Negev (North of Nahal Paran). Sifriat Poalim, Merhavia (in Hebrew)
Danin A (1996) Plants of desert dunes. Springer Verlag, Berlin
Danin A (2004) Distribution atlas of plants in the *Flora Palaestina* area. The Israel Academy of Sciences and Humanities, Jerusalem
Darwin C (1877) The different forms of flowers on plants of the same species. John Murray, London
Darwin C (1881) The power of movement in plants. D. Appleton, New York; Da Capo Press reprint edition 1966, New York
Daskin JH, Stalmans M, Pringle RM (2016) Ecological legacies of civil war: 35-year increase in savanna tree cover following wholesale large-mammal declines. J Ecol 104:79–89
Dawkins M (1971) Shifts of 'attention' in chicks during feeding. Anim Behav 19:575–582
Dawkins R (1982) The extended phenotype. Oxford University Press, Oxford
de Almeida A, do Couto HTZ, de Almeida AF (2010a) Are camouflaged seeds less attacked by wild birds? Sci Agric 67:170–175
de Almeida A, do Couto HTZ, de Almeida AF (2010b) Camouflaging of seeds treated with pesticides mitigates the mortality of wild birds in wheat and rice crops. Sci Agric 67:176–182
De Bona S, Valkonen JK, López-Sepulcre A, Mappes J (2015) Predator mimicry, not conspicuousness, explains the efficacy of butterfly eyespots. Proc R Soc B 282:20150202
De Costa C (2002) St Anthony's fire and living ligatures: a short history of ergometrine. Lancet 359:1768–1770
De Moraes CM, Lewis WJ, Paré PW, Alborn HT, Tumlinson JH (1998) Herbivore-infested plants selectively attract parasitoids. Nature 393:570–573
De Moraes CM, Mescher MC, Tumlinson JH (2001) Caterpillar-induced nocturnal plant volatiles repel conspecific females. Nature 410:577–580
de Wert L, Mahon K, Ruxton GD (2012) Protection by association: evidence for aposematic commensalism. Biol J Linn Soc 106:81–89

Delcourt PA, Delcourt HR (1993) Paleoclimates, paleovegetation, and paleofloras during the Late Quaternary. In: Flora of North America North of Mexico, vol 1. Oxford University Press, New York, pp 71–94
Dell B, Jones S, Wilson SA (1987) Phosphorus nutrition of jarrah (*Eucalyptus marginata*) seedlings. Plant Soil 97:369–379
Detto C (1905) Blütenbiologische Untersuchungen, I. Über die Bedeutung der Insektenähnlichkeit der Ophrysblute bebst bemerkungen über die Möhrenblütte bei *Daucus carota*. Flora 94:287–329
D'Horta FM, Kirwan GM, Buzzetti D (2012) Gaudy juvenile plumages of cinereous mourner (*Laniocera hypopyrra*) and Brazilian laniisoma (*Laniisoma elegans*). Wilson J Ornithol 124:429–435
Diamond JM (1990) Biological effects of ghosts. Nature 345:769–770
Diamond J (2005) Collapse. How societies choose to fail or succeed. Viking, New York
Diamond J, Bond AB (2013) Concealing coloration in animals. Harvard University Press, Cambridge
Dias GG, Moreira GRP, Ferreira BG, Isaias RMS (2013) Why do the galls induced by *Calophya duvauae* Scott on *Schinus polygamus* (Cav.) Cabrera (Anacardiaceae) change colors? Biochem Syst Ecol 48:111–122
Dicke M, Sabelis MW (1988) How plants obtain predatory mites as bodyguards. Neth J Zool 38:148–165
Dicke M, Sabelis MW, Takabayashi J, Bruin J, Posthumus MA (1990) Plant strategies of manipulating predator-prey interactions through allelochemicals: prospects for application in pest control. J Chem Ecol 16:3091–3118
Dirzo R (2002) Lying leaves. Science 297:1119
Dirzo R, Harper JL (1982) Experimental studies on slug-plant interactions IV. The performance of cyanogenic and acyanogenic morphs of Trifolium repens in the field. J Ecol 70:119–138
Dixon AFG (1998) Aphid ecology. An optimization approach, 2nd edn. Chapman and Hall, London
Dixon AFG, Hopkins GW (2010) Temperature, seasonal development and distribution of insects with particular reference to aphids. In: Kindlmann P, Dixon AFG, Michaud JP (eds) Aphid biodiversity under environmental change. Springer, Dordrecht, pp 129–147
Dobson HEM, Bergström G (2000) The ecology and evolution of pollen odors. Plant Syst Evol 222:63–87
Dodson C (1962) The importance of pollination in the evolution of the orchids of tropical America. Am Orchid Soc Bull 31:525–534, 641–649, 731–735
Dodson C, Frymire G (1961) Natural pollination of orchids. Mo Bot Gard Bull 49:133–152
Doku C, Brady J (1989) Landing site preferences of *Glossina morsitans morsitans* Westwood (Diptera: Glossinidae) in the laboratory: avoidance of horizontal features? Bull Entomol Res 79:521–528
Domínguez K, Jiménez M-L (2005) Mating and self-burying behavior of *Homalonychus theologus* Chamberlin (Araneae, Homalonychidae) in Baja California sur. J Arachnol 33:167–174
Dominy NJ, Lucas PE (2001) Ecological importance of trichromatic vision to primates. Nature 410:363–366
Dominy ND, Lucas PW, Ramsden W, Riba-Hernandez P, Stoner KE, Turner IM (2002) Why are young leaves red? Oikos 98:163–176
Döring TF, Archetti M, Hardie J (2009) Autumn leaves seen through herbivore eyes. Proc R Soc B 276:121–127
Dukas R (2001) Effects of perceived danger on flower choice by bees. Ecol Lett 4:327–333
Dukas R, Ellner S (1993) Information processing and prey detection. Ecology 74:1337–1346
Dukas R, Kamil AC (2000) The cost of limited attention in blue jays. Behav Ecol 11:502–506
Duke S (1994) Glandular trichomes – a focal point of chemical and structural interactions. Int J Plant Sci 155:617–620
Durnal EW (2010) Crime scene investigation (as seen on TV). Forensic Sci Int 199:1–5
Dussourd DE (1995) Entrapment of aphids and whiteflies in lettuce latex. Ecol Popul Biol 88:163–172

Dussourd DE, Eisner T (1987) Vein-cutting behavior: insect counterploy to the latex defense of plants. Science 237:898–901

Eadie MJ (2004) Ergot of rye – the first specific for migraine. J Clinic Neurosci 11:4–7

Ebeling W (1971) Sorptive dusts for pest control. Annu Rev Entomol 16:123–158

Edmunds M (1974) Defence in animals. A survey of anti-predator defences. Longman Group Ltd., Harlow

Edmunds M (2000) Why are there good and poor mimics? Biol J Linn Soc 70:459–466

Edmunds M, Grayson J (1991) Camouflage and selective predation in caterpillars of the poplar and eyed hawkmoths (*Laothoe populi* and *Smerinthus ocellata*). Biol J Linn Soc 42:467–480

Egri Á, Blahó M, Kriska G, Farkas R, Gyurkovszky M, Åkesson S, Horváth G (2012) Polarotactic tabanids find striped patterns with brightness and/or polarization modulation least attractive: an advantage of zebra stripes. J Exp Biol 215:736–745

Ehleringer J, Björkman O, Mooney HA (1976) Leaf pubescence: effects on absorptance and photosynthesis in a desert shrub. Science 192:376–377

Ehleringer JR, Ullmann I, Lange OL, Farquhar GD, Cowan IR, Schulze E-D, Ziegler H (1986) Mistletoes: a hypothesis concerning morphological and chemical avoidance of herbivory. Oecologia 70:234–237

Eisikowitch D (1980) The role of dark flowers in the pollination of certain Umbelliferae. J Nat Hist 14:737–742

Eisner T (1964) Catnip: its Raison d'Être. Science 146:1318–1320

Eisner T (1981) Leaf folding in a sensitive plant: a defensive thorn-exposure mechanism? Proc Natl Acad Sci USA 78:402–404

Eisner T (1985) A fly that mimics jumping spiders. Psyche 92:103–104

Eisner T, Eisner M (2000) Defensive use of a fecal thatch by a beetle larva (*Hemisphaerota cyanea*). Proc Natl Acad Sci USA 97:2632–2636

Eisner T, Grant RP (1981) Toxicity, odor aversion, and "olfactory aposematism". Science 213:476

Eisner T, Eisner M, Hoebeke ER (1998) When defense backfires: detrimental effect of a plant's protective trichomes on an insect beneficial to the plant. Proc Natl Acad Sci USA 95:4410–4414

Eisner T, Eisner M, Siegler M (2005) Secret weapons. Defenses of insects, spiders, scorpions, and other many-legged creatures. Harvard University Press, Cambridge

Elmi AA, West CP (1995) Endophyte infection effects on stomatal conductance, osmotic adjustment, and drought recovery of tall fescue. New Phytol 131:61–67

Embar K, Kotler BP, Mukherjee S (2011) Risk management in optimal foragers: the effect of sightlines and predator type on patch use, time allocation, and vigilance in gerbils. Oikos 120:1657–1666

Endara MJ, Coley PD (2011) The resource availability hypothesis revisited: a meta-analysis. Funct Ecol 25:389–398

Endler JA (1978) A predator's view of animal color patterns. Evol Biol 11:319–364

Endler JA (1981) An overview of the relationships between mimicry and crypsis. Biol J Linn Soc 16:25–31

Endler JA (1984) Progressive background matching in moths, and a quantitative measure of crypsis. Biol J Linn Soc 22:187–231

Endler JA (1986) Natural selection in the wild. Princeton University Press, Princeton

Endler JA, Mappes J (2004) Predator mixes and the conspicuousness of aposematic signals. Am Nat 163:532–547

Enquist M, Arak A (1994) Symmetry, beauty and evolution. Nature 372:169–172

Ergonul O, Erbay A, Eren S, Dokuzoguz B (2003) Analysis of the case fatality rate of tetanus among adults in a tertiary hospital in Turkey. Eur J Clin Microbiol Infect Dis 22:188–190

Euw JV, Fishelson L, Parsons JA, Reichstein T, Rothschild M (1967) Cardenolides (heart poisons) in a grasshopper feeding on milkweeds. Nature 214:35–39

Evans CRH (1987) Oral ulceration after contact with the houseplant *Dieffenbachia*. Br Dent J 162:467–468

Evenari M, Shanan L, Tadmor N (1982) The Negev. The challenge of a desert, 2nd edn. Harvard University Press, Cambridge
Facelli JM (1993) Experimental evaluation of the foliar flag hypothesis using fruits of *Rhus glabra* (L.). Oecologia 93:70–72
Fadzly N, Burns KC (2010) Hiding from the ghost of herbivory past: evidence for crypsis in an insular tree species. Int J Plant Sci 171:828–833
Fadzly N, Jack C, Schaefer HM, Burns KC (2009) Ontogenetic colour changes in an insular tree species: signalling to extinct browsing birds? New Phytol 184:495–501
Faegri K, van der Pijl L (1979) The principles of pollination ecology, 3rd edn. Pergamon Press, Oxford
Faeth SH, Connor EF, Simberloff D (1981) Early leaf abscission: a neglected source of mortality for folivores. Am Nat 117:409–415
Fahn A (1979) Secretory tissues in plants. Academic Press, London
Fahn A (1988) Secretory tissues in vascular plants. New Phytol 108:229–257
Fahn A (1990) Plant anatomy, 4th edn. Pergamon Press, Oxford
Fahn A, Cutler DF (1992) Xerophytes. Gebrüder Borntraeger, Berlin
Fahn A, Shimony C (1996) Glandular trichomes of *Fagonia* L. (Zygophyllaceae) species: structure, development and secreted materials. Ann Bot 77:25–34
Fahn A, Zohary M (1955) On the pericarpial structure of the legumen, its evolution and relation to dehiscence. Phytomorphology 5:99–111
Farmer EE (2014) Leaf defence. Oxford University Press, Oxford
Fedriani JM, Boulay R (2006) Foraging by fearful frugivores: combined effect of fruit ripening and predation risk. Funct Ecol 20:1070–1079
Feeny O (1970) Seasonal changes in oak leaf tannins and nutrients as a cause of spring feeding by winter moth caterpillars. Ecology 51:565–581
Feeny P (1976) Plant apparency and chemical defense. Recent Adv Phytochem 10:1–40
Feild TS, Lee DW, Holbrook NM (2001) Why leaves turn red in autumn. the role of anthocyanins in senescing leaves of red-osier dogwood. Plant Physiol 127:566–574
Feinbrun-Dothan N (1978) Flora Palaestina, vol III. The Israel Academy of Sciences and Humanities, Jerusalem
Feinbrun-Dothan N, Danin A (1991) Analytical flora of Eretz-Israel. Cana Publishing House Ltd., Jerusalem (in Hebrew)
Feinsinger P (1978) Ecological interactions between plants and hummingbirds in a successional tropical community. Ecol Monogr 48:269–287
Fendt M (2006) Exposure to urine of canids and felids, but not of herbivores, induces defensive behavior in laboratory rats. J Chem Ecol 32:2617–2627
Field J (1992) Guild structure in solitary spider-hunting wasps (Hymenoptera: Pompilidae) compared with null model predictions. Ecol Entomol 17:198–208
Finch S, Collier RH (2000) Host plant selection by insects: a theory based on "appropriate/inappropriate landings" by pest insects of cruciferous plants. Entomol Exp Appl 96:91–102
Finch S, Jones TH (1989) An analysis of the deterrent effect of Aphids on cabbage root fly (*Delia radicum*) egg-laying. Ecol Entomol 14:387–391
Fineblum WL, Rausher MD (1997) Do floral pigmentation genes also influence resistance to enemies? The *W* locus in *Ipomoea purpurea*. Ecology 78:1646–1654
Fink LS (1995) Foodplant effects on colour morphs of *Eumorpha fasciata* caterpillar (Lepidoptera: Sphingidae). Biol J Linn Soc 56:423–437
Forbes P (2009) Dazzled and deceived: mimicry and camouflage. Yale University Press, New Haven
Ford AT, Goheen JR, Otieno TO, Bidner L, Isbell LA, Palmer TM, Ward D, Woodroffe R, Pringle RM (2014) Large carnivores make savanna tree communities less thorny. Science 346:346–349
Forsman A, Herrström J (2004) Asymmetry in size, shape, and color impairs the protective value of conspicuous color patterns. Behav Ecol 15:141–147

Forsman A, Merilaita S (1999) Fearful symmetry: pattern size and asymmetry affects aposematic signal efficacy. Evol Ecol 13:131–140
Forsman A, Merilaita S (2003) Fearful symmetry? Intra-individual comparisons of asymmetry in cryptic vs. signalling colour patterns in butterflies. Evol Ecol 17:491–507
Franceschi VR, Horner HT Jr (1980) Calcium oxalate crystals in plants. Bot Rev 46:361–427
Franceschi VR, Nakata PA (2005) Calcium oxalate in plants: formation and function. Annu Rev Plant Biol 56:41–71
Frankie GW, Baker HG, Opler PA (1974) Comparative phenological studies of trees in tropical wet and dry forests in the lowlands of Costa Rica. J Ecol 62:881–919
Franks NR, Hooper J, Webb C, Dornhaus A (2005) Tomb evaders: house-hunting hygiene in ants. Biol Lett 1:190–192
Frederickson ME, Greene MJ, Gordon DM (2005) 'Devil's gardens' bedevilled by ants. Nature 437:495–496
Fu HY, Chen SJ, Chen RF, Ding WH, Kuo-Huang LL, Huang RN (2006) Identification of oxalic acid and tartaric acids as major persistent pain-inducing toxins in the stinging hairs of the Nettle, *Urtica thunbergiana*. Ann Bot 98:57–65
Fuentes M (1995) The effect of unripe fruits on ripe fruit removal by birds in *Pistacia terebinthus*: flag or handicap? Oecologia 101:55–58
Fukui A (2001) Indirect interactions mediated by leaf shelters in animal-plant communities. Popul Ecol 43:31–40
Furth DG (2004) Fun with flea beetle feces. Chrysomela 43:10
Furuta K (1986) Host preference and population dynamics in an autumnal population of the maple aphid, *Periphyllus californiensis* Shinji (Homoptera, Aphididae). J Appl Entomol 102:93–100
Futuyma DJ, Agrawal AA (2009) Macroevolution and the biological diversity of plants and herbivores. Proc Natl Acad Sci USA 106:18054–18061
Gagliano M, Renton M, Depczynski M, Mancuso S (2014) Experience teaches plants to learn faster and forget slower in environments where it matters. Oecologia 175:63–72
Gal A, Brumfeld V, Weiner S, Addadi L, Oron D (2012) Certain biominerals in leaves function as light scatters. Adv Mater 24:OP77–OP83
Gamberale-Stille G, Tullberg BS (2001) Fruit or aposematic insect? Context-dependent colour preferences in domestic chicks. Proc R Soc Lond B 268:2525–2529
Ganders FR, Griffiths AJF, Carey K (1980) Natural selection for spotted leaves: parallel morph ratio variation in three species of annual plants. Can J Bot 58:689–693
Gardner DG (1994) Injury to the oral mucous membranes caused by the common houseplant, diffenbachia. A review. Oral Surg Oral Med Oral Pathol 78:631–633
Gaskett AC, Herberstein ME (2010) Colour mimicry and sexual deception by Tongue orchids (*Cryptostylis*). Naturwiss 97:97–102
Gates DM (1980) Biophysical ecology. Springer-Verlag, New York
Geburek T, Robitschek K, Milasowszky N, Schadauer K (2007) Different cone colours pay off: lessons learnt from European larch (*Larix decidua*) and Norway spruce (*Picea abies*). Can J Bot 85:132–140
Gentry HS (1982) Agaves of continental North America. University of Arizona Press, Tucson
Gerchman Y, Dodek I, Petichov R, Yerushalmi Y, Lerner A, Keasar T (2012) Beyond pollinator attraction: extra-floral displays deter herbivores in a Mediterranean annual plant. Evol Ecol 26:499–512
Gerchman Y, Lev-Yadun S, Inbar M (2013) Red gall pigmentation: cytokinin stimulation is not everything. Arthropod Plant Interact 7:335–337
Gianoli E, Carrasco-Urra F (2014) Leaf mimicry in a climbing plant protects against herbivory. Curr Biol 24:984–987
Gibson DO (1980) The role of escape in mimicry and polymorphism: I. The response of captive birds to artificial prey. Biol J Linn Soc 14:201–214
Gibson G (1992) Do tsetse flies 'see' zebras? A field study of the visual response of tsetse to striped targets. Physiol Entomol 17:141–147
Gibson AC, Nobel PS (1986) The cactus primer. Harvard University Press, Cambridge

Gilbert LE (1980) Ecological consequences of a coevolved mutualism between butterflies and plants. In: Gilbert LE, Raven PH (eds) Coevolution of animals and plants. University of Texas Press, Austin, pp 210–240

Gilbert LE (1982) The coevolution of a butterfly and a vine. Sci Am 247(8):102–107

Gillaspy G, Ben-David H, Gruissem W (2003) Fruits: a developmental perspective. Plant Cell 5:1439–1451

Gittleman JL, Harvey PH (1980) Why are distasteful prey not cryptic? Nature 286:149–150

Givnish TJ (1990) Leaf mottling: relation to growth form and leaf phenology and possible role as camouflage. Funct Ecol 4:463–474

Givnish TJ, Sytsma KJ, Smith JF, Hahn WJ (1994) Thorn-like prickles and heterophylly in *Cyanea*: adaptations to extinct avian browsers on Hawaii? Proc Natl Acad Sci USA 91:2810–2814

Gleadow RM, Woodrow IE (2002) Constraints on effectiveness of cyanogenic glycosides in herbivore defense. J Chem Ecol 28:1301–1313

Glinwood R, Pettersson J (2000) Movement by mating females of a host alternating aphid: a response to leaf fall. Oikos 90:43–49

Gloger CL (1833) Das Abändern der Vögel durch Einfluss der Klima's. A. Schulz, Breslau

Glover BJ, Whitney MJ (2010) Structural colour and iridescence in plants: the poorly studied relations of pigment colour. Ann Bot 105:505–511

Gong B, Zhang G (2014) Interactions between plants and herbivores: a review of plant defense. Acta Ecol Sin 34:325–336

González-Teuber M, Heil M (2015) Comparative anatomy and physiology of myrmecophytes: ecological and evolutionary perspectives. Res Rep Biodivers Stud 4:21–32

Gotelli NJ, Ulrich W (2012) Statistical challenges in null model analysis. Oikos 121:171–180

Goto T, Wicklow DT, Ito Y (1996) Aflatoxin and cyclopiazonic acid production by a sclerotium-producing *Aspergillus tamarii* strain. Appl Environ Microbiol 62:4036–4038

Goubitz S, Werger M, Ne'eman G (2003) Germination response to fire related factors of seeds from non-serotinous and serotinous cones. Plant Ecol 169:195–204

Goubitz S, Nathan R, Roitemberg R, Shmida A, Ne'eman G (2004) Canopy seed bank structure in relation to: fire, tree size and density. Plant Ecol 173:191–201

Gould KS (2004) Nature's Swiss army knife: the diverse protective roles of anthocyanins in leaves. J Biomed Biotechnol 2004(5):314–320

Gould KS, Lee DW, Callow JA (eds) (2002a) Anthocyanins in leaves. Adv Bot Res 37

Gould KS, Neill SO, Vogelmann TC (2002b) A unified explanation for anthocyanins in leaves? Adv Bot Res 37:167–192

Goulson D, McGuire K, Munro EE, Adamson S, Colliar L, Park KJ, Tinsley MC, Gilburn AS (2009) Functional significance of the dark central floret of *Daucus carota* (Apiaceae) L.; is it an insect mimic? Plant Species Biol 24:77–82

Gowda JH (1996) Spines of *Acacia tortilis*: what do they defend and how? Oikos 77:279–284

Gowda JH, Palo RT (2003) Age-related changes in defensive traits of *Acacia tortilis* Hayne. Afr J Ecol 41:218–223

Grafen A (1990) Biological signals as handicaps. J Theor Biol 144:517–546

Graham A (1993) History of North American vegetation – Cretaceous (Maastrichtian) – Tertiary. In: Flora of North America, vol 1. Oxford University Press, New York, pp 57–70

Grammatikopoulos G, Karabourniotis G, Kyparissis A, Petropoulou Y, Manetas Y (1994) Leaf hairs of olive (*Olea europaea*) prevent stomatal closure by ultraviolet-B radiation. Aust J Plant Physiol 21:293–301

Grant V, Grant KA (1965) Flower pollination in the phlox family. Columbia University Press, New York

Grant BS, Wiseman LL (2002) Recent history of melanism in American peppered moths. J Hered 93:86–90

Grayson J, Edmunds M (1989) The causes of colour change in caterpillars of the poplar and eyed hawkmoths (*Laothoe populi* and *Smerinthus ocellata*). Biol J Linn Soc 37:263–279

Green E (1989) A diet-induced developmental polymorphism in a caterpillar. Science 243:643–646

Green TR, Ryan CA (1972) Wound-induced proteinase inhibitor in plant leaves: a possible defense mechanism against insects. Science 175:776–777
Greenwood RM, Atkinson IAE (1977) Evolution of divaricating plants in New Zealand in relation to Moa browsing. Proc N Z Ecol Soc 24:21–33
Greig-Smith PW (1986) Bicolored fruit displays and frugivorous birds: the importance of fruit quality to dispersers and bird predators. Am Nat 127:246–251
Gressitt JL, Sedlacek J, Szent-Ivany JJH (1965) Flora and fauna on backs of large Papuan moss-forest weevils. Science 150:1833–1835
Grey-Wilson C (1988) The genus Cyclamen. The Royal Botanic Gardens, Kew, in association with Christopher Helm and Timber Press, Kew
Gronquist M, Bezzerides A, Attygalle A, Meinwald J, Eisner M, Eisner T (2001) Attractive and defensive functions of the ultraviolet pigments of a flower (*Hypericum calycinum*). Proc Natl Acad Sci USA 98:13745–13750
Groover A (2016) Gravitropisms and reaction woods of forest trees – evolution, functions and mechanisms. New Phytol 211:790–802
Grose MJ (2007) 'Green above, paler below': descriptions in the literature of the colour in trees from southwest Australia. J R Soc West Aust 90:179–194
Grose MJ (2014) Reading the colours of plants at a finer scale. J Landsc Archit 9:42–47
Grotewold E (2006) The genetics and biochemistry of floral pigments. Annu Rev Plant Biol 57:761–780
Grubb PJ (1992) A positive distrust in simplicity – lessons from plant defences and from competition among plants and among animals. J Ecol 80:585–610
Guevara R, Dirzo R (1999) Consumption of macro-fungi by invertebrates in a Mexican tropical cloud forest: do fruit body characteristics matter? J Trop Ecol 15:603–617
Guilford T (1994) "Go-slow" signalling and the problem of automimicry. J Theor Biol 170:311–316
Guilford T, Dawkins MS (1993) Are warning colors handicaps? Evolution 47:400–416
Guilford T, Nicol C, Rothschild M, Moore BP (1987) The biological roles of pyrazines: evidence for a warning odour function. Biol J Linn Soc 31:113–128
Guimarães PR Jr, Galetti M, Jordano P (2008) Seed dispersal anachronisms: rethinking the fruits extinct megafauna ate. PLoS One 3:e1745
Gunnarsson B (2007) Bird predation on spiders: ecological mechanisms and evolutionary consequences. J Arachnol 35:509–529
Gupta D, Bleakley B, Gupta RK (2008) Dragon's blood: botany, chemistry and therapeutic uses. J Ethnopharmacol 115:361–380
Guthrie RD, Petocz RG (1970) Weapon automimicry among animals. Am Nat 104:585–588
Gutman M, Seligman NG (1979) Grazing management of Mediterranean foot-hill range in the Upper Jordan River Valley. J Range Manag 32:86–92
Gutman M, Seligman NG, Noy-Meir I (1990) Herbage production of Mediterranean grassland under seasonal and yearlong grazing systems. J Range Manag 43:64–68
Gutman M, Holzer Z, Baram H, Noy-Meir I, Seligman NG (1999) Heavy stocking and early-season deferment of grazing on Mediterranean-type grassland. J Range Manage 52:590–599
Gutterman Y (1978) Seed coat permeability as a function of photoperiodical treatments of the mother plants during seed maturation in the desert annual plant: *Trigonella arabica*, Del. J Arid Environ 1:141–144
Gutterman Y, Chauser-Volfson E (2000) The distribution of the phenolic metabolites barbaloin, aloeresin and aloenin as a peripheral defense strategy in the succulent leaf parts of *Aloe arborescens*. Biochem Syst Ecol 28:825–838
Gutterman Y, Evenari M (1972) The influence of daylength on seed coat colour, an index of water permeability of the desert annual *Ononis sicula* Guss. J Ecol 60:713–719
Haberlandt G (1914) Physiological plant anatomy (trans: Montagu Drummond). MacMillan and Co Ltd, London
Hagel JM, Yeung EC, Facchini PJ (2008) Got milk? The secret life of laticifers. Trends Plant Sci 13:631–639

Hagen SB, Folstad I, Jakobsen SW (2003) Autumn colouration and herbivore resistance in mountain birch (*Betula pubescens*). Ecol Lett 6:807–811
Hagen SB, Debeausse S, Yoccoz NG, Folstad I (2004) Autumn coloration as a signal of tree condition. Proc R Soc Lond B 271(Suppl):S184–S185
Haim A, Izhaki I (1994) Changes in rodent community during recovery from fire: relevance to conservation. Biodivers Conserv 3:573–585
Haim A, Izhaki I, Golan A (1996) Rodent species diversity in pine forests recovering from fire. Isr J Zool 42:353–359
Hairston NG, Smith FE, Slobodkin LB (1960) Community structure, population control, and competition. Am Nat 94:421–425
Halpern M, Raats D, Lev-Yadun S (2007a) Plant biological warfare: thorns inject pathogenic bacteria into herbivores. Environ Microbiol 9:584–592
Halpern M, Raats D, Lev-Yadun S (2007b) The potential anti-herbivory role of microorganisms on plant thorns. Plant Signal Behav 2:503–504
Halpern M, Waissler A, Dror A, Lev-Yadun S (2011) Biological warfare of the spiny plant: introducing pathogenic microorganisms into herbivore's tissues. Adv Appl Microbiol 74:97–116
Hämäläinen L, Valkonen J, Mappes J, Rojas B (2015) Visual illusions in predator-prey interactions: birds find moving patterned prey harder to catch. Anim Cogn 18:1059–1068
Hämet-Ahti L, Palmén A, Alanko P, Tigersted PMA (1992) Woody flora of Finland. Yliopistopaino, Helsinki
Hamilton WJ III (1973) Life's color code. McGraw-Hill, New York
Hamilton WD, Brown SP (2001) Autumn tree colours as a handicap signal. Proc R Soc Lond B 268:1489–1493
Hancock AM, Witonsky DB, Alkorta-Aranburu G, Beall CM, Gebremedhin A, Sukernik R, Utermann G, Pritchard JK, Coop G (2011) Adaptations to climate-mediated selective pressures in humans. PLoS Genet 7:e1001375
Hanley ME, Lamont BB, Armbruster WS (2009) Pollination and plant defence traits co-vary in Western Australian *Hakeas*. New Phytol 182:251–260
Hansen I, Brimer L, Mølgaard P (2003) Herbivore-deterring secondary compounds in heterophyllous woody species of the Mascarene Islands. Perspect Plant Ecol Evol Syst 6:187–203
Hansen DM, Olesen JM, Mione T, Johnson SD, Müller CB (2007) Coloured nectar: distribution, ecology, and evolution of an enigmatic floral trait. Biol Rev 82:83–111
Hara N (1957) Study of the variegated leaves, with special reference to those caused by air spaces. Jap J Bot 16:86–101
Harborne JB (1982) Introduction to ecological biochemistry. Academic Press, London
Harper JL (1977) Population biology of plants. Academic Press, London
Harper DGC (2006) Maynard Smith: amplifying the reasons for signal reliability. J Theor Biol 239:203–209
Hartley SE (1998) The chemical composition of plant galls: are levels of nutrients and secondary compounds controlled by the gall-former? Oecologia 113:492–501
Hartley SE, Lawton JH (1987) Effects of different types of damage on the chemistry of birch foliage, and the responses of birch feeding insects. Oecologia 74:432–437
Harvey PH, Paxton RJ (1981) The evolution of aposematic coloration. Oikos 37:391–396
Hatier J-HB, Gould KS (2008) Foliar anthocyanins as modulators of stress signals. J Theor Biol 253:625–627
Hawkins BA (1994) Pattern and process in host-parasitoid interactions. Cambridge University Press, Cambridge
Hawlena D, Schmitz OJ (2010) Herbivore physiological response to predation risk and implications for ecosystem nutrient dynamics. Proc Natl Acad Sci USA 107:15503–15507
Heath JJ, Cipollini DF, Stireman JO III (2013) The role of cartenoids and their derivatives in mediating interactions between insects and their environment. Arthropod Plant Interact 7:1–20
Hegna RH, Nokelainen O, Hegna JR, Mappes J (2013) To quiver or to shiver: increased melanization benefits thermoregulation, but reduces warning signal efficacy in the wood tiger moth. Proc R Soc B 280:20122812

Heil M (2008) Indirect defence via tritrophic interactions. New Phytol 178:41–61
Heil M, Koch T, Hilpert A, Fiala B, Boland W, Linsenmair KE (2001) Extrafloral nectar production of the ant-associated plant, *Macaranga tanarius*, is an induced, indirect, defensive response elicited by jasmonic acid. Proc Natl Acad Sci USA 98:1083–1088
Heinrich B (1975) Bee flowers: a hypothesis on flower variety and blooming times. Evolution 29:325–334
Heinrich B (1979) Foraging strategies of caterpillars. Leaf damage and possible predator avoidance strategies. Oecologia 42:325–337
Heinrich B (1993) How avian predators constrain caterpillar foraging. In: Stamp NE, Casey TM (eds) Caterpillars: ecological and evolutionary constraints on foraging. Chapman & Hall, New York, pp 224–247
Heinrich B, Collins SL (1983) Caterpillar leaf damage, and the game of hide-and-seek with birds. Ecology 64:592–602
Heithaus ER (1974) The role of plant pollinator interaction in determining community structure. Ann Mo Bot Gard 61:675–691
Henkin Z, Seligman NG, Kafkafi U, Noy-Meir I (1998) 'Effective growing days': a simple predictive model of the response of herbaceous plant growth in a Mediterranean ecosystem to variation in rainfall and phosphorus availability. J Ecol 86:137–148
Herms DA, Mattson WJ (1992) The dilemma of plants: to grow or defend. Q Rev Biol 67:283–335
Herrera CM (1982) Defense of ripe fruit from pests: its significance in relation to plant-disperser interactions. Am Nat 120:218–241
Herrera CM (1984) A study of avian frugivores, bird-dispersed plants, and their interaction in Mediterranean scrublands. Ecol Monogr 54:1–23
Herrera CM (1985a) Aposematic insects as six-legged fruits: incidental short-circuiting of their defense by frugivorous birds. Am Nat 126:286–293
Herrera CM (1985b) Determinants of plant-animal coevolution: the case of mutualistic dispersal of seeds by vertebrates. Oikos 44:132–141
Herrera J (1991) The reproductive biology of a riparian Mediterranean shrub, *Nerium oleander* L. (Apocynaceae). Bot J Linn Soc 106:147–172
Herrera CM (2002) Seed dispersal by vertebrates. In: Herrera CM, Pellmyr O (eds) Plant-animal interactions. An evolutionary approach. Blackwell, Oxford, pp 185–208
Herrera CM, Medrano M, Rey PJ, Sánchez-Lafuente AM, Garcia MB, Guitián J, Manzaneda AJ (2002) Interaction of pollinators and herbivores on plant fitness suggests a pathway for correlated evolution of mutualism- and antagonism-related traits. Proc Natl Acad Sci USA 99:16823–16828
Hewitt G (2000) The genetic legacy of the Quaternary ice ages. Nature 405:907–913
Higginson AD, Gilbert FS, Reader T, Barnard CJ (2007) Reduction of visitation rates by honeybees (*Apis mellifera*) to individual inflorescences of lavender (Lavandula stoechas) upon removal of coloured accessory bracts (Hymenoptera: Apidae). Entomol Gener 29:165–178
Hilker M, Meiners T (2011) Plants and insect eggs: how do they affect each other? Phytochemistry 72:1612–1623
Hill ME (2006) The effect of aposematic coloration on the food preference of *Aphelocoma coerulescens*, the Florida scrub jay. Bios 77:97–106
Hill DA, Lucas PW, Cheng PY (1995) Bite forces by Japanese macaques (*Macaca fuscata yakui*) on Yakushima island, Japan to open aphid-induced galls on *Distylium racemosum* (Hamamelidaceae). J Zool 237:57–63
Hinton HE (1973) Natural deception. In: Gregory RL, Gombrich EH (eds) Illusion in nature and art. Duckworth, London, pp 97–159
Hlivko JT, Rypstra AL (2003) Spiders reduce herbivory: nonlethal effects of spiders on the consumption of soybean leaves by beetle pests. Ann Entomol Soc Am 96:914–919
Hoch WA, Zeldin EL, McCown BH (2001) Physiological significance of anthocyanins during autumnal leaf senescence. Tree Physiol 21:1–8

Hoch WA, Singsaas EL, McCown BH (2003) Resorption protection. Anthocyanins facilitate nutrient recovery in autumn by shielding leaves from potentially damaging light levels. Plant Physiol 133:1296–1305
Hocking B (1964) Fire melanism in some African grasshoppers. Evolution 18:332–335
Hocking B (1968) Insect-flower associations in the high Arctic with special reference to nectar. Oikos 19:359–388
Hodes RM, Teferedegne B (1990) Tetanus in Ethiopia: analysis of 55 cases from Addis Ababa. East Afr Med J 67:887–893
Hoekstra HE (2006) Genetics, development and evolution of adaptive pigmentation in vertebrates. Heredity 97:222–234
Hoey BK, Crowe KR, Jones VM, Polans NO (1966) A phylogenetic analysis of *Pisum* based on morphological characters, and allozyme and RAPD markers. Theor Appl Genet 92:92–100
Holen ØH, Svennungsen TO (2012) Aposematism and the handicap principle. Am Nat 180:629–641
Holmes RT, Schultz JC, Nothnagle P (1979) Bird predation on forest insects: an exclosure experiment. Science 206:462–463
Holopainen JK (2008) Importance of olfactory and visual signals of autumn leaves in the coevolution of aphids and trees. BioEssays 30:889–896
Holopainen JK, Peltonen P (2002) Bright autumn colours of deciduous trees attract aphids: nutrient retranslocation hypothesis. Oikos 99:184–188
Holopainen JK, Semiz G, Blande JD (2009) Life-history strategies affect aphid preference for yellowing leaves. Biol Lett 5:603–605
Holopainen JK, Heijari J, Oksanen E, Alessio GA (2010) Leaf volatile emissiones of *Betula pendula* during autumn coloration and leaf fall. J Chem Ecol 36:1068–1075
Honkavaara J, Koivula M, Korpimäki E, Siitari H, Viitala J (2002) Ultraviolet vision and foraging in terrestrial vertebrates. Oikos 98:505–511
Horn HS (1971) The adaptive geometry of trees. Princeton University Press, Princeton
Hornell J (1941) Fishing-poisons. Man 41:126–128
Horovitz A (1976) Edaphic factors and flower colour distribution in the *Anemoneae* (Ranunculaceae). Plant Syst Evol 126:229–242
Horovitz A, Galil J, Zohary D (1975) Biological flora of Israel. 6. *Anemone coronaria* L. Isr J Bot 24:26–41
Howarth B, Edmunds M (2000) The phenology of Syrphidae (Diptera): are they Batesian mimics of Hymenoptera? Biol J Linn Soc 71:437–457
Howe HF, Westley LC (1988) Ecological relationships of plants and animals. Oxford University Press, New York
Howell CJ, Kelly D, Turnbull MH (2002) Moa ghosts exorcised? New Zealand's divaricate shrubs avoid photoinhibition. Funct Ecol 16:232–240
Hughes NM (2011) Winter leaf reddening in 'evergreen' species. New Phytol 190:573–581
Hughes NM, Lev-Yadun S (2015) Red/purple leaf margin coloration: potential ecological and physiological functions. Environ Exp Bot 119:27–39
Hughes NM, Smith WK (2007) Seasonal photosynthesis and anthocyanin production in 10 broadleaf evergreen species. Funct Plant Biol 34:1072–1079
Hughes NM, Smith WK, Gould KS (2010) Red (anthocyanic) leaf margins do not correspond to increased phenolic content in New Zealand *Veronica* spp. Ann Bot 105:647–654
Hughes AE, Troscianko J, Stevens M (2014) Motion dazzle and the effects of target patterning on capture success. BMC Evol Biol 14:201
Hughes AE, Magor-Elliott RS, Stevens M (2015) The role of stripe orientation in target capture success. Front Zool 12:17
Hunter AF (2000) Gregariousness and repellent defences in the survival of phytophagous insects. Oikos 91:213–224
Huxley CR, Cutler DF (1991) Ant-plant interactions. Oxford University Press, London
Imbrie J, Palmer-Imbrie K (1979) Ice ages. Solving the mystery. Enslow Publishers, Hillside

Inbar M, Lev-Yadun S (2005) Conspicuous and aposematic spines in the animal kingdom. Naturwiss 92:170–172

Inbar M, Eshel A, Wool D (1995) Interspecific competition among phloem-feeding insects mediated by induced host-plant sinks. Ecology 76:1506–1515

Inbar M, Doostdar H, Mayer RT (1999) Effects of sessile Whitefly nymphs (Homoptera: Aleyrodidae) on leaf-chewing larvae (Lepidoptera: Noctuidae). Environ Entomol 28:353–357

Inbar M, Wink M, Wool D (2004) The evolution of host plant manipulation by insects: molecular and ecological evidence from gall-forming aphids on *Pistacia*. Mol Phylogenet Evol 32:504–511

Inbar M, Izhaki I, Lupo I, Silanikove N, Glasser T, Gerchman Y, Perevolotsky A, Lev-Yadun S (2010a) Why do many galls have conspicuous colors? A new hypothesis. Arthropod Plant Interact 4:1–6

Inbar M, Izhaki I, Lupo I, Silanikove N, Glasser T, Gerchman Y, Perevolotsky A, Lev-Yadun S (2010b) Conspicuous gall colors. A response to T.C.R. White. Arthropod Plant Interact 4:151–152

Irwin RE, Strauss SY, Storz S, Emerson A, Guibert G (2003) The role of herbivores in the maintenance of a flower color polymorphism in wild radish. Ecology 84:1733–1743

Irwin RE, Adler LS, Brody AK (2004) The dual role of floral traits: pollinator attraction and plant defense. Ecology 85:1503–1511

Isaksen T, Folstad I (2014) Flowers for bees, autumn colours for whom? An experimental test with autumn colouration in mountain birches (*Betula pubescens* ssp. *czerepanovii*) as a signal against the moth *Epirrita autumnata*. Entomol Exp Appl 152:141–147

Ishiwata S, Hanada S, Hayashi F, Yamasaki T, Uemura Y, Aoki N, Hayashi M, Nozaki T, Fukuda H, Kishida Y, Hayashi N, Shinohara A, Shinonaga S (2005) Insect larvae of Japan. Gakken, Tokyo (in Japanese)

Izhaki I (1986) Seed dispersal by birds in east Mediterranean scrub. Ph.D. dissertation. The Hebrew University of Jerusalem, Jerusalem (in Hebrew with English summary)

Izhaki I (1998) The relationships between fruit ripeness, wasp seed predation, and avian fruit removal in *Pistacia palaestina*. Isr J Plant Sci 46:273–278

Izhaki I (2002) Emodin – a secondary metabolite with multiple ecological functions in higher plants. New Phytol 155:205–217

Izhaki I, Idelovich B, Laster R, Ofer Y (2009) The impact of macro- vs. micro-environmental factors on the structure of ant communities inhabiting East-Mediterranean Aleppo pine forests. Isr J Entomol 39:129–146

Jacobs GH (1993) The distribution and nature of colour vision among the mammals. Biol Rev 68:413–471

Jaenisch R, Bird A (2003) Epigenetic regulation of gene expression: how the genome integrates intrinsic and environmental signals. Nat Genet 33:245–254

Jain G, Gould KS (2015) Are betalain pigments the functional homologues of anthocyanins in plants? Environ Exp Bot 119:48–53

Janson CH (1987) Bird consumption of bicolored fruit displays. Am Nat 130:788–792

Janzen DH (1966) Coevolution of mutualism between ants and acacias in Central America. Evolution 20:249–275

Janzen DH (1970) Herbivores and the number of tree species in tropical forests. Am Nat 104:501–528

Janzen DH (1976) Why bamboos wait so long to flower. Annu Rev Ecol Syst 7:347–391

Janzen DH (1977) Why fruits rot, seeds mold, and meat spoils. Am Nat 111:691–713

Janzen DH (1979) New horizons in the biology of plant defenses. In: Rosenthal GA, Janzen DH (eds) Herbivores their interaction with secondary plant metabolites. Academic Press, Orlando, pp 331–350

Janzen DH (1986) Chihuahuan Desert nopaleras: defaunated big mammal vegetation. Annu Rev Ecol Syst 17:595–636

Janzen DH, Martin PS (1982) Neotropical anachronisms: the fruits the gomphotheres ate. Science 215:19–27

Janzen DH, Hallwachs W, Burns JM (2010) A tropical horde of counterfeit predator eyes. Proc Natl Acad Sci USA 107:11659–11665

Jayasuriya KMGG, Baskin JM, Geneve RL, Baskin CC, Chien C-T (2008) Physical dormancy in seeds of the holoparasitic angiosperm *Cuscuta australis* (Convolvulaceae, Cuscuteae): dormancy-breaking requirements, anatomy of the water gap and sensitivity cycling. Ann Bot 102:39–48

Jayasuriya KMGG, Athugala YS, Wijayasinghe MM, Baskin JM, Baskin CC, Mahadevan N (2015) The crypsis hypothesis: a stenopic view of the selective factors in the evolution of physical dormancy in seeds. Seed Sci Res 25:127–137

Jensen GB, Hansen BM, Eilenberg J, Mahillon J (2003) The hidden lifestyles of *Bacillus cereus* and relatives. Environ Microbiol 5:631–640

Jensen EL, Dill LM, Cahill JF Jr (2011) Applying behavioral-ecological theory to plant defense: light-dependent movement in *Mimosa pudica* suggests a trade-off between predation risk and energetic reward. Am Nat 177:377–381

Jersáková J, Johnson SD, Kindlmann P (2006) Mechanisms and evolution of deceptive pollination in orchids. Biol Rev 81:219–235

Joel DM, Juniper BE, Dafni A (1985) Ultraviolet patterns in the traps of carnivorous plants. New Phytol 101:585–593

Joern A (2002) Context-dependent foraging and enemy-free space: Grasshopper sparrows (*Ammodramus savannarum*) searching for grasshoppers (Acrididae). Ecoscience 9:231–240

Johnsen S (2001) Hidden in plain sight: the ecology and physiology of organismal transparency. Biol Bull 201:301–318

Johnson HB (1975) Plant pubescence: an ecological perspective. Bot Rev 41:233–258

Johnson CN (2009) Ecological consequences of Late Quaternary extinctions of megafauna. Proc R Soc B 276:2509–2519

Johnson SD, Midgley JJ (1997) Fly pollination of *Gorteria diffusa* (Asteraceae), and a possible mimetic function for dark spots on the capitulum. Am J Bot 84:429–436

Johnson SD, Midgley JJ (2001) Pollination by monkey beetles (Scarabaeidae: Hopliini): do color and dark centers of flowers influence alighting behavior? Environ Entomol 30:861–868

Johnson SD, Hargreaves AL, Brown M (2006) Dark, bitter-tasting nectar functions as a filter of flower visitors in a bird-pollinated plant. Ecology 87:2709–2716

Johnstone RA (1995) Sexual selection, honest advertisement and the handicap principle: reviewing the evidence. Biol Rev 70:1–65

Jolivet P (1998) Interrelationship between insects and plants. CRC Press, Boca Raton

Jones CG, Lawton JH (1991) Plant chemistry and insect species richness of British umbellifers. J Anim Ecol 60:767–777

Jones KA, Krebs JR, Whittingham MJ (2006) Interaction between seed crypsis and habitat structure influence patch choice in a granivorous bird, the chaffinch *Fringilla coelebs*. J Avian Biol 37:413–418

Jordano P (1982) Migrant birds are the main seed dispersers of blackberries in southern Spain. Oikos 38:183–193

Jordano P (1984) Seed weight variation and differential avian dispersal in blackberries *Rubus ulmifolius*. Oikos 43:149–153

Jordano P (1989) Pre-dispersal biology of *Pistacia lentiscus* (Anacardiaceae): cumulative effects on seed removal by birds. Oikos 55:375–386

Juniper BE (1994) Flamboyant flushes: a reinterpretation of non-green flush colours in leaves. In: International Dendrology Society Yearbook. International Dendrology Society Yearbook, London, pp 49–57

Juniper BE, Robins RJ, Joel DM (1989) The carnivorous plants. Academic Press, London

Junker R, Chung AYC, Blüthgen N (2007) Interaction between flowers, ants and pollinators: additional evidence for floral repellence against ants. Ecol Res 22:665–670

Jura-Morawiec J, Tulik M (2015) Morpho-anatomical basis of dragon's blood secretion in *Dracaena*. Flora 213:1–5

Jürgens N (1996) Psammophorous plants and other adaptations to desert ecosystems with high incidence of sandstorms. Feddes Repert 107:345–357

Jürgens A, Shuttleworth A (2016) Carrion and dung mimicry in plants. In: Benbow ME, Tomberlin JK, Tarone AM (eds) Carrion ecology, evolution, and their applications. CRC Press, Boca Raton, pp 361–386

Justus M, Witte L, Hartmann T (1997) Levels and tissue distribution of loline alkaloids in endophyte-infected *Festuca pratensis*. Phytochemistry 44:51–57

Kamilar JM, Bradley BJ (2011) Interspecific variation in primate coat colour supports Gloger's rule. J Biogeogr 38:2270–2277

Kant MR, Jonckheere W, Knegt B, Lemos F, Liu J, Schimmel BCJ, Villarroel CA, Ataide LMS, Dermauw W, Glas JJ, Egas M, Janssen A, Van Leeuwen T, Schuurink RC, Sabelis MW, Alba JM (2015) Mechanisms and ecological consequences of plant defence induction and suppression in herbivore communities. Ann Bot 115:1015–1051

Kaplan I, Lynch ME, Dively GP, Denno RF (2007) Leafhopper-induced plant resistance enhances predation risk in a phytophagous beetle. Oecologia 152:665–675

Kappers IF, Aharoni A, van Herpen TWJM, Luckerhoff LLP, Dicke M, Bouwmeester HJ (2005) Genetic engineering of terpenoid metabolism attracts bodyguards to *Arabidopsis*. Science 309:2070–2072

Karageorgou P, Manetas Y (2006) The importance of being red when young: anthocyanins and the protection of young leaves of *Quercus coccifera* from insect herbivory and excess light. Tree Physiol 26:613–621

Karageorgou P, Buschmann C, Manetas Y (2008) Red leaf color as a warning signal against insect herbivory: honest or mimetic? Flora 203:648–652

Karban R (2007) Deciduous leaf drop reduces insect herbivory. Oecologia 153:81–88

Karban R (2015) Plant sensing & communication. University of Chicago Press, Chicago

Karban R, Agrawal AA (2002) Herbivore offense. Annu Rev Ecol Syst 33:641–664

Karban R, Baldwin IT (1997) Induced responses to herbivory. University of Chicago Press, Chicago

Karban R, Shiojiri K, Ishizaki S, Wetzel WC, Evans RY (2013) Kin recognition affects plant communication and defence. Proc R Soc B 280:20123062

Karlsson M, Caesar S, Ahnesjö J, Forsman A (2008) Dynamics of colour polymorphism in a changing environment: fire melanism and then what? Oecologia 154:715–724

Karpestam E, Merilaita S, Forsman A (2014) Natural level of colour polymorphism reduce performance of visual predators searching for camouflaged prey. Biol J Linn Soc 112:546–555

Katewa SS, Guria BD, Jain A (2001) Ethnomedicinal and obnoxious grasses of Rajasthan, India. J Ethnopharmacol 76:293–297

Katz O (2015) Silica phytoliths in Angiosperms: phylogeny and early evolutionary history. New Phytol 208:642–646

Kauppinen J, Mappes J (2003) Why are wasps so intimidating: field experiments on hunting dragonflies (Odonata: *Aeshna grandis*). Anim Behav 66:505–511

Kavanagh PH (2015) Herbivory and the evolution of divaricate plants: structural defences lost on an offshore island. Austral Ecol 40:206–211

Kavanagh PH, Shaw RC, Burns KC (2016) Potential aposematism in an insular tree species: are signals dishonest early in ontogeny? Biol J Linn Soc 118:951–958

Kaye H, Mackintosh NJ, Rothschild M, Moore BP (1989) Odour of pyrazine potentiates an association between environmental cues and unpalatable taste. Anim Behav 37:563–568

Keasar T, Kishinevsky M, Shmida A, Gerchman Y, Chinkov N, Koplovich A, Katzir G (2013) Plant-derived visual signals may protect beetle herbivores from bird predators. Behav Ecol Sociobiol 67:1613–1622

Kelber A (2001) Receptor based models for spontaneous colour choices in flies and butterflies. Entomol Exp Appl 99:231–244

Kelber A, Vorobyev M, Osorio D (2003) Animal colour vision – behavioural tests and physiological concepts. Biol Rev 78:81–118

Kelly D, Sork VL (2002) Mast seeding in perennial plants: why, how, where? Annu Rev Ecol Syst 33:427–447

Kemp DJ, Herberstein ME, Fleishman LJ, Endler JA, Bennett ATD, Dyer AG, Hart NS, Marshall J, Whiting MJ (2015) An integrative framework for the appraisal of coloration in nature. Am Nat 185:705–724

Kendrick JW, Tucker J, Peoples SA (1955) Nitrate poisoning in cattle due to ingestion of variegated thistle, *Silybum marianum*. J Am Vet Med Assoc 126:53–56

Keskitalo J, Bergquist G, Gardeström P, Jansson S (2005) A cellular timetable of autumn senescence. Plant Physiol 139:1635–1648

Kessler A, Baldwin IT (2001) Defensive function of herbivore induced plant volatile emissions in nature. Science 291:2141–2144

Kettlewell B (1973) The evolution of melanism. Clarendon Press, Oxford

Kikuchi DW, Pfennig DW (2010) High-model abundance may permit the gradual evolution of Batesian mimicry: an experimental test. Proc R Soc B 277:1041–1048

Kim TN, Underwood N (2015) Plant neighborhood effects on herbivory: damage is both density and frequency dependent. Ecology 96:1431–1437

Kimoto S, Takizawa H (1994) Leaf beetles (Chrysomelidae) of Japan. Tokai University Press, Tokyo

King LE, Douglas-Hamilton I, Vollrath F (2007) African elephants run from the sound of disturbed bees. Curr Biol 17:R832–R833

King LE, Lawrence A, Douglas-Hamilton I, Vollrath F (2009) Beehive fence deters crop-raiding elephants. Afr J Ecol 47:131–137

King LE, Lawrence A, Douglas-Hamilton I, Vollrath F (2011) Beehive fences as effective deterrents for crop-raiding elephants: field trials in northern Kenya. Afr J Ecol 49:431–439

Kingdon J (1984) The zebra's stripes: an aid to group cohesion. In: MacDonald D (ed) The encyclopedia of mammals. Equinox, Oxford, pp 486–487

Klooster MR, Clark D, Culley TA (2009) Cryptic bracts facilitate herbivore avoidance in the mycoheterotrophic plant *Monotropsis odorata* (Ericaceae). Am J Bot 96:2197–2205

Knight RS, Siegfried WR (1983) Inter-relationships between type, size and color of fruits and dispersal in Southern African trees. Oecologia 56:405–412

Knight AP, Walter RG (2001) A guide to plant poisoning of animals in North America. Teton NewMedia, Jackson

Knudsen JT, Eriksson R, Gershenzon J, Ståhl B (2006) Diversity and distribution of floral scent. Bot Rev 72:1–120

Koike T (1990) Autumn coloring, photosynthetic performance and leaf development of deciduous broad-leaved trees in relation to forest succession. Tree Physiol 7:21–32

Koller D (2011) The restless plant. Harvard University Press, Cambridge

Komárek S (1998) Mimicry, aposematism and related phenomena in animals and plants: bibliography 1800–1990. Vesmir, Prague

Konno K (2011) Plant latex and other exudates as plant defense systems: roles of various defense chemicals and proteins contained therein. Phytochemistry 72:1510–1530

Konoplyova A, Petropoulou Y, Yiotis C, Psaras GK, Manetas Y (2008) The fine structure and photosynthetic cost of structural leaf variegation. Flora 203:653–662

Koski MH, Ashman T-L (2015) Floral pigmentation patterns provide an example of Gloger's rule in plants. Nat Plants 1:14007

Kostal V, Finch S (1996) Preference of the cabbage root fly, *Delia radicum* (L.), for coloured traps: influence of sex and physiological status of the flies, trap background and experimental design. Physiol Entomol 21:123–130

Kozlowski TT (1971) Growth and development of trees, vol I. Academic Press, New York

Kozlowski TT, Pallardy SG (1997) Physiology of woody plants, 2nd edn. Academic Press, San Diego

Kozlowski TT, Kramer PJ, Pallardy SG (1991) The physiological ecology of woody plants. Academic Press, San Diego

Krimmel BA, Pearse IS (2013) Sticky plant traps insects to enhance indirect defence. Ecol Lett 16:219–224

Krueger WC, Laycock WA, Price DA (1974) Relationships of taste, smell, sight, and touch to forage selection. J Range Manag 27:258–262

Krulik GA (1980) Light transmission in window-leaved plants. Can J Bot 58:1591–1600

Kruuk H (1972) The spotted hyena. A study of predation and social behavior. University of Chicago Press, Chicago

Kullenberg B (1950) Investigation on the pollination of *Ophrys* species. Oikos 2:1–19

Kullenberg B (1956) On the scents and colours of *Ophrys* flowers and their specific pollinators among the aculeate Hymenoptera. Svensk Bot Tidskr 50:25–46

Kullenberg B (1961) Studies in *Ophrys* Pollination. Zool Bidrag Uppsala 34:1–340

Kunimi Y (1983) Spiders inhabiting the colonial-webs of the fall webworm, *Hyphantria cunea* Drury (Lepidoptera: Arctiidae). Appl Entomol Zool 18:81–89

Kursar TA, Coley PD (1991) Nitrogen content and expansion rate of young leaves of rainforest species: implications for herbivory. Biotropica 23:141–150

Kursar TA, Coley PD (1992) Delayed greening in tropical leaves: an antiherbivory defense? Biotropica 24:256–262

Kursar TA, Coley PD (2003) Convergence in defense syndromes of young leaves in tropical rainforests. Biochem Syst Ecol 31:929–949

Kytridis V-P, Karageorgou P, Levizou E, Manetas Y (2008) Intra-species variation in transient accumulation of leaf anthocyanins in *Cistus creticus* during winter: evidence that anthocyanins may compensate for an inherent photosynthetic and photoprotective inferiority of the red-leaf phenotype. J Plant Physiol 165:952–959

Labandeira CC (1998) Early history of arthropod and vascular plant associations. Annu Rev Earth Planet Sci 26:329–377

Lachmann ML, Számadó S, Bergstrom CT (2001) Cost and conflict in animal signals and human language. Proc Natl Acad Sci USA 98:13189–13194

Lack A (1976) Competition for pollinators and evolution in *Centaurea*. New Phytol 77:787–792

Lamborn E, Ollerton J (2000) Experimental assessment of the functional morphology of inflorescences of *Daucus carota* (Apiaceae): testing the 'fly catcher effect'. Funct Ecol 14:445–454

Lamont BB (1993) Injury-induced cyanogenesis in vegetative and reproductive parts of two *Grevillea* species and their F1 hybrid. Ann Bot 71:537–542

Langenheim JH (2003) Plant resins. Chemistry, evolution, ecology, and ethnobotany. Timber Press, Portland

Langford SD, Boor PJ (1996) Oleander toxicity: an examination of human and animal toxic exposures. Toxicology 109:1–13

Langridge KV (2006) Symmetrical crypsis and asymmetrical signalling in the cuttlefish *Sepia officinalis*. Proc R Soc B 273:959–967

Lanner RM (1998) Seed dispersal in *Pinus*. In: Richardson DM (ed) Ecology and biogeography of *Pinus*. Cambridge University Press, Cambridge, pp 281–295

Lapointe SL (2000) Particle film deters oviposition by *Diaprepes abbreviatus* (Coleoptera: Curculionidae). J Econ Entomol 93:1459–1463

La Rocca N, Pupillo P, Puppi G, Rascio N (2014) *Erythronium dens-canis* L. (Liliaceae): an unusual case of change of leaf mottling. Plant Physiol Biochem 74:108–117

Launchbaugh KL, Provenza FD (1993) Can plants practice mimicry to avoid grazing by mammalian herbivores? Oikos 66:501–504

Laundré JW (2014) How large predators manage the cost of hunting. Science 346:33

Lavranos JJ (1974) Notes on the succulent flora of north east Africa and southern Arabia. Part II. Cactus Succul J US 46:125–130

Lawler IR, Foley WJ, Eschler BM, Pass DM, Handasyde K (1998) Intraspecific variation in *Eucalyptus* secondary metabolites determines food intake by folivorous marsupials. Oecologia 116:160–169

Lawler I, Eschler B, Schliebs DM, Foley WJ (1999) Relationship between chemical functional groups on *Eucalyptus* secondary metabolites and their effectiveness as marsupial antifeedants. J Chem Ecol 25:2561–2573
Laycock WA (1978) Coevolution of poisonous plants and large herbivores on rangelands. J Range Manag 31:335–342
Leal IR, Fischer E, Kost C, Tabarelli M, Wirth R (2006) Ant protection against herbivores and nectar thieves in *Passiflora coccinea* flowers. Ecoscience 13:431–438
Lee DW (2002) Anthocyanins in autumn leaf senescence. Adv Bot Res 37:147–165
Lee D (2007) Nature's palette. The science of plant color. University of Chicago Press, Chicago
Lee D (2012) The mystery of seasonal color change. Arnoldia 70(2):2–10
Lee DW, Collins TM (2001) Phylogenetic and ontogenetic influences on the distribution of anthocyanins and betacyanins in leaves of tropical plants. Int J Plant Sci 162:1141–1153
Lee DW, Gould KS (2002a) Anthocyanins in leaves and other vegetative organs: an introduction. Adv Bot Res 37:1–16
Lee DW, Gould KS (2002b) Why leaves turn red. Am Sci 90:524–531
Lee DW, Gould KS (2009) Three birds with one stone: moas, heteroblasty and the New Zealand flora. New Phytol 184:282–284
Lee DW, Brammeler S, Smith AP (1987) The selective advantages of anthocyanins in developing leaves of mango and cacao. Biotropica 19:40–49
Lee DW, O'Keefe J, Holbrook NM, Feild TS (2003) Pigment dynamics and autumn leaf senescence in a New England deciduous forest, eastern USA. Ecol Res 18:677–694
Lehtilä K, Strauss SY (1997) Leaf damage by herbivores affects attractiveness to pollinators in wild radish, *Raphanus raphanistrum*. Oecologia 111:396–403
Leonard RJ, McArthur C, Hochuli DF (2016) Exposure to wind alters insect herbivore behaviour in larvae of *Uraba lugens* (Lepidoptera: Nolidae). Aust Entomol 55:242–246
Levin DA (1972) Competition for pollinator service: a stimulus for the evolution of autogamy. Evolution 26:668–669
Levin DA (1973) The role of trichomes in plant defense. Q Rev Biol 48:3–15
Levin DA, Anderson WW (1970) Competition for pollinators between simultaneously flowering species. Am Nat 104:455–461
Lev-Yadun S (1994) Induction of sclereid differentiation in the pith of *Arabidopsis thaliana* (L.) Heynh. J Exp Bot 45:1845–1849
Lev-Yadun S (1997) Fibres and fibre-sclereids in wild-type *Arabidopsis thaliana*. Ann Bot 80:125–129
Lev-Yadun S (2001) Aposematic (warning) coloration associated with thorns in higher plants. J Theor Biol 210:385–388
Lev-Yadun S (2003a) Why do some thorny plants resemble green zebras? J Theor Biol 244:483–489
Lev-Yadun S (2003b) Weapon (thorn) automimicry and mimicry of aposematic colorful thorns in plants. J Theor Biol 244:183–188
Lev-Yadun S (2005) Shade avoidance and Zahavi's handicap principle in dense plant populations. Biol J Linn Soc 84:313–319
Lev-Yadun S (2006a) Defensive coloration in plants: a review of current ideas about anti-herbivore coloration strategies. In: Teixeira da Silva JA (ed) Floriculture, ornamental and plant biotechnology: advances and topical issues, vol IV. Global Science Books, London, pp 292–299
Lev-Yadun S (2006b) Defensive functions of white coloration in coastal and dune plants. Isr J Plant Sci 54:317–325
Lev-Yadun S (2009a) Aposematic (warning) coloration in plants. In: Baluska F (ed) Plant-environment interactions. From sensory plant biology to active plant behavior. Springer-Verlag, Berlin, pp 167–202
Lev-Yadun S (2009b) Müllerian and Batesian mimicry rings of white-variegated aposematic spiny and thorny plants: a hypothesis. Isr J Plant Sci 57:107–116
Lev-Yadun S (2009c) Müllerian mimicry in aposematic spiny plants. Plant Signal Behav 4:482–483

Lev-Yadun S (2009d) Ant mimicry by *Passiflora* flowers? Isr J Entomol 39:159–163

Lev-Yadun S (2010) The shared and separate roles of aposematic (warning) coloration and the co-evolutionary hypothesis in defending autumn leaves. Plant Signal Behav 5:937–939

Lev-Yadun S (2011) Fearful symmetry in aposematic plants. Plant Signal Behav 6:1739–1740

Lev-Yadun S (2013a) Theoretical and functional complexity of white variegation of unripe fleshy fruits. Plant Signal Behav 8:e25851

Lev-Yadun S (2013b) The enigmatic fast leaflet rotation in *Desmodium motorium*: butterfly mimicry for defense? Plant Signal Behav 8:e24473

Lev-Yadun S (2013c) Does chemical aposematic (warning) signaling occur between host plants and their potential parasitic plants? Plant Signal Behav 8:e24907

Lev-Yadun S (2014a) The proposed anti-herbivory roles of white leaf variegation. Prog Bot 76:241–269

Lev-Yadun S (2014b) Defensive masquerade by plants. Biol J Linn Soc 113:1162–1166

Lev-Yadun S (2014c) Potential defence from herbivory by dazzle effects and trickery coloration of leaf variegation. Biol J Linn Soc 111:692–697

Lev-Yadun S (2014d) Why is latex usually white and only sometimes yellow, orange or red? Simultaneous visual and chemical plant defense. Chemoecology 24:215–218

Lev-Yadun S (2014e) Do plants use visual carrion-based aposematism to deter herbivores? Biochemist 36(5):36–39

Lev-Yadun S (2015a) Partly transparent young legume pods: do they mimic caterpillars for defense and simultaneously enable better photosynthesis? Plant Signal Behav 10:e1048941

Lev-Yadun S (2015b) Gloger's rule in plants: the species and ecosystem levels. Plant Signal Behav 10:e1040968

Lev-Yadun S, Aloni R (1995) Differentiation of the ray system in woody plants. Bot Rev 61:45–88

Lev-Yadun S, Gould KS (2007) What do red and yellow autumn leaves signal? Bot Rev 73:279–289

Lev-Yadun S, Gould KS (2009) Role of anthocyanins in plant defense. In: Gould KS, Davies KM, Winefield C (eds) Life's colorful solutions: the biosynthesis, functions, and applications of anthocyanins. Springer-Verlag, Berlin, pp 21–48

Lev-Yadun S, Gutman M (2013) Carrion odor and cattle grazing: evidence for plant defense by carrion odor. Commun Integr Biol 6:e26111

Lev-Yadun S, Halpern M (2007) Ergot (*Claviceps purpurea*) – an aposematic fungus. Symbiosis J 43:105–108

Lev-Yadun S, Halpern M (2008) External and internal spines in plants insert pathogenic microorganisms into herbivore's tissues for defense. In: Van Dijk T (ed) Microbial ecology research trends. Nova Scientific Publishers, Inc., New York, pp 155–168

Lev-Yadun S, Holopainen JK (2009) Why red-dominated American autumn leaves and yellow-dominated autumn leaves in Northern Europe? New Phytol 183:506–512

Lev-Yadun S, Holopainen JK (2011) How red is the red autumn leaf herring and did it lose its red color? Plant Signal Behav 6:1879–1880

Lev-Yadun S, Inbar M (2002) Defensive ant, aphid and caterpillar mimicry in plants. Biol J Linn Soc 77:393–398

Lev-Yadun S, Keasar T (2012) Prerequisites for evolution: variation and selection in yellow autumn birch leaves. New Phytol 195:282–284

Lev-Yadun S, Mirsky N (2007) False satiation: the probable antiherbivory strategy of *Hoodia gordonii*. Funct Plant Sci Biotechnol 1:56–57

Lev-Yadun S, Ne'eman G (2004) When may green plants be aposematic? Biol J Linn Soc 81:413–416

Lev-Yadun S, Ne'eman G (2006) Color changes in old aposematic thorns, spines, and prickles. Isr J Plant Sci 54:327–333

Lev-Yadun S, Ne'eman G (2012) Does bee or wasp mimicry by orchid flowers also deter herbivores? Arthropod Plant Interact 6:327–332

Lev-Yadun S, Ne'eman G (2013) Bimodal colour pattern of individual *Pinus halepensis* Mill. seeds: a new type of crypsis. Biol J Linn Soc 109:271–278

Lev-Yadun S, Sederoff R (2000) Pines as model gymnosperms to study evolution, wood formation and perennial growth. J Plant Growth Regul 19:290–305
Lev-Yadun S, Sederoff R (2001) Grafting for transgene containment. Nat Biotechnol 19:1104
Lev-Yadun S, Gopher A, Abbo S (2000) The cradle of agriculture. Science 288:1602–1603
Lev-Yadun S, Dafni A, Inbar M, Izhaki I, Ne'eman G (2002) Colour patterns in vegetative parts of plants deserve more research attention. Trends Plant Sci 7:59–60
Lev-Yadun S, Dafni A, Flaishman MA, Inbar M, Izhaki I, Katzir G, Ne'eman G (2004a) Plant coloration undermines herbivorous insect camouflage. BioEssays 26:1126–1130
Lev-Yadun S, Flaishman MA, Atzmon N (2004b) Nonchimeric variegated mutation in *Cupressus sempervirens* L. Int J Plant Sci 165:257–261
Lev-Yadun S, Ne'eman G, Izhaki I (2009a) Unripe red fruits may be aposematic. Plant Signal Behav 4:836–841
Lev-Yadun S, Ne'eman G, Shanas U (2009b) A sheep in wolf's clothing: do carrion and dung odours of flowers not only attract pollinators but also deter herbivores? BioEssays 31:84–88
Lev-Yadun S, Katzir G, Ne'eman G (2009c) *Rheum palaestinum* (desert rhubarb), a self-irrigating desert plant. Naturwissenschaften 96:393–397
Lev-Yadun S, Yamazaki K, Holopainen JK, Sinkkonen A (2012) Spring versus autumn leaf colours: evidence for different selective agents and evolution in various species and floras. Flora 207:80–85
Li T, Holopainen JK, Kokko H, Tervahauta AI, Blande JD (2012) Herbivore-induced aspen volatiles temporally regulate two different indirect defences in neighbouring plants. Funct Ecol 26:1176–1185
Lichter-Marck IH, Wylde M, Aaron E, Oliver JC, Singer MS (2015) The struggle for safety: effectiveness of caterpillar defenses against bird predation. Oikos 124:525–533
Lillywhite HB, Friedman G, Ford N (1977) Color matching and perch selection by lizards in recently burned chaparral. Copeia 1977:115–121
Lima SL, Dill LM (1990) Behavioral decisions made under the risk of predation: a review and prospectus. Can J Zool 68:619–640
Lindström L, Alatalo RV, Mappes J (1999) Reactions of hand-reared and wild-caught predators toward waringly colored, gregarious, and conspicuous prey. Behav Ecol 10:317–322
Lindström L, Rowe C, Guilford T (2001) Pyrazine odour makes visually conspicuous prey aversive. Proc R Soc Lond B 268:159–162
Linnen CR, Poh Y-P, Peterson BK, Barrett RDH, Larson JG, Jensen JD, Hoekstra HE (2013) Adaptive evolution of multiple traits through multiple mutations at a single gene. Science 229:1312–1316
Lloyd JC, Zakhleniuk OV, Raines CA (2001) Identification of mutants in phosphorus metabolism. Ann Appl Biol 138:111–115
Londoño GA, García DA, Martínez MAS (2015) Morphological and behavioral evidence of Batesian mimicry in nestlings of a lowland Amazonian bird. Am Nat 185:135–141
LoPresti EF (2014) Chenopod salt bladders deter insect herbivores. Oecologia 174:921–930
LoPresti EF, Karban R (2016) Chewing sandpaper: grit, plant apparency and plant defense in sand-entrapping plants. Ecology 97:826–833
LoPresti EF, Pearse IS, Charles GK (2015) The siren song of a sticky plant: Columbines provision mutualist arthropods by attracting and killing passerby insects. Ecology 96:2862–2869
Lorenz K (1979) Ergot on cereal grains. Crit Rev Food Sci Nutr 11:311–354
Louda SM (1982) Inflorescence spiders: a cost/benefit analysis for the host plant, *Haplopappus venetus* Blake (Asteraceae). Oecologia 55:185–191
Lucas PW, Darvell BW, Lee PKD, Yuen PDB, Choong MF (1998) Colour cues for leaf food selection by long-tailed macaques (*Macaca fascicularis*) with a new suggestion for the evolution of trichromatic colour vision. Folia Primatol 69:139–152
Lucas PW, Turner IM, Dominy NJ, Yamashita N (2000) Mechanical defenses to herbivory. Ann Bot 86:913–920
Lüttge U (1997) Physiological ecology of tropical plants. Springer-Verlag, Berlin

Madden D, Young TP (1992) Symbiotic ants as an alternative defense against giraffe herbivory in spinescent *Acacia drepanolobium*. Oecologia 91:235–238

Maddocks SA, Church SC, Cuthill IC (2001) The effects of the light environment on prey choice by zebra finches. J Exp Biol 204:2509–2515

Mainland IL (2003) Dental microwear in grazing and browsing Gotland sheep (*Ovis aries*) and its implications for dietary reconstruction. J Archaeol Sci 30:1513–1527

Majerus MEN (1998) Melanism. Evolution in action. Oxford University Press, Oxford

Majerus MEN, Brunton CFA, Stalker J (2000) A bird's eye view of the peppered moth. J Evol Biol 13:155–159

Mallet J, Joron M (1999) Evolution of diversity in warning color and mimicry: polymorphisms, shifting balance, and speciation. Annu Rev Ecol Syst 30:201–233

Manchester SR (1999) Biogeographical relationships of North American Tertiary floras. Ann Mo Bot Gard 86:472–522

Manetas Y (2003) The importance of being hairy: the adverse effects of hair removal on stem photosynthesis of *Verbascum speciosum* are due to solar UV-B radiation. New Phytol 158:503–508

Manetas Y (2006) Why some leaves are anthocyanic and why most anthocyanic leaves are red? Flora 201:163–177

Mantle PG (1969) The role of alkaloids in the poisoning of mammals by sclerotia of *Claviceps* spp. J Stored Prod Res 5:237–244

Mäntylä E, Klemola T, Sirkiä P, Laaksonen T (2008) Low light reflectance may explain the attraction of birds to defoliated trees. Behav Ecol 19:325–330

Manzur MI, Courtney SP (1984) Influence of insect damage in fruits of hawthorn on bird foraging and seed dispersal. Oikos 43:265–270

Mappes J, Kaitala A (1995) Host-plant selection and predation risk for offspring of the parent bug. Ecology 76:2668–2670

Mappes J, Marples N, Endler JA (2005) The complex business of survival by aposematism. Trends Ecol Evol 20:598–603

Mappes J, Kokko H, Ojala K, Lindström L (2014) Seasonal changes in predator community switch the direction of selection for prey defences. Nat Commun 5:5016

Marchand D, McNeil JN (2004) Avoidance of intraspecific competition via host modification in a grazing, fruit-eating insect. Anim Behav 67:397–402

Markwick NP, Poulton J, Espley RV, Rowan DD, McGhie TK, Wadasinghe G, Wohlers M, Jia Y, Allan AC (2013) Red-foliaged apples affect the establishment, growth, and development of the light brown apple moth, *Epiphyas postvittana*. Entomol Exp Appl 146:261–275

Marshall NJ, Messenger JB (1996) Colour-blind camouflage. Nature 382:408–409

Martínez-Abraín A (2008) Statistical significance and biological relevance: a call for a more cautious interpretation of results in ecology. Acta Oecol 34:9–11

Maskato Y, Talal S, Keasar T, Gefen E (2014) Red foliage color reliably indicates low host quality and increased metabolic load for development of an herbivorous insect. Arthropod Plant Interact 8:285–292

Massei G, Cotterill JV, Coats JC, Bryning G, Cowan DP (2007) Can Batesian mimicry help plants to deter herbivory? Pest Manag Sci 63:559–563

Massey FP, Hartley SE (2006) Experimental demonstration of the antiherbivore effects of silica in grasses: impacts on foliage digestibility and vole growth rates. Proc R Soc Lond B 273:2299–2304

Mather MH, Roitberg BD (1987) A sheep in wolf's clothing: tephritid flies mimic spider predators. Science 236:308–310

Matile P (2000) Biochemistry of Indian summer: physiology of autumnal leaf coloration. Exp Gerontol 35:145–158

Matossian MK (1989) Poisons of the past. Molds, epidemics, and history. Yale University Press, New Haven

Mattatia J (1977) Amphicarpy and variability in Pisum fulvum. Bot Notiser 130:27–34

Mauseth JD (2006) Structure-function relationships in highly modified shoots of Cactaceae. Ann Bot 98:901–926

Maw MG (1981) *Rhamnus cathartica* L., common or European buckthorn (Rhamnaceae). Biological control program against insects and weeds in Canada 1969–1980. Commonwealth Agricultural Bureau, London, pp 185–189

Maynard Smith J (1976) Sexual selection and the handicap principle. J Theor Biol 57:239–242

Maynard Smith J, Harper D (2003) Animal signals. Oxford University Press, Oxford

McKey D, Waterman PG, Mbi CN, Gartlan JS, Struhsaker TT (1978) Phenolic content of vegetation in two African rain forests: ecological implications. Science 202:61–64

McNaughton SJ, Tarrants JL (1983) Grass leaf silicification: natural selection for an inducible defense against herbivores. Proc Natl Acad Sci USA 80:790–791

McNaughton SJ, Tarrants JL, McNaughton MM, Davis RH (1985) Silica as a defense against herbivory and a growth promotor in African grasses. Ecology 66:528–535

Meharg C, Meharg AA (2015) Silicon, the silver bullet for mitigating biotic and abiotic stress, and improving grain quality, in rice? Environ Exp Bot 120:8–17

Melville R, Wrigley FA (1969) Fenestration in the leaves of *Monstera* and its bearing on the morphogenesis and colour patterns of leaves. Bot J Linn Soc 62:1–16

Mendelssohn H (1963) On the biology of the venomous snakes of Israel. Part I. Isr J Zool 12:143–170

Menzies IJ, Youard LW, Lord JM, Carpenter KL, van Klink JW, Perry NB, Schaefer HM, Gould KS (2016) Leaf colour polymorphisms: a balance between plant defence and photosynthesis. J Ecol 104:104–113

Merilaita S (1998) Crypsis through disruptive coloration in an isopod. Proc R Soc Lond B 265:1059–1064

Merilaita S (2003) Visual background complexity facilitates the evolution of camouflage. Evolution 57:1248–1254

Merilaita S, Lind J (2005) Background-matching and disruptive coloration, and the evolution of cryptic coloration. Proc R Soc Lond B 272:665–670

Merilaita S, Tuomi J, Jormalainen V (1999) Optimization of cryptic coloration in heterogeneous habitat. Biol J Linn Soc 67:151–161

Merilaita S, Lyytinen A, Mappes J (2001) Selection for cryptic coloration in a visually heterogeneous habitat. Proc R Soc Lond B 268:1925–1929

Merzlyak MN, Chivkunova OB, Solovchenko AE, Naqvi KR (2008) Light absorption by anthocyanins in juvenile, stressed, and senescing leaves. J Exp Bot 59:3903–3911

Meunier JD, Colin F (eds) (2001) Phytoliths: application in earth sciences and human history. A.A. Balkema Publishers, Lisse

Meyers MA, Lin AYM, Lin YS, Olevsky EA, Georgalis S (2008) The cutting edge: sharp biological materials. JOM J Miner Met Mater Soc 60:19–24

Midgley JJ (2004) Why are spines of African Acacia species white? Afr J Range Forage Sci 21:211–212

Midgley JJ, Botha MA, Balfour D (2001) Patterns of thorn length, density, type and colour in African Acacias. Afr J Range Forage Sci 18:59–61

Midgley JJ, White JDM, Johnson SD, Bronner GN (2015) Faecal mimicry by seeds ensures dispersal by dung beetles. Nat Plants 1:15141

Midgley JJ, Abbas H, Armel MP (2016) Further evidence that in African acacia, white is a warning colour to herbivores; the white pseudo-galls of *Vachellia seyal*. Afr J Range Forage Sci 33:127–129

Milchunas DG, Noy-Meir I (2002) Grazing refuges, external avoidance of herbivory and plant diversity. Oikos 99:113–130

Milewski AV, Young TP, Madden D (1991) Thorns as induced defenses: experimental evidence. Oecologia 86:70–75

Milne RI (2004) Phylogeny and biogeography of *Rhododendron* subsection *Pontica*, a group with a tertiary relict distribution. Mol Phylogenet Evol 33:389–401

Milne RI, Abbott RJ (2002) The origin and evolution of Tertiary relict floras. Adv Bot Res 38:281–314

Milton SJ (1991) Plant spinescence in arid southern Africa: does moisture mediate selection by mammals? Oecologia 87:279–287

Mirocha CJ, Pathre SV, Schauerhamer B, Christensen CM (1976) Natural occurrence of *Fusarium* toxins in feedstuff. Appl Environ Microbiol 32:553–556

Mitchell RJ, Flanagan RJ, Brown BJ, Waser NM, Karron JD (2009) New frontiers in competition for pollination. Ann Bot 103:1403–1413

Mithöfer A, Boland W (2012) Plant defense against herbivores: chemical aspects. Annu Rev Plant Biol 63:431–450

Mogi T, Takahashi H, Katsuyama T, Ishii H, Ohta K, Kigawa S, Sakio H, Nakagawa S, Yoshiyama H (2000) Woody plants of Japan: Choripetalae 2. Yama-kei, Tokyo

Moore R, Langenkamp M (1991) Tissue partitioning during leaf development in ornamentally-grown *Frithia pulchra* (Mesembryanthemaceae), a 'window plant'. Ann Bot 67:279–283

Moore BP, Brown WV, Rothschild M (1990) Methylalkylpyrazines in aposematic insects, their hostplants and mimics. Chemoecology 1:43–51

Morais-Filho JC, Romero GQ (2010) Plant glandular trichomes mediate protective mutualism in a spider-plant system. Ecol Entomol 35:485–494

Moran N, Hamilton WD (1980) Low nutritive quality as defence against herbivores. J Theor Biol 86:247–254

Moran JA, Booth WE, Charles JK (1999) Aspects of pitcher morphology and spectral characteristics of six Bornean *Nepenthes* pitcher plant species: implications for prey capture. Ann Bot 83:521–528

Moran JA, Clarke C, Gowen BE (2012) The use of light in prey capture by the tropical pitcher plant *Nepenthes aristolochioides*. Plant Signal Behav 7:957–960

Morden-Moore AL, Willson MF (1982) On the ecological significance of fruit color in *Prunus serotina* and *Rubus occidentalis*: field experiments. Can J Bot 60:1554–1560

Morrell R (1969) Play snake for safety. Animals 12:154–155

Morris D (1990) Animal watching. A field guide to animal behaviour. Jonathan Cape, London

Morton TC, Vencl FV (1998) Larval beetles form a defense from recycled host-plant chemicals discharged as fecal wastes. J Chem Ecol 24:765–785

Mosquin T (1971) Competition for pollinators as a stimulus for the evolution of flowering time. Oikos 22:398–402

Motoyoshi I, Nishida S, Sharan L, Adelson EH (2007) Image statistics and the perception of surface qualities. Nature 447:206–209

Moya D, Espelta JM, Verkai I, López-Serrano F, De Las HJ (2007) Tree density and site quality influence on *Pinus halepensis* Mill. Reproductive characteristics after large fires. Ann For Sci 64:649–656

Moya D, Saracino A, Salvatore R, Lovreglio R, de Las Heras JV (2008) Anatomic basis and insulation of serotinous cones in *Pinus halepensis* Mill. Trees Struct Funct 22:511–519

Mudge JF (2013) Explicit consideration of critical effect sizes and costs of errors can improve decision-making in plant science. New Phytol 199:876–878

Muir CD (2013) How did the Swiss cheese plant get its holes? Am Nat 181:273–281

Mukherjee R, Kodandaramaiah U (2015) What makes eyespots intimidating-the importance of pairedness. BMC Evol Biol 15:34

Müller F (1879) *Ituna* and *Thyridia*; a remarkable case of mimicry in butterflies. Proc Entomol Soc Lond 1879:xx–xxix

Myczko Ł, Skórka P, Dylewski Ł, Sparks TH, Tryjanowski P (2015) Color mimicry of empty seeds influences the probability of predation by birds. Ecosphere 6:177

Myers JH, Bazely D (1991) Thorns, spines, prickles, and hairs: are they stimulated by herbivory and do they deter herbivores? In: Tallamy DW, Raupp MJ (eds) Phytochemical induction by herbivores. John Wiley & Sons Inc, New York, pp 325–344

Nahas L, Gonzaga MO, Del-Claro K (2012) Emergent impacts of ant and spider interactions: herbivory reduction in a tropical savanna tree. Biotropica 44:498–505

Nakata PA (2003) Advances in our understanding of calcium oxalate crystal formation and function in plants. Plant Sci 164:901–909

Nassar O, Lev-Yadun S (2009) How prickly is a prickly pear? Isr J Plant Sci 57:117–124

Nathan R, Ne'eman G (2000) Serotiny, seed dispersal and seed predation. In: Ne'eman G, Trabaud L (eds) Ecology, biogeography and management of *Pinus halepensis* and *Pinus brutia* forest ecosystems in the Mediterranean basin. Backhuys Publishers, Leiden, pp 105–118

Nathan R, Ne'eman G (2004) Spatiotemporal dynamics of recruitment in Aleppo pine (*Pinus halepensis* Miller). Plant Ecol 171:123–137

Nathan R, Safriel UN, Noy-Meir I, Schiller G (1999) Seed release without fire in *Pinus halepensis*, a Mediterranean serotinous wind-dispersed tree. J Ecol 87:659–669

Nathan R, Safriel UN, Noy-Meir I, Schiller G (2000) Spatiotemporal variation in seed dispersal and recruitment near and far from *Pinus halepensis* trees. Ecology 81:2156–2169

Neal PR, Dafni A, Giurfa M (1998) Floral symmetry and its role in plant-pollinator systems: terminology, distribution, and hypotheses. Annu Rev Ecol Syst 29:345–373

Ne'eman G, Izhaki I (1999) The effect of stand age and microhabitat on soil seed banks in Mediterranean Aleppo pine forests after fire. Plant Ecol 144:115–125

Ne'eman G, Goubitz S, Nathan R (2004) Reproductive traits of *Pinus halepensis* in the light of fire – a critical review. Plant Ecol 171:69–79

Neinhuis C, Müller-Doblies U, Müller-Doblies D (1996) *Psammophora* and other sand-coated plants from southern Africa. Feddes Repert 107:549–555

Neish PG, Drinnan AN, Ladiges PY (1995) Anatomy of leaf-margin lenticels in *Eucalyptus denticulata* and three other eucalypts. Aust J Bot 43:211–221

Nelson ZJ, Jackson RR, Li D, Barrion AT, Edwards GB (2006) Innate aversion to ants (Hymenoptera: Formicidae) and ant mimics: experimental findings from mantises (Mantodea). Biol J Linn Soc 88:23–32

Nersesian CL, Banks PB, McArthur C (2011) Titrating the cost of plant toxins against predators: determining the tipping point for foraging herbivores. J Anim Ecol 80:753–760

Newark T, Miller J (2007) Camouflage. Thames & Hudson, New York

Nicotra AB, Hofmann M, Siebke K, Ball MC (2003) Spatial patterning of pigmentation in evergreen leaves in response to freezing stress. Plant Cell Environ 26:1893–1904

Nicotra AB, Leigh A, Boyce CK, Jones CS, Niklas KJ, Royer DL, Tsukaya H (2011) The evolution and functional significance of leaf shape in the angiosperms. Funct Plant Biol 38:535–552

Niemelä P (1979) Topographical delimination of *Oporinia*-damages: Experimental evidence of the effect of winter temperature. Rep Kevo Subarct Res Sta 15:33–36

Niemelä P, Tuomi J (1987) Does the leaf morphology of some plants mimic caterpillar damage? Oikos 50:256–257

Niemelä P, Tuomi J, Siren S (1984) Selective herbivory on mosaic leaves of variegated *Acer pseudoplatanus*. Experientia 40:1433–1434

Nikiforou C, Zeliou K, Kytridis VP, Kyzeridou A, Manetas Y (2010) Are red leaf phenotypes more or less fit? The case of winter leaf reddening in *Cistus creticus*. Environ Exp Bot 67:509–514

Nikiforou C, Nikolopoulos D, Manetas Y (2011) The winter-red-leaf syndrome in *Pistacia lentiscus*: evidence that the anthocyanic phenotype suffers from nitrogen deficiency, low carboxylation efficiency and high risk of photoinhibition. J Plant Physiol 168:2184–2187

Nishida R (2002) Sequestration of defensive substances from plants by Lepidoptera. Annu Rev Entomol 47:57–92

Niskanen M, Mappes J (2005) Significance of the dorsal zigzag pattern of *Vipera latastei gaditana* against avian predators. J Anim Ecol 74:1091–1101

Niu Y, Sun H (2014) Alpine scree plants benefit from cryptic coloration with limited cost. Plant Singnal Behav 9:e29698

Niu Y, Chen G, Peng D-L, Song B, Yang Y, Li Z-M, Sun H (2014) Grey leaves in an alpine plant: a cryptic colouration to avoid attack? New Phytol 203:953–963

Nobel PS (1978) Surface temperature of cacti – influences of environmental and morphological factors. Ecology 59:986–996

Nobel PS (1994) Remarkable agaves and cacti. Oxford University Press, New York

Nolte DL, Mason JR, Epple G, Aronov E, Campbell DL (1994) Why are predator urines aversive to prey? J Chem Ecol 20:1505–1516

Noy-Meir I (1990) The effect of grazing on the abundance of wild wheat, barley and oat in Israel. Biol Conserv 51:299–310

Numata S, Kachi N, Okuda T, Manokaran N (2004) Delayed greening, leaf expansion, and damage to sympatric *Shorea* species in a lowland rain forest. J Plant Res 117:19–25

Nyffeler M (2000) Ecological impact of spider predation: a critical assessment of Bristowe's and Turnbull's estimates. Bull Br Arachnol Soc 11:367–373

Nyffeler M, Sterling WL, Dean DA (1994) How spiders make a living. Environ Entomol 23:1357–1367

Nyman T, Julkunen-Titto RR (2000) Manipulation of the phenolic chemistry of willows by gall-inducing sawflies. Proc Natl Acad Sci USA 97:13184–13187

Nystrand O, Granström A (1997) Post-dispersal predation on *Pinus sylvestris* seeds by *Fringilla* spp: ground substrate affects selection for seed color. Oecologia 110:353–359

Oberbauer SF, Starr G (2002) The role of anthocyanins for photosynthesis of Alaskan arctic evergreens during snowmelt. Adv Bot Res 37:129–145

Obeso JR (1997) The induction of spinescence in European holly leaves by browsing ungulates. Plant Ecol 129:149–156

Ochoa-López S, Villamil N, Zedillo-Avelleyra P, Boege K (2015) Plant defence as a complex and changing phenotype throughout ontogeny. Ann Bot 116:797–806

Oelschlägel B, Gorb S, Wanke S, Neinhuis C (2009) Structure and biomechanics of trapping flower trichomes and their role in the pollination biology of *Aristolochia* plants (Aristolochiaceae). New Phytol 184:988–1002

Oliveira AP, Silva LR, Ferreres F, de Pinho PG, Valentão P, Silva BM, Pereira JA, Andrade PB (2010) Chemical assessment and *in vitro* antioxidant capacity of *Ficus carica* latex. J Agric Food Chem 58:3393–3398

Omacinl M, Chaneton EJ, Ghersa CM, Müller CB (2001) Symbiotic fungal endophytes control insect host-parasite interaction webs. Nature 409:78–81

Ono H (ed) (2009) The spiders of Japan: with keys and Genera and illustrations of the species. Tokai University Press, Tokyo

Opitz SEW, Müller C (2009) Plant chemistry and insect sequestration. Chemoecol 19:117–154

Ori M, Ikeda H (1998) Spider venoms and spider toxins. J Toxicol – Toxin Rev 17:405–426

Orshan G (1963) Seasonal dimorphism of desert and Mediterranean chamaephytes and its significance as a factor in their water economy. In: Rutter AJ, Whitehead FH (eds) The water relations of plants. John Wiley & Sons Inc, New York, pp 206–222

Orshan G (1986) The deserts of the Middle East. In: Evenari M, Noy-Meir I, Goodall DW (eds) Hot deserts and arid shrublands. Elsevier, Amsterdam, pp 1–28

Ortolani A (1999) Spots, stripes, tail tips and dark eyes: predicting the function of carnivore colour patterns using the comparative method. Biol J Linn Soc 67:433–476

Osterloh J, Herold S, Pond S (1982) Oleander interference in the digoxin radioimmunoassay in a fatal ingestion. JAMA 247:1596–1597

Ougham HJ, Morris P, Thomas H (2005) The colors of autumn leaves as symptoms of cellular recycling and defenses against environmental stresses. Curr Top Dev Biol 66:135–160

Ougham H, Thomas H, Archetti M (2008) The adaptive value of leaf colour. New Phytol 179:9–13

Owen D (1982) Camouflage and mimicry. University of Chicago Press, Chicago

Owen-Smith N (2015) Mechanisms of coexistence in diverse herbivore-carnivore assemblages: demographic, temporal and spatial heterogeneities affecting prey vulnerability. Oikos 124:1417–1426

Pannell JR (2014) Leaf mimicry: chameleon-like leaves in a Patagonian vine. Curr Biol 24:R357–R359

Papaj DR, Newsom GM (2005) A within-species warning function for an aposematic signal. Proc R Soc Lond B 272:2519–2523

Parsons DJ, Moldenke AR (1975) Convergence in vegetation structure along analogous climatic gradients in California and Chile. Ecology 56:950–957

Pascual FB, McGinley EL, Zanardi LR, Cortese MM, Murphy TV (2003) Tetanus surveillance, United States, 1998–2000. 52, SS03:1–8. Available from: http://www.cdc.gov/mmwr/preview/mmwrhtml/ss5203a1.htm

Pasteur G (1982) A classificatory review of mimicry systems. Annu Rev Ecol Syst 13:169–199

Paul ND, Gwynn-Jones D (2003) Ecological roles of solar UV radiation: towards an integrated approach. Trends Ecol Evol 18:48–55

Paulsen TR, Colville L, Kranner I, Daws MI, Högstedt G, Vandvik V, Thompson K (2013) Physical dormancy in seeds: a game of hide and seek? New Phytol 198:496–503

Paulsen TR, Högstedt G, Thompson K, Vandvik V, Eliassen S (2014) Conditions favouring hard seededness as a dispersal and predator escape strategy. J Ecol 102:1475–1484

Paulsen TR, Colville L, Daws MI, Eliassen S, Högstedt G, Kranner I, Thompson K, Vandvik V (2015) The crypsis hypothesis explained: a reply to Jayasuriya et al. (2015). Seed Sci Res 25:402–408

Payne GA, Brown MP (1998) Genetics and physiology of aflatoxin biosynthesis. Annu Rev Phytopathol 36:329–362

Pažoutová S (2006) Website at the Institute of Microbiology, Academy of Sciences of the Czech Republic ("Ergot infopage, what is *Claviceps*", the table)

Pažoutová S, Parbery DP (1999) The taxonomy and phylogeny of Claviceps. In: Křen V, Cvak L (eds) Ergot the genus *Claviceps*. Harwood Academic Publishers, Australia, pp 57–77

Pearn J (1987) Oleander poisoning. In: Covacevich J, Davie P, Pearn J (eds) Toxic plants & animals; a guide for Australia. Queensland Museum, Brisbane, pp 37–49

Peira N, Golkar A, Larsson M, Wiens S (2010) What you fear will appear: detection of schematic spiders in spider fear. Exp Psychol 57:470–475

Pellmyr O, Thien LB (1986) Insect reproduction and floral fragrances: keys to the evolution of the angiosperms? Taxon 35:76–85

Penacchio O, Lovell PG, Cuthill IC, Ruxton GD, Harris JM (2015a) Three-dimensional camouflage: exploiting photons to conceal form. Am Nat 186:553–563

Penacchio O, Cuthill IC, Lovell PG, Ruxton GD, Harris JM (2015b) Orientation to the sun by animals and its interaction with crypsis. Funct Ecol 29:1165–1177

Pener MP (1991) Locust phase polymorphism and its endocrine relations. Adv Insect Physiol 23:1–79

Penney HD, Hassall C, Skevington JH, Abbott KR, Sherratt TN (2012) A comparative analysis of the evolution of imperfect mimicry. Nature 483:461–466

Perevolotsky A, Haimov Y (1991) Structural response of Mediterranean woodland species to disturbance: evidence of different defense strategies. Isr J Bot 40:305–313

Perevolotsky A, Seligman N (1998) Role of grazing in Mediterranean rangeland ecosystems. Inversion of a paradigm. Bioscience 48:1007–1017

Perevolotsky A, Schwartz-Tzachor R, Yonathan R, Ne'eman G (2011) Geophytes-herbivore interactions: reproduction and population dynamics of *Anemone coronaria* L. Plant Ecol 212:563–571

Petanidou T, Ellis WN, Margaris NS, Vokou D (1995) Constraints on flowering phenology in a phryganic (East Mediterranean Shrub) community. Am J Bot 82:607–620

Pfeister JA (1999) Behavioral strategies for coping with poisonous plants. Idaho for wildlife range. Exp Sta Bull 70:45–59

Pfister JA, Müller-Schwarze D, Balph DF (1990) Effects of predator fecal odors on feed selection by sheep and cattle. J Chem Ecol 16:573–583

Pleasants JM (1980) Competition for bumblebee pollinators in rocky mountain plant communities. Ecology 61:1446–1459

Plowright RC, Owen RE (1980) The evolutionary significance of bumble bee color patterns: a mimetic interpretation. Evolution 34:622–637

Polte S, Reinhold K (2013) The function of the wild carrot's dark central floret: attract, guide or deter? Plant Species Biol 28:81–86

Pope D (2005) 73 north. The battle of the Barents sea. McBooks Press, Ithaca

Porter JK (1994) Chemical constituents of grass endophytes. In: Bacon CW, White JF (eds) Biotechnology of endophytic fungi of grasses. CRC Press, Boca Raton, pp 103–123

Porter SS (2013) Adaptive divergence in seed color camouflage in contrasting soil environments. New Phytol 197:1311–1320

Potter DA, Kimmerer TW (1988) Do holly leaf spines really deter herbivory? Oecologia 75:216–221

Potterat O (2010) Goji (*Lycium barbarum* and *L. chinense*): phytochemistry, pharmacology and safety in the perspective of traditional uses and recent popularity. Planta Med 76:7–19

Poulton EB (1890) The colours of animals. Their meaning and use. Especially considered in the case of insects, 2nd edn. Kegan Paul, Trench, Trübner & Co Ltd, London

Preisser EL (2009) The physiology of predator stress in free-ranging prey. J Anim Ecol 78:1103–1105

Prescott LM, Harley JP, Klein DA (2005) The fungi (Eumycota), slime molds and water molds. In: Prescott LM, Harley JP, Klein DA (eds) Microbiology, 5th edn. McGraw-Hill, New York, pp 537–552

Preston-Mafham R, Preston-Mafham K (1994) Cacti the illustrated dictionary. Timber Press, Portland

Price PW (1997) Insect ecology, 3rd edn. John Wiley & Sons Inc, New York

Price PW, Bouton CE, Gross P, McPheron BA, Thompson JN, Weis AE (1980) Interactions among three throphic levels: influence of plants on interactions between insect herbivores and natural enemies. Annu Rev Ecol Syst 11:41–65

Price PW, Fernandes WG, Waring GL (1987) Adaptive nature of insect galls. Environ Entomol 16:15–24

Prokopy RJ (1972) Evidence for a marking pheromone deterring repeated oviposition in apple-maggot flies. Environ Entomol 1:326–332

Prokopy RJ, Owens ED (1983) Visual detection of plants by herbivorous insects. Annu Rev Entomol 28:337–364

Provan J, Bennett KD (2008) Phylogeographic insights into cryptic glacial refugia. Trends Ecol Evol 23:564–571

Provenza FD, Kimball BA, Villalba JJ (2000) Roles of odor, taste, and toxicity in the food preferences of lambs: implications for mimicry in plants. Oikos 88:424–432

Prudic KL, Skemp AK, Papaj DR (2007) Aposematic coloration, luminance contrast, and the benefits of conspicuousness. Behav Ecol 18:41–46

Prychild CJ, Rudall PJ, Gregory M (2004) Systematics and biology of silica bodies in Monocotyledons. Bot Rev 69:377–440

Purcell JE (1980) Influence of siphonophore behavior upon their natural diets: evidence for aggressive mimicry. Science 209:1045–1047

Purser B (2003) Jungle bugs: masters of camouflage and mimicry. Firefly Books, Toronto

Queenborough SA, Metz MR, Valencia R, Wright SJ (2013) Demographic consequences of chromatic leaf defence in tropical tree communities: do red young leaves increase growth and survival? Ann Bot 112:677–684

Raese JT (2002) Phosphorous deficiency symptoms in leaves of apple and pear trees as influenced by available soil phosphorous. Commun Soil Sci Plant Anal 33:461–477

Rafferty C (1999) Dietary selection of macropods in south western Australia with particular reference to plant physiological defences. Honours thesis, Curtin University of Technology, Perth

Raghothama KG (1999) Phosphate acquisition. Annu Rev Plant Physiol Plant Mol Biol 50:665–693

Ralphs MH, Jones WA, Pfister JA (1997) Damage from the larkspur mirid deters cattle grazing of larkspur. J Range Manag 50:371–373

Ramachandran VS, Tyler CW, Gregory RL, Rogers-Ramachandran D, Duensing S, Pillsbury C, Ramachandran C (1996) Rapid adaptive camouflage in tropical flounders. Nature 379:815–818

Raman A, Schaefer CW, Withers TM (2005) Biology, ecology, and evolution of gall-inducing arthropods. Science Publishers, Enfield
Ramirez CC, Lavandero B, Archetti M (2008) Coevolution and the adaptive value of autumn tree colours: colour preference and growth rates of a southern beech aphid. J Evol Biol 21:49–56
Rand K, Bar E, Ben-Ari M, Lewinsohn E, Inbar M (2014) The mono – and sesquiterpene content of aphid-induced galls on *Pistacia palaestina* is not a simple reflection of their composition in intact leaves. J Chem Ecol 40:632–642
Raper KB, Fennell DI (1965) The genus *Aspergillus*. Robert E. Krieger Publishing Company, Huntington
Rapp G Jr, Mulholland SC (eds) (1992) Phytolith systematics. Emerging issues. Plenum Press, New York
Rausher MD (1978) Search image for leaf shape in a butterfly. Science 200:1071–1073
Rausher MD (1979) Egg recognition: its advantage to a butterfly. Anim Behav 27:1034–1040
Rebollo S, Milchunas DG, Noy-Meir I, Chapman PL (2002) The role of spiny plant refuge in structuring grazed shortgrass steppe plant communities. Oikos 98:53–64
Redfern M (2011) Plant galls. Collins, Hong Kong
Reeves JL (2011) Vision should not be overlooked as an important sensory modality for finding host plants. Environ Entomol 40:855–863
Rehill BJ, Schultz JC (2012) *Hormaphis hamamelidis* fundatrices benefit by manipulating phenolic metabolism of their host. J Chem Ecol 38:496–498
Renoult JP, Valido A, Jordano P, Schaefer HM (2014) Adaptation of flower and fruit colours to multiple, distinct mutualists. New Phytol 201:678–686
Renwick JAA (2002) The chemical world of crucivores: lures, treats and traps. Entomol Exp Appl 104:35–42
Reynolds GW (1969) The Aloes of South Africa. A.A. Balkema, Cape Town
Reznicek G, Jurenitsch J, Robien W, Kubelka W (1989) Saponins in *Cyclamen* species: configuration of cyclamiretin C and structure of isocyclamin. Phytochemistry 28:825–828
Richards PW (1996) The tropical rain forest an ecological study, 2nd edn. Cambridge University Press, Cambridge
Richmond KE, Sussman M (2003) Got silicon? The non-essential beneficial plant nutrient. Curr Opin Plant Biol 6:268–272
Ridley HN (1930) The dispersal of plants throughout the world. L. Reeve & Co Ltd, Ashford
Ripple WJ, Beschta RL (2004) Wolves and the ecology of fear: can predation risk structure ecosystems? Bioscience 54:755–766
Rivero-Lynch AP, Brown VK, Lawton JH (1996) The impact of leaf shape on the feeding preference of insect herbivores: experimental and field studies with *Capsella* and *Phyllotreta*. Philos Trans R Soc Lond B 351:1671–1677
Robbins RK (1980) The lycaened "false head" hypothesis: historical review and quantitative analysis. J Lepidopteran Soc 34:194–208
Robbins RK (1981) The 'false head' hypothesis: predation and wing pattern variation of lycaenid butterflies. Am Nat 118:770–775
Robbins CT, Hanley TA, Hagerman AE, Hjeljord O, Baker DL, Schwartz CC, Mautz WW (1987) Role of tannins in defending plants against ruminants: reduction in protein availability. Ecology 68:98–107
Rolfe RA (1910) The bee-orchis. Orch Rev 10:261
Rolshausen G, Schaefer HM (2007) Do aphids paint the tree red (or yellow) – can herbivore resistance or photoprotection explain colourful leaves in autumn? Plant Ecol 191:77–84
Rolston MP (1978) Water impermeable seed dormancy. Bot Rev 44:365–396
Romero GQ, Souza JC, Vasconcellos-Neto J (2008) Anti-herbivore protection by mutualistic spiders and the role of plant glandular trichomes. Ecology 89:3105–3115
Ronel M, Lev-Yadun S (2009) Spiny plants in the archaeological record of Israel. J Arid Environ 73:754–761
Ronel M, Lev-Yadun S (2012) The spiny, thorny and prickly plants in the flora of Israel. Bot J Linn Soc 168:344–352

Ronel M, Malkiel H, Lev-Yadun S (2007) Quantitative characterization of the thorn system of the common shrubs *Sarcopoterium spinosum* and *Calicotome villosa*. Isr J Plant Sci 55:63–72

Ronel M, Khateeb S, Lev-Yadun S (2009) Protective spiny modules in thistles of the Asteraceae in Israel. J Torrey Bot Soc 136:46–56

Ronel M, Ne'eman G, Lev-Yadun S (2010) Spiny east-Mediterranean plant species flower later and in a drier season than non-spiny species. Flora 205:276–281

Rosenblum EB (2006) Convergent evolution and divergent selection: Lizards at the white sands ecotone. Am Nat 167:1–15

Roslin T, Syrjälä H, Roland J, Harrison PJ, Fownes S, Matter SF (2008) Caterpillars on the run – induced defenses create spatial patterns in host plant damage. Ecography 31:335–347

Rostás M, Maag D, Ikegami M, Inbar M (2013) Gall volatiles defend aphids against a browsing mammal. BMC Evol Biol 13:193

Rota J, Wagner DL (2006) Predator mimicry: metalmark moths mimic their jumping spider predators. PLoS One 1:e45

Rothschild M (1972) Some observations on the relationship between plants, toxic insects and birds. In: Harborne JB (ed) Phytochemical ecology. Academic Press, London, pp 1–12

Rothschild M (1973) Secondary plant substances and warning colouration in insects. In: van Emden HF (ed) Insect/plant relationships. Symposia of the Royal Entomological Society of London: Number six. Blackwell Scientific Publications, Oxford, pp 59–83

Rothschild M (1974) Modified stipules of *Passiflora* which resemble horned caterpillars. Proc R Entomol Soc Lond 39:16

Rothschild M (1980) Remarks on carotenoids in the evolution of signals. In: Gilbert LE, Raven PH (eds) Coevolution of animals and plants. University of Texas Press, Austin, pp 20–51

Rothschild M (1984) Aide mémoire mimicry. Ecol Entomol 9:311–319

Rothschild M (1986) The red smell of danger. New Sci 111(4):34–36

Rothschild M, Moore B (1987) Pyrazines as alerting signals in toxic plants and insects. In: Labeyrie V, Fabres G, Lachaise D (eds) Insects – plants. Dr W. Junk Publishers, Dordrecht, pp 97–101

Rothschild M, Schoonhoven LM (1977) Assessment of egg load by *Pieris brassicae* (Lepidoptera: Pieridae). Nature 266:352–355

Rothschild M, Moore BP, Brown WV (1984) Pyrazines as warning odour components in the Monarch butterfly, *Danaus plexippus*, and in moths of the genera *Zygaena* and *Amata* (Lepidoptera). Biol J Linn Soc 23:375–380

Rowe C, Halpin C (2013) Why are warning displays multimodal? Behav Ecol Sociobiol 67:1425–1439

Rowland HM, Ihalainen E, Lindström L, Mappes J, Speed M (2007) Co-mimics have a mutualistic relationship despite unequal defences. Nature 448:64–67

Rowland HM, Mappes J, Ruxton GD, Speed MP (2010) Mimicry between unequally defended prey can be parasitic: evidence for quasi-Batesian mimicry. Ecol Lett 13:1494–1502

Roy BA, Widmer A (1999) Floral mimicry: a fascinating yet poorly understood phenomenon. Trends Plant Sci 4:325–330

Roze JA (1996) Coral snakes of the Americas: biology, identification, and venoms. Krieger Publishing Company, Malabar

Rubino DL, McCarthy BC (2004) Presence of aposematic (warning) coloration in vascular plants of southeastern Ohio. J Torrey Bot Soc 131:252–256

Ruhren S, Handel SN (1999) Jumping spiders (Salticidae) enhance the seed production of a plant with extrafloral nectaries. Oecologia 119:227–230

Ruiz N, Ward D, Saltz D (2002) Calcium oxalate crystals in leaves of *Pancratium sickenbergeri*: constitutive or induced defence? Funct Ecol 16:99–105

Russell BG, Banks PB (2007) Do Australian small mammals respond to native and introduced predator odours? Austral Ecol 32:277–286

Russo RA (2007) Field guide to plant galls of California and other western states. University of California Press, Berkeley

Ruxton GD (2002) The possible fitness benefits of striped coat coloration for zebra. Mammal Rev 32:237–244

Ruxton GD (2009) Non-visual crypsis: a review of the empirical evidence for camouflage to senses other than vision. Philos Trans R Soc B 364:549–557

Ruxton GD (2014) Empirically exploring why latex might be white: a comment on Lev-Yadun 2014. Chemoecology 24:219–220

Ruxton GD, Stevens M (2015) The evolutionary ecology of decorating behaviour. Biol Lett 11:20150325

Ruxton GD, Sherratt TN, Speed MP (2004) Avoiding attack. The evolutionary ecology of crypsis, warning signals & mimicry. Oxford University Press, Oxford

Ruxton GD, Wilkinson DM, Schaefer HM, Sherratt TN (2015) Why fruit rots: theoretical support for Janzen's theory of microbe-macrobe competition. Proc R Soc B 281:20133320

Ryan CA (1990) Protease inhibitors in plants: genes for improving defenses against insects and pathogens. Annu Rev Phytopathol 28:425–449

Rypstra AL, Buddle CM (2013) Spider silk reduces insect herbivory. Biol Lett 9:20120948

Sajeva M, Costanzo M (1994) Succulents. The illustrated dictionary. Timber Press, Portland

Sakai WS, Hanson M, Jones RC (1972) Raphids with barbs and grooves in *Xanthosoma sagittifolium* (Araceae). Science 178:314–315

Salvatore R, Moya D, Pulido L, Lovreglio R, Lo'pez-Serrano FR, De las Heras J, Leone V (2010) Morphological and anatomical differences in Aleppo pine seeds from serotinous and non-serotinous cones. New For 39:329–341

Sanson GD, Kerr SA, Gross KA (2007) Do silica phytoliths really wear mammalian teeth? J Archaeol Sci 34:526–531

Santer RD (2013) Motion dazzle: a locust's eye view. Biol Lett 9:20130811

Saracino A, Pacella R, Leone V, Borghetti M (1997) Seed dispersal and changing seed characteristics in a *Pinus halepensis* Mill. forest after fire. Plant Ecol 130:13–19

Saracino A, D'Alessandro CM, Borghetti M (2004) Seed colour and post-fire bird predation in a Mediterranean pine forest. Acta Oecol 26:191–196

Sato Y, Kudoh H (2016) Associational effects against a leaf beetle mediate a minority advantage in defense and growth between hairy and glabrous plants. Evol Ecol 30:137–154

Savage JM, Slowinski JB (1992) The colouration of the venomous coral snakes (family Elapidae) and their mimics (families Aniliidae and Colubridae). Biol J Linn Soc 45:235–254

Scarchuk J, Lent JM (1965) The structure of mottled-leaf summer squash. J Hered 56:167–168

Schaberg PG, van den Berg AK, Murakami PF, Shane JB, Donnelly JR (2003) Factors influencing red expression in autumn foliage of sugar maple trees. Tree Physiol 23:325–333

Schaberg PG, Murakami PF, Turner MR, Heitz HK, Hawley GJ (2008) Association of red coloration with senescence of sugar maple leaves in autumn. Trees Struct Funct 22:573–578

Schaedelin FC, Taborsky M (2009) Extended phenotypes as signals. Biol Rev 84:293–313

Schaefer HM (2011) Why fruits go to the dark side. Acta Oecol 37:604–610

Schaefer HM, Gould KS (2007) Modelling the evolution of leaf colouration with binary assumptions is barking up the wrong tree. J Theor Biol 249:638–639

Schaefer HM, Rolshausen G (2006a) Plants on red alert: do insects pay attention? BioEssays 28:65–71

Schaefer HM, Rolshausen G (2006b) Do alternative hypotheses evoke red alert among co-evolutionists? BioEssays 28:960–961

Schaefer HM, Rolshausen G (2007a) Aphids do not attend to leaf colour as visual signal, but to the handicap of reproductive investment. Biol Lett 3:1–4

Schaefer HM, Rolshausen G (2007b) How to investigate a putative signal? Stick to the right method when assessing the response of a receiver. Biol Lett 3:152–153

Schaefer HM, Ruxton GD (2008) Fatal attraction: carnivorous plants roll out the red carpet to lure insects. Biol Lett 4:153–155

Schaefer HM, Ruxton GD (2009) Deception in plants: mimicry or perceptual exploitation? Trends Ecol Evol 24:676–685

Schaefer HM, Ruxton GD (2011) Plant-animal communication. Oxford University Press, New York

Schaefer HM, Ruxton GD (2014) Fenestration: a window of opportunity for carnivorous plants. Biol Lett 10:20140134

Schaefer HM, Schaefer V (2007) The evolution of visual fruit signals: concepts and constraints. In: Dennis AJ, Schupp EW, Green R, Wescott DW (eds) Seed dispersal: theory and its application in a changing world. CAB International, Wallingford, pp 59–77

Schaefer HM, Schmidt V (2004) Detectability and content as opposing signal characteristics in fruits. Proc R Soc Lond B 271(Suppl):S370–S373

Schaefer M, Stobbe N (2006) Disruptive coloration provides camouflage independent of background matching. Proc R Soc Lond B 273:2427–2432

Schaefer HM, Wilkinson DM (2004) Red leaves, insects and coevolution: a red herring? Trends Ecol Evol 19:616–618

Schaefer HM, Schaefer V, Levey DJ (2004) How plant-animal interactions signal new insights in communication. Trends Ecol Evol 19:577–584

Schaefer HM, Levey DJ, Schaefer V, Avery ML (2006) The role of chromatic and achromatic signals for fruit detection by birds. Behav Ecol 17:784–789

Schaefer HM, Schaefer V, Vorobyev M (2007) Are fruit colors adapted to consumer vision and birds equally efficient in detecting colorful signals? Am Nat 169(Suppl):S159–S169

Schaefer HM, Rentzsch M, Breuer M (2008) Anthocyanins reduce fungal growth in fruits. Nat Prod Commun 3:1267–1272

Schaefer HM, Valido A, Jordano P (2014) Birds see the true colours of fruits to live off the fat of the land. Proc R Soc B 281:20132516

Schiestl FP (2005) On the success of a swindle: pollination by deception in orchids. Naturwiss 92:255–264

Schiestl FP, Ayasse M (2001) Post-pollination emission of a repellent compound in a sexually deceptive orchid: a new mechanism for maximising reproductive success? Oecologia 126:531–534

Schiestl FP, Ayasse M, Paulus HF, Löfstedt C, Hansson BS, Ibarra F, Francke W (2000) Sex pheromone mimicry in the early spider orchid (*Ophrys sphegodes*): patterns of hydrocarbons as the key mechanism for pollination by sexual deception. J Comp Physiol A 186:567–574

Schiller G (1978) Factors involved in natural regeneration of Aleppo pine. Ph.D. thesis, Tel Aviv University, Tel Aviv (in Hebrew with English summary)

Schmidt-Entling MH, Siegenthaler E (2009) Herbivore release through cascading risk effects. Biol Lett 5:773–776

Schmitz OJ (2008) Herbivory from individuals to ecosystems. Annu Rev Ecol Evol Syst 39:133–152

Schoener TW (1987) Leaf pubescence in buttonwood: community variation in a putative defense against defoliation. Proc Natl Acad Sci USA 84:7992–7995

Schoener TW (1988) Leaf damage in island buttonwood, *Conocarpus erectus*: correlations with pubescence, island area, isolation and the distribution of major carnivores. Oikos 53:253–266

Schoener TW, Spiller DA (1987) Effect of lizards on spider populations: manipulative reconstruction of a natural experiment. Science 236:949–952

Schonrogge K, Walker P, Crawely MJ (1999) Complex life cycles in *Andricus kollari* (Hymenoptera: Cynipidae) and their impact on associated parasitoids and inquilines species. Oikos 84:293–301

Schultz BB (1992) Insect herbivores as potential causes of mortality and adaptation in gallforming insects. Oecologia 90:297–299

Schultz JC, Baldwin IT (1982) Oak leaf quality declines in response to defoliation by gypsy moth larvae. Science 217:149–151

Schuman MC, Baldwin IT (2016) The layers of plant responses to insect herbivores. Annu Rev Entomol 61:373–394

Scott-Samuel NE, Baddeley R, Palmer CE, Cuthill IC (2011) Dazzle camouflage affects speed perception. PLoS One 6:e20233

Seligman N, Rosensaft Z, Tadmor N, Katzenelson J, Naveh Z (1959) Natural pasture of Israel. Vegetation, carrying capacity and improvement. Sifriat Poalim, Merhavia (in Hebrew)

Sendoya SF, Freitas AVL, Oliveira PS (2009) Egg-laying butterflies distinguish predaceous ants by sight. Am Nat 174:134–140

Shapiro AM (1981a) Egg-mimics of *Streptanthus* (Cruciferae) deter oviposition by *Pieris sisymbrii* (Lepidoptera: Pieridae). Oecologia 48:142–143

Shapiro AM (1981b) The pierid red-egg syndrome. Am Nat 117:276–294

Shaw GR (1914) The genus *Pinus*, Publication # 5. The Arnold Arboretum, Cambridge

Shears J (1988) The use of a sand-coat in relation to feeding and dial activity in the sepiolid squid *Euprymna scolopes*. Malacologia 29:121–133

Sherburne JA (1972) Effects of seasonal changes in the abundance and chemistry of the fleshy fruits of northeastern woody shrubs on patterns of exploitation by frugivorous birds. Ph.D. thesis, Cornell University

Sheriff MJ, Krebs CJ, Boonstra R (2009) The sensitive hare: sublethal effects of predator stress on reproduction in snowshoe hares. J Anim Ecol 78:1249–1258

Sherratt TN (2007) Mimicry on the edge. Nature 448:34–36

Sherratt TN, Wilkinson DM, Bain RS (2005) Explaining Dioscorides' "double difference": why are some mushrooms poisonous, and do they signal their unprofitability? Am Nat 166:767–775

Sherzer W (1896) Pebble mimicry in Philippine Island beans. Bot Gaz 21:235–237

Shi Y, Yokoyama S (2003) Molecular analysis of the evolutionary significance of ultraviolet vision in vertebrates. Proc Natl Acad Sci USA 100:8308–8313

Shifriss O (1981) Do *Cucurbita* plants with silvery leaves escape virus infection? Cucurbit Genet Cooperative Rep 4:42–43

Shimizu N, Hirota S, Morita H (2001) Plant invader 600 species. Zenkoku-noson-kyoiku-kyokai, Tokyo

Shimizu T, Ohtani K, Hirakawa H, Ohshima K, Yamashita A, Shiba T, Ogasawara N, Hattori M, Kuhara S, Hayashi H (2002) Complete genome sequence of *Clostridium perfringens*, an anaerobic flesh-eater. Proc Natl Acad Sci U S A 99:996–1001

Shimshi D (1979/1980) Two ecotypes of *Iris atrofusca* Bak. and their relations to man-modified habitats. Isr J Bot 28:80–86

Shiojiri K, Ozawa R, Kugimiya S, Uefune M, van Wijk M, Sabelis MW, Takabayashi J (2010) Herbivore-specific, density-dependent induction of plant volatiles: honest or "cry wolf" signal? PLoS One 5:e12161

Shmida A (1981) Mediterranean vegetation in California and Israel: similarities and differences. Isr J Bot 30:105–123

Shmida A, Darom D (1986) Handbook of wildflowers of Israel. Desert flora. Keter Publishing House Ltd, Jerusalem (in Hebrew)

Shorthouse JD, Rohfritsch O (1992) The biology of insect-induced galls. Oxford University Press, Oxford

Showler AT (2002) Effects of kaolin-based particle film application on boll weevil (Coleoptera: Curculionidae) injury to cotton. J Econ Entomol 95:754–762

Showler AT (2003) Effects of kaolin particle film on beet armyworm, *Spodoptera exigua* (Hübner) (Lepidoptera: Noctuidae), oviposition, larval feeding and development on cotton, *Gossypium hirsutum* L. Agric Ecosyst Environ 95:265–271

Silfver T, Sinkkonen A, Oksanen E, Rousi M (2015) Early shoot growth termination in Betula pendula is associated with the number of overwintering aphid eggs on boreal birches. Evol Ecol 29:157–167

Sillén-Tullberg B, Brtant EH (1983) The evolution of aposematic coloration in distasteful prey: an individual selection model. Evolution 37:993–1000

Simons P (1992) The action plant. Movement and nervous behaviour in plants. Blackwell, Oxford

Sinkkonen A (2006a) Sexual reproduction advances autumn leaf colours in mountain birch (*Betula pubescens* ssp. *czerepanovii*). J Evol Biol 19:1722–1724

Sinkkonen A (2006b) Do autumn leaf colours serve as a reproductive insurance against sucking herbivores? Oikos 113:557–562

Sinkkonen A (2008) Red reveals branch die-back in Norway maple *Acer plantoides*. Ann Bot 102:361–366

Sinkkonen A (2009) Ultraviolet leaf pigments as a main component in autumn colour show: a constructive comment on Archetti et al. Trends Ecol Evol 24:236–237

Sinkkonen A, Somerkoski E, Paaso U, Holopainen JK, Rousi M, Mikola J (2012) Genotypic variation in yellow autumn leaf colours explains aphid load in silver birch. New Phytol 195:461–469

Sipura M (1999) Tritrophic interactions: willows, herbivorous insects and insectivorous birds. Oecologia 121:537–545

Sivinski J, Stowe M (1980) A kleptoparasitic cecidomyiid and other flies associated with spiders. Psyche 87:337–348

Skelhorn J (2015) Masquerade. Curr Biol 25:R643–R644

Skelhorn J, Ruxton GD (2011) Context-dependent misclassification of masquerading prey. Evol Ecol 25:751–761

Skelhorn J, Ruxton GD (2013) Size-dependent microhabitat selection by masquerading prey. Behav Ecol 24:89–97

Skelhorn J, Rowland HM, Ruxton GD (2010a) The evolution and ecology of masquerade. Biol J Linn Soc 99:1–8

Skelhorn J, Rowland HM, Speed MP, Ruxton GD (2010b) Masquerade: camouflage without crypsis. Science 327:51

Skelhorn J, Rowland HM, Speed MP, De Wert L, Quinn L, Delf J, Ruxton GD (2010c) Size-dependent misclassification of masqueradinf prey. Behav Ecol 21:1344–1348

Skelhorn J, Rowland HM, Delf J, Speed MP, Ruxton GD (2011) Density-dependent predation influences the evolution and behavior of masqueradinf prey. Proc Natl Acad Sci USA 108:6532–6536

Slocum RD, Lawrey JD (1976) Viability of the epizoic lichen flora carried and dispersed by green lacewing (*Nodita pavida*) larvae. Can J Bot 54:1827–1831

Smith SM (1975) Innate recognition of coral snake pattern by a possible avian predator. Science 187:759–760

Smith SM (1977) Coral-snake pattern recognition and stimulus generalization by naïve great kiskadees (Aves: Tyrannidae). Nature 265:535–536

Smith AP (1986) Ecology of leaf color polymorphism in a tropical forest species: habitat segregation and herbivory. Oecologia 69:283–287

Smits BL, Peterson WJ (1942) Carotenoids of telial galls of *Gymnosporangium juniperi-virginianae lk*. Science 96:210–211

Snow B, Snow D (1988) Birds and berries. A study of an ecological interaction. T. & A. D. Poyser, Calton

Soltau U, Dötterl S, Liede-Schumann S (2009) Leaf variegation in *Caladium steudneriifolium* (Araceae): a case of mimicry? Evol Ecol 23:503–512

Soltis DE, Morris AB, McLachlan JS, Manos PS, Soltis PS (2006) Comparative phylogeography of unglaciated eastern North America. Mol Ecol 15:4261–4293

Spaethe J, Chittka L (2003) Interindividual variation of eye optics and single object resolution in bumblebees. J Exp Biol 206:3447–3453

Speed MP, Ruxton GD (2005) Warning displays in spiny animals: one (more) evolutionary route to aposematism. Evolution 59:2499–2508

Stafford HA (1994) Anthocyanins and betalains: evolution of the mutually exclusive pathways. Plant Sci 101:91–98

Stamp NE, Wilkens RT (1993) On the cryptic side of life: being unapparent to enemies and the consequences for foraging and growth caterpillars. In: Stamp NE, Casey TM (eds) Caterpillars:

ecological and evolutionary constraints on foraging. Chapman & Hall, New York, pp 283–330
Stanley RM II (1998) To fool a glass eye. Camouflage versus photoreconnaissance in World War II. Smithsonian Institute Press, Washington D.C
Stant MY (1967) Anatomy of the Butomaceae. J Linn Soc Bot 60:31–60
Starrett A (1993) Adaptive resemblance: a unifying concept for mimicry and crypsis. Biol J Linn Soc 48:299–317
Stavenga DG, Hardie RC (eds) (1989) Facets of vision. Springer Verlag, Berlin
Stebbins GL (1974) Flowering plants. Evolution above the species level. Harvard University Press, Cambridge
Stephenson AG (1981) Toxic nectar deters nectar thieves of *Catalpa speciosa*. Am Midl Nat 105:381–383
Stevens M (2007) Predator perception and the interrelation between different forms of protective coloration. Proc R Soc B 274:1457–1464
Stevens M (2013) Sensory ecology, behaviour, & evolution. Oxford University Press, Oxford
Stevens M (2016) Cheats and deceits. How animals and plants exploit and mislead. Oxford University Press, Oxford
Stevens M, Merilaita S (2009) Animal camouflage: current issues and new perspectives. Philos Trans R Soc B 364:423–427
Stevens M, Merilaita S (eds) (2011) Animal camouflage. Mechanisms and function. Cambridge University Press, Cambridge
Stevens M, Ruxton GD (2014) Do animal eyespots really mimic eyes? Curr Zool 60:26–36
Stevens M, Castor-Perry SA, Price JRF (2009) The protective value of conspicuous signals is not impared by shape, size, or position asymmetry. Behav Ecol 20:96–102
Stevens M, Hopkins E, Hinde W, Adcock A, Connolly Y, Troscianko T, Cuthill IC (2007) Field experiments on the effectiveness of 'eyespots' as predator deterrents. Anim Behav 74:1215–1227
Stevens M, Yule DH, Ruxton GD (2008a) Dazzle coloration and prey movement. Proc R Soc B 275:2639–2643
Stevens M, Hardman CJ, Stubbins CL (2008b) Conspicuousness, not eye mimicry, makes "eyespots" effective antipredator signals. Behav Ecol 19:525–531
Stevens M, Searle WTL, Seymour JE, Marshall KLA, Ruxton GD (2011) Motion dazzle and camouflage as distinct anti-predator defenses. BMC Biol 9:81
Stiles EW (1982) Fruit flags: two hypotheses. Am Nat 120:500–509
Stone BC (1979) Protective coloration of young leaves in certain Malaysian palms. Biotropica 11:126
Stone GN, Schonrogge K (2003) The adaptive significance of insect gall morphology. Trends Ecol Evol 18:512–522
Stone GN, Schonrogge K, Atkinson RJ, Bellido D, Pujade-Villar J (2002) The population biology of oak gall wasps (Hymenoptera: Cynipidae). Annu Rev Entomol 47:633–668
Stoner CJ, Bininda-Emonds ORP, Caro T (2003a) The adaptive significance of coloration in lagomorphs. Biol J Linn Soc 79:309–328
Stoner CJ, Caro TM, Graham CM (2003b) Ecological and behavioral correlates of coloration in artiodactyls: systematic analyses of conventional hypotheses. Behav Ecol 14:823–840
Stournaras KE, Lo E, Böhning-Gaese K, Cazetta E, Dehling DM, Schleuning M, Stoddard MC, Donoghue MJ, Prum RO, Schaefer HM (2013) How colorful are fruits? Limited color diversity in fleshy fruits on local and global scales. New Phytol 198:617–629
Strathdee AT, Bale JS (1998) Life on the edge: insect ecology in arctic environments. Annu Rev Entomol 43:85–106
Strathdee AT, Howling GG, Bale JS (1995) Cold hardiness of overwintering aphid eggs. J Insect Physiol 41:654–657
Strauss SY, Cacho NI (2013) Nowhere to run, nowhere to hide: the importance of enemies and apparency in adaptation to harsh soil environments. Am Nat 182:E1–E14

Strauss SY, Irwin RE (2004) Optimal defence theory and flower petal colour predict variation in the secondary chemistry of wild radish. J Ecol 92:132–141

Strauss SY, Whittall JB (2006) Non-pollinator agents of selection on floral traits. In: Harder LD, Barrett SCH (eds) The ecology and evolution of flowers. Oxford University Press, Oxford, pp 120–138

Strauss SY, Cacho NI, Schwartz MW, Schwartz AC, Burns KC (2015) Apparency revisited. Entomol Exp Appl 157:74–85

Styrsky JD, Eubanks MD (2007) Ecological consequences of interactions between ants and honeydew-producing insects. Proc R Soc B 274:151–164

Sugiura S, Yamazaki K (2006) Consequences of scavenging behaviour in a plant bug associated with a glandular plant. Biol J Linn Soc 88:593–602

Sumner P, Mollon JD (2000a) Chromaticity as a signal of ripeness in fruits taken by primates. J Exp Biol 203:1987–2000

Sumner P, Mollon JD (2000b) Catarrhine photopigments are optimized for detecting targets against a foliar background. J Exp Biol 203:1963–1986

Summers K, Speed MP, Blount JD, Stuckert AMM (2015) Are aposematic signals honest? A review. J Evol Biol 28:1583–1599

Suzuki TK, Tomita S, Sezutsu H (2014) Gradual and contingent evolutionary emergence of leaf mimicry in butterfly wing patterns. BMC Ecol Biol 14:229

Takabayashi J, Dicke M (1996) Plant-carnivore mutualism through herbivore-induced carnivore attractants. Trends Plant Sci 1:109–113

Takhtajan A (1991) Evolutionary trends in flowering plants. Columbia University Press, New York

Talbert CM, Holch AE (1957) A study of the lobing of sun and shade leaves. Ecology 38:655–658

Taulavuori K, Pihlajaniemi H, Huttunen S, Taulavuori E (2011) Reddish spring colouring of deciduous leaves: a sign of ecotype? Trees Struct Funct 25:231–236

Tautz J, Markl H (1978) Caterpillars detect flying wasps by hairs sensitive to airborne vibration. Behav Ecol Sociobiol 4:101–110

Tautz J, Rostás M (2008) Honeybee buzz attenuates plant damage by caterpillars. Curr Biol 18:R1125–R1126

Taylor K (2009) Biological flora of the British isles: *Urtica dioica* L. J Ecol 97:1436–1458

Tellez P, Rojas E, Van Bael S (2016) Red coloration in young tropical leaves associated with reduced fungal pathogen damage. Biotropica 48:150–153

Tenberge KB (1999) Biology and life strategy of the ergot fungi. In: Křen V, Cvak L (eds) Ergot the genus *Claviceps*. Harwood Academic Publishers, Australia, pp 25–56

Tenow O, Nilssen A (1990) Egg cold hardiness and topoclimatic limitations to outbreaks of *Epirrita autumnata* in northern Fennoscandia. J Appl Ecol 27:723–734

Terlouw EMC, Boissy A, Blinet P (1998) Behavioural responses of cattle to the odours of blood and urine from conspecifics and to the odour of faeces from carnivores. Appl Anim Behav Sci 57:9–21

Tewksbury JJ, Nabhan GP (2001) Directed deterrence by capsaicin in chillies. Nature 412:403–404

Thayer GH (1918) Concealing-coloration in the animal kingdom. An exposition of the laws of disguise through color and pattern: being a summary of Abbott H. Thayer's disclosures. Macmillan, New York

Theimer TC, Bateman GC (1992) Patterns of prickly-pear herbivory by collared peccaries. J Wildl Manag 56:234–240

Théry M, Gomez D (2010) Insect colours and visual appearance in the eyes of their predators. Adv Insect Physiol 38:267–353

Thirgood JV (1981) Man and the Mediterranean forest. A history of resource depletion. Academic Press, London

Thornhill R (1975) Scorpionflies as kleptoparasites of web-building spiders. Nature 258:709–711

Thurston EL, Lersten NR (1969) The morphology and toxicology of plant stinging hairs. Bot Rev 35:393–412

Tiffney BH (1985) Perspectives on the origin of the floristic similarity between eastern Asia and eastern North America. J Arnold Arbor 66:73–94

Timell TE (1986) Compression wood in gymnosperms. Springer-Verlag, Berlin

Tng DYP, Janos DP, Jordan GJ, Weber E, Bowman DMJS (2014) Phosphorus limits *Eucalyptus grandis* seedling growth in an unburnt rain forest soil. Front Plant Sci 5:527

Tomlinson PB (1990) The structural biology of palms. Clarendon Press, Oxford

Tooker JF, Rohr JR, Abrahamson WF, De Moraes CM (2008) Gall insects can avoid and alter indirect plant defenses. New Phytol 178:657–671

Toussoun TA, Nelson PE (1968) A pictorial guide to the identification of *Fusarium* species according to the taxonomic system of Snyder and Hansen. Pennsylvania State University Press, University Park

Trabaud L (2000) Post-fire regeneration of *Pinus halepensis* forests in the west Mediterranean. In: Ne'eman G, Trabaud L (eds) Ecology, biogeography and management of *Pinus halepensis* and *Pinus brutia* forest ecosystems in the Mediterranean basin. Backhuys Publishers, Leiden, pp 257–268

Traveset A (1994) Cumulative effects on the reproductive output of *Pistacia terebinthus* (Anacardiaceae). Oikos 71:152–162

Traveset A, Willson MF (1998) Ecology of the fruit-colour polymorphism in *Rubus spectabilis*. Evol Ecol 12:331–345

Troscianko T, Benton CP, Lovell PG, Tolhurst DJ, Pizlo Z (2009) Camouflage and visual perception. Philos Trans R Soc B 364:449–461

Trullas SC, van Wyk JH, Spotila JR (2007) Thermal melanism in ectotherms. J Therm Biol 32:235–245

Trumbo ST, Huang Z-Y, Robinson GE (1997) Division of labor between undertaker specialists and other middle-aged workers in honey bee colonies. Behav Ecol Sociobiol 41:151–163

Tsahar E, Friedman J, Izhaki I (2002) Impact on fruit removal and seed predation of a secondary metabolite, emodin, in *Rhamnus alaternus* fruit pulp. Oikos 99:290–299

Tsahar E, Izhaki I, Lev-Yadun S, Bar-Oz G (2009) Distribution and extinction of ungulates during the Holocene of the southern Levant. PLoS One 4:e5316

Tsukaya H, Okada H, Mohamed M (2004) A novel feature of structural variegation in leaves of the tropical plant *Schismatoglottis calyptrata*. J Plant Res 117:477–480

Tullberg BS, Merilaita S, Wiklund C (2005) Aposematism and crypsis combined as a result of distance dependence: functional versatility of the colour pattern in the swallowtail butterfly larva. Proc R Soc Lond B 272:1315–1321

Tuomi J, Augner M (1993) Synergistic selection of unpalatability in plants. Evolution 47:668–672

Turini A, Veselý P, Fuchs R (2016) Five species of passerine bird differ in their ability to detect Batesian mimics. Biol J Linn Soc 117:832–841

Ulmer T, MacDougal JM (2004) Passiflora. Passionflowers of the world. Timber Press, Portland

Unagul P, Wongsa P, Kittakoop P, Intamas S, Srikitikulchai P, Tanticharoen M (2005) Production of red pigments by the insect pathogenic fungus *Cordyceps unilateralis* BCC 1869. J Ind Microbiol Biotechnol 32:135–140

Ungar PS, Teaford MF, Glander KE, Pastor RF (1995) Dust accumulation in the canopy: a potential cause of dental microwear in primates. Am J Phys Anthropol 97:93–99

Unsicker SB, Kunert G, Gershenzon J (2009) Protective perfumes: the role of vegetative volatiles in plant defense against herbivores. Curr Opin Plant Biol 12:479–485

Valeix M, Fritz H, Loveridge AJ, Davidson Z, Hunt JE, Murindagomo F, Macdonald DW (2009) Does the risk of encountering lions influence African herbivore behavior at waterholes? Behav Ecol Sociobiol 63:1483–1494

Valkonen J, Niskanen M, Björklund M, Mappes J (2011) Disruption or aposematism? Significance of dorsal zigzag pattern of European vipers. Evol Ecol 25:1047–1063

van Baalen M, Jansen VAA (2001) Dangerous liaisons: the ecology of private interest and common good. Oikos 95:211–224

Van Bael SA, Brawn JD, Robinson SK (2003) Birds defend trees from herbivores in a Neotropical forest canopy. Proc Natl Acad Sci USA 100:8304–8307

van der Pijl L (1982) Principles of dispersal in higher plants, 3rd edn. Springer-Verlag, Berlin
van der Pijl L, Dodson CH (1966) Orchid flowers. Their pollination and evolution. University of Miami Press, Coral Gables
van Dongen PWJ, de Groot ANJA (1995) History of ergot alkaloids from ergotism to ergometrine. Eur J Obstet Gynecol Reprod Biol 60:109–116
van Hezewijk BH, Roland J (2003) Gall size determines the structure of the *Rabdophaga strobiloides* host-parasitoid community. Ecol Entomol 28:593–603
van Kleunen M, Nänni I, Donaldson JS, Manning JC (2007) The role of beetle marks and flower colour on visitation by monkey beetles (Hopliini) in the greater Cape floral region, South Africa. Ann Bot 100:1483–1489
van Poecke RMP, Posthumus MA, Dicke M (2001) Herbivore-induced volatile production by *Arabidopsis thaliana* leads to attraction of the parasitoid *Cotesia rubecula*: chemical, behavioral, and gene-expression analysis. J Chem Ecol 27:1911–1928
Vencl FV, Morton TC, Mumma RO, Schultz JC (1999) Shield defense of a larval tortoise beetle. J Chem Ecol 25:549–566
Vereecken NJ, Schiestl FP (2009) On the roles of colour and scent in a specialized floral mimicry system. Ann Bot 104:1077–1084
Vereecken NJ, Teng J, Kullenberg G, Bergström G (2009) Bertil Kullenberg (1913–2007) and his *Ophrys* orchids. J Eur Orch 41:3–18
Vermeij GJ (2015) Plants that lead: do some surface features direct enemy traffic on leaves and stems? Biol J Linn Soc 116:288–294
Vetter RS (2013) Arachnophobic entomologists: when two more legs makes a big difference. Am Entomol 59:168–175
Vogel S (2012) The life of a leaf. Chicago University Press, Chicago
Vogel ER, Neitz M, Dominy NG (2006) Effect of color vision phenotype on the foraging of wild white-faced capuchins, *Cebus capucinus*. Behav Ecol 18:292–297
Vogelmann TC (1993) Plant tissue optics. Annu Rev Plant Physiol Plant Mol Biol 44:231–251
Vollrath F, Douglas-Hamilton I (2002) African bees to control African elephants. Naturwiss 89:508–511
von Helversen B, Schooler LJ, Czienskowski U (2013) Are stripes beneficial? Dazzle camouflage influences perceived speed and hit rates. PLoS One 8:e61173
Waage JK (1981) How the zebra got its stripes – biting flies as selective agents in the evolution of zebra coloration. J Ent Soc Sth Afr 44:351–358
Wagner GJ, Wang E, Shepherd RW (2004) New approaches for studying and exploiting an old protuberance, the plant trichome. Ann Bot 93:3–11
Waisel Y (1972) Biology of halophytes. Academic Press, New York
Waisel Y, Litav M, Agami M (1975) Coastal plants of Israel. Division of Ecology, Department of Botany, Tel Aviv University, Petah Tiqva (in Hebrew)
Waisel Y, Cohen S, Binyamini N (1977) Poison plants of Israel. Division of Ecology Ltd., Petach Tiqwa (in Hebrew)
Waldbauer GP (1988) Aposematism and Batesian mimicry. Measuring mimetic advantage in natural habitats. Evol Biol 22:227–259
Wäli PP, Wäli PR, Saikkonen K, Tuomi J (2013) Is the pathogenic ergot fungus a conditional defensive mutualist for its host grass? PLoS One 8:e69249
Walker A, Hoeck HN, Perez L (1978) Microwear of mammalian teeth as an indicator of diet. Science 201:908–910
Wallace AR (1889) Darwinism. Macmillan and Co, London
Walters DR, McRoberts N, Fitt BDL (2008) Are green islands red herrings? Significance of green islands in plant interactions with pathogens and pests. Biol Rev 83:79–102
Wang Y, Stass A, Horst WJ (2004) Apoplastic binding of aluminum is involved in silicon-induced amelioration of aluminum toxicity in maize. Plant Physiol 136:3762–3770
Warren J (2015) Is wind-mediated passive leaf movement an effective form of herbivore defence? Plant Ecol Evol 148:52–56

Warren J, James P (2008) Do flowers wave to attract pollinators? A case study with *Silene maritima*. J Evol Biol 21:1024–1029

Waser NM (1978) Competition for hummingbird pollination and sequential flowering in two Colorado wildflowers. Ecology 59:934–944

Waters CN, Zalasiewicz J, Summerhayes C, Barnosky AD, Poirier C, Gałuszka A, Cearreta A, Edgeworth M, Ellis EC, Ellis M, Jeandel C, Leinfelder R, McNeill JR, Richter D dB, Steffen W, Syvitski J, Vidas D, Wagreich M, Williams M, Zhisheng A, Grinevald J, Odada E, Oreskes N, Wolfe AP (2016) The Anthropocene is functionally and stratigraphically distinct from the Holocene. Science 351:aad2622-1–aad2622-10

Weeden NF (2007) Genetic changes accompanying the domestication of *Pisum sativum*: is there a common genetic basis to the 'domestication syndrome' for legumes? Ann Bot 100:1017–1025

Weis AE, Walton R, Crego CL (1988) Reactive plant tissue sites and the population biology of gall makers. Annu Rev Entomol 33:467–486

Weiss MR (1991) Floral colour changes as cues for pollinators. Nature 354:227–229

Weiss MR (1995) Floral colour change: a widespread functional convergence. Am J Bot 82:167–195

Weiss MR (2006) Defecation behavior and ecology of insects. Annu Rev Entomol 51:635–661

Weiss MR, Lamont BB (1997) Floral colour change and insect pollination: a dynamic relationship. Isr J Plant Sci 45:185–199

Wen J (1999) Evolution of Eastern Asian and Eastern North American disjunct distributions in flowering plants. Annu Rev Ecol Syst 30:421–455

Werker E (2000) Trichome diversity and development. Adv Bot Res 31:1–35

Werner EE, Peacor SD (2003) A review of trait-mediated indirect interactions in ecological communities. Ecology 84:1083–1100

Westmoreland D, Muntan C (1996) The influence of dark central florets on insect attraction and fruit production in Queen Anne's Lace (*Daucus carota* L.). Am Midl Nat 135:122–129

Wheeler KGR (2004) A natural history of nettles. Trafford, Victoria

Wheelwright NT, Janson CH (1985) Colors of fruit displays of bird-dispersed plants in two tropical forests. Am Nat 126:777–799

White PS (1988) Prickle distribution in *Aralia spinosa* (Araliaceae). Am J Bot 75:282–285

White TCR (2009) Catching a red herring: autumn colours and aphids. Oikos 118:1610–1612

White TCR (2010) Why do many galls have conspicuous colours? An alternative hypothesis revisited. Arthropod Plant Interact 4:149–150

Whitehouse ME (1997) The benefits of stealing from a predator: foraging rates, predation risk, and intraspecific aggression in the kleptoparasitic spider *Argyrodes antipodiana*. Behav Ecol 8:663–667

Whitfield J (2002) Plants feign infestation. In: Nature science update. Nature News Service, November 14, 2002

Whitney KD (2004) Experimental evidence that both parties benefit in a facultative plant-spider mutualism. Ecology 85:1642–1650

Wickler W (1968) Mimicry in plants and animals. Weidenfeld and Nicolson, London

Wiens D (1978) Mimicry in plants. Evol Biol 11:365–403

Wilkinson N (1969) A brush with life. Seeley Service & Co Ltd, London

Wilkinson DM, Sherratt TN, Phillip DM, Wratten SD, Dixon AFG, Young AJ (2002) The adaptive significance of autumn leaf colours. Oikos 99:402–407

Willey JM, Sherwood LM, Woolverton CJ (2008) Prescot, Harley, and Klein's Microbiology, 7th edn. McGraw Hill, New York

Williams D (2001) Naval camouflage 1914–1945. A complete visual reference. Naval Institute Press, Annapolis

Williams KS, Gilbert LE (1981) Insects as selective agents on plant vegetative morphology: egg mimicry reduces egg laying by butterflies. Science 212:467–469

Williamson GB (1982) Plant mimicry: evolutionary constraints. Biol J Linn Soc 18:49–58

Willis KJ, van Andel TH (2004) Trees or no trees? The environments of central and eastern Europe during the last Glaciation. Quat Sci Rev 23:2369–2387

Willson MF (1983) Plant reproductive ecology. Wiley, New York

Willson MF, Hoppes WG (1986) Foliar 'flags' for avian frugivores: signal or serendipity? In: Estrada A, Fleming TH (eds) Frugivores and seed dispersal. Dr W. Junk, Dordrecht, pp 55–69

Willson MF, Melampy MN (1983) The effect of bicolored fruit displays on fruit removal by avian frugivores. Oikos 41:27–31

Willson MF, Thompson JN (1982) Phenology and ecology of color in bird-dispersed fruits, or why some fruits are red when they are "green". Can J Bot 60:701–713

Willson MF, Whelan CJ (1990) The evolution of fruit color in fleshy-fruited plants. Am Nat 136:790–809

Wimp GM, Whitham TG (2001) Biodiversity consequences of predation and host plant hybridization on an aphid-ant mutualism. Ecology 82:440–452

Winter K, Smith JAC (1996) Crassulacean acid metabolism. Biochemistry, ecophysiology and evolution. Springer-Verlag, Berlin

Wise DH (1993) Spiders in ecological webs. Cambridge University Press, Cambridge

Wold EN, Marquis RJ (1997) Induced defense in white oak: effects on herbivores and consequences for the plant. Ecology 78:1356–1369

Wong AR, Srivastava DS (2010) Red abaxial coloring reduces herbivory in *Columnea consanguinea*. Ecotropica 16:93–99

Wool D (2004) Galling aphids: specialization, biological complexity, and variation. Annu Rev Entomol 49:175–192

Woolfson A, Rothschild M (1990) Speculating about pyrazines. Proc R Soc Lond B 242:113–119

Wüster W, Allum CSE, Bjargardóttir IB, Bailey KL, Dawson KJ, Guenioui J, Lewis J, McGurk J, Moore AG, Niskanen M, Pollard CP (2004) Do aposematic and Batesian mimicry require bright colours? A test, using European viper markings. Proc R Soc Lond B 271:2495–2499

Yager KG, Schaefer HM, Gould KS (2016) Shared leaf shape in *Alseuosmia pusilla* and *Pseudowintera colorata*: a case of Batesian mimicry? Botany 94:555–564

Yamaguchi H, Tanaka H, Hasegawa M, Tokuda M, Asami T, Suzuki Y (2012) Phytohormones and willow gall induction by a gall-inducing sawfly. New Phytol 196:586–595

Yamazaki K (2008a) Autumn leaf colouration: a new hypothesis involving plant-ant mutualism via aphids. Naturwiss 95:671–676

Yamazaki K (2008b) Colors of young and old spring leaves as a potential signal for ant-tended hemipterans. Plant Signal Behav 3:984–985

Yamazaki K (2010) Leaf mines as visual defensive signals to herbivores. Oikos 119:796–801

Yamazaki K (2011) Gone with the wind: trembling leaves may deter herbivory. Biol J Linn Soc 104:738–747

Yamazaki K (2014) Slug caterpillars of *Parasa lepida* (Cramer, 1799) (Lepidoptera: Limacodidae) become stuck on rose prickles. Pan Pac Entomol 90:221–225

Yamazaki K (2016) Caterpillar mimicry by plant galls as a visual defense against herbivores. J Theor Biol 404:10–14

Yamazaki K, Lev-Yadun S (2014) Dark axils and nodes in various plant species may serve as defensive mimicry of beetles and beetle faces. J Nat Hist 48:691–698

Yamazaki K, Lev-Yadun S (2015) Dense white trichome production by plants as possible mimicry of arthropod silk or fungal hyphae that deter herbivory. J Theor Biol 364:1–6

Yano S (1994) Flower nectar of an autogamous perennial *Rorippa indica* as an indirect defense mechanism against herbivorous insects. Res Popul Ecol 36:63–71

Yano S (1997) Silique burst of *Cardamine scutata* (Cruciferae) as a physical inducible defense against seed predatory caterpillars. Res Popul Ecol 39:95–100

Yano S (2012) Cooperative web sharing against predators promotes group living in spider mites. Behav Ecol Sociobiol 66:845–853

Yeo PF (1972) Miscellaneous notes on pollination and pollinators. J Nat Hist 6:667–686

Yokoi T, Fujisaki K (2009) Hesitation behaviour of hoverflies *Sphaerophoria* spp. to avoid ambush by crab spiders. Naturwiss 96:195–200

Yong JW, Wong SC, Letham DS, Hocart CH, Farquhar GD (2000) Effects of elevated CO_2 and nitrogen nutrition on cytokinins in the xylem sap and leaves of cotton. Plant Physiol 124:767–780

Young TP, Okello BD (1998) Relaxation of an induced defense after exclusion of herbivores: spines on *Acacia drepanolobium*. Oecologia 115:508–513

Young TP, Stanton ML, Christian CE (2003) Effects of natural and simulated herbivory on spine lengths of *Acacia drepanolobium* in Kenya. Oikos 101:171–179

Zachos J, Pagani M, Sloan L, Thomass E, Billups K (2001) Trends, rhythms, and aberrations in global climate 65 Ma to present. Science 292:686–693

Zahavi A (1975) Mate selection – a selection for a handicap. J Theor Biol 53:205–214

Zahavi A (1977) The cost of honesty (further remarks on the handicap principle). J Theor Biol 67:603–605

Zahavi A (1991) On the definition of sexual selection, Fisher's model, and the evolution of waste and of signals in general. Anim Behav 42:501–503

Zahavi A, Zahavi A (1997) The handicap principle. A missing piece of Darwin's puzzle. Oxford University Press, New York

Zalucki MP, Brower LP, Malcolm SB (1990) Oviposition by *Danaus plexippus* in relation to cardenolide content of three *Asclepias* species in the southeastern U.S.A. Ecol Entomol 15:231–240

Zamora R, Gómez JM (1993) Vertebrate herbivores as predators of insect herbivores: an asymmetrical interaction mediated by size differences. Oikos 66:223–228

Zhan S, Zhang W, Niitepõld K, Hsu J, Haeger JF, Zalucki MP, Altizer S, de Roode JC, Reppert SM, Kronforst MR (2014) The genetics of monarch butterfly migration and warning colouration. Nature 514:317–321

Zohary M (1962) Plant life of Palestine. Israel and Jordan. Ronald Press, New York

Zohary M (1973) Geobotanical foundations of the Middle East. Gustav Fischer, Stuttgart

Zohary M (1980) Vegetal landscapes of Israel. Am Oved Publishers Ltd, Tel Aviv (in Hebrew)

Zohary M (1983) Man and vegetation in the Middle East. In: Holzner W, Werger MJA, Ikusima I (eds) Man's impact on vegetation. Dr W. Junk BV Publishers, The Hague, pp 287–295

Zohary D, Hopf M, Weiss E (2012) Domestication of plants in the old world, 4th edn. Clarendon, Oxford

Zolnerowich G (1992) A unique *Amycle* numph (Homoptera: Fulgoridae) that mimics jumping spiders (Araneae: Salticidae). J NY Entomol Soc 100:498–502

The manufacturer's authorised representative in the EU is Springer Nature Customer Service Centre GmbH, Europaplatz 3, 69115 Heidelberg, Germany. If you have any concerns regarding our products, please contact ProductSafety@springernature.com

Printed and bound by CPI Group (UK) Ltd, Croydon, CR0 4YY
25/09/2026
02222851-0003